Digital Learning and Teaching in Chemistry

Advances in Chemistry Education Series

Titles in the Series:
1: Professional Development of Chemistry Teachers: Theory and Practice
2: Argumentation in Chemistry Education: Research, Policy and Practice
3: The Nature of the Chemical Concept: Re-constructing Chemical Knowledge in Teaching and Learning
4: Creative Chemists: Strategies for Teaching and Learning
5: Engaging Learners with Chemistry: Projects to Stimulate Interest and Participation
6: The Johnstone Triangle: The Key to Understanding Chemistry
7: Problems and Problem Solving in Chemistry Education
8: Teaching and Learning in the School Chemistry Laboratory
9: Nanochemistry for Chemistry Educators
10: Student Reasoning in Organic Chemistry: Research Advances and Evidence-based Instructional Practices
11: Digital Learning and Teaching in Chemistry

How to obtain future titles on publication:
A standing order plan is available for this series. A standing order will bring delivery of each new volume immediately on publication.

For further information please contact:
Book Sales Department, Royal Society of Chemistry, Thomas Graham House, Science Park, Milton Road, Cambridge, CB4 0WF, UK
Telephone: +44 (0)1223 420066, Fax: +44 (0)1223 420247
Email: booksales@rsc.org
Visit our website at books.rsc.org

Digital Learning and Teaching in Chemistry

Edited by

Yehudit Judy Dori
Technion – Israel Institute of Technology, Israel
The Samuel Neaman Institute for National Policy, Israel
Email: yjdori@technion.ac.il

Courtney Ngai
Colorado State University, USA
Email: Courtney.ngai@colostate.edu

and

Gabriela Szteinberg
College Office, College of Arts & Sciences, Washington University in St. Louis, USA
Email: gsztein@wustl.edu

ROYAL SOCIETY OF **CHEMISTRY**

Advances in Chemistry Education Series No. 11

Print ISBN: 978-1-83916-523-8
PDF ISBN: 978-1-83916-794-2
EPUB ISBN: 978-1-83916-795-9
Print ISSN: 2056-9335
Electronic ISSN: 2056-9343

A catalogue record for this book is available from the British Library

The Royal Society of Chemistry is a charity, registered in England and Wales, Number 207890, and a company incorporated in England by Royal Charter (Registered No. RC000524), registered office: Burlington House, Piccadilly, London W1J 0BA, UK, Telephone: +44 (0) 20 7437 8656.

For further information see our website at www.rsc.org

Foreword by Keith S. Taber

The Royal Society of Chemistry's '*Advances in Chemistry Education*' series provides books that address key issues in the teaching and learning of chemistry and offer readers research-informed accounts of topics to both inform good teaching practice and support further research in the field. In that regard, I am very pleased to see the publication of this timely book edited by Yehudit Judy Dori, Courtney Ngai, and Gabriela Szteinberg.

Chemistry education is now a well-established research field and has had time to develop multiple foci and themes. At its heart, however, are the related and complex processes of student learning of chemistry and how this can best be supported through teaching. This is the core focus of the '*Advances in Chemistry Education*' book series, and '*Digital Learning and Teaching in Chemistry*' is clearly 'on-topic' in that regard. I was encouraged to learn from the introductory chapter that the most frequent words identified from the text were 'students' and 'learning', with 'teachers' following on close behind.

The notion of 'digital learning and teaching' is of course at one level a misnomer—learning itself is no more digital or analogue than it ever was. Yet this terminology does reflect something of a revolution in the extent and power of the educational technology now available to support teaching and learning. Digital tools—computers of various kinds (including mobile phones and tablets), the internet, simulations, virtual learning environments, automatic data-logging, editable textbooks and student notebooks, Wikis, and so forth—have in a few decades completely changed the nature of the technology available (or at least, potentially available) in the classroom. Indeed, the change has been so extensive that the 'classroom' no longer needs to be a shared *physical* space but can be a manifold of connected spaces.

The history of this revolution has not always been smooth. Initially the hardware was expensive at a time when it was largely unproven—and so it often

Advances in Chemistry Education Series No. 11
Digital Learning and Teaching in Chemistry
Edited by Yehudit Judy Dori, Courtney Ngai and Gabriela Szteinberg

Published by the Royal Society of Chemistry, www.rsc.org

had to be used in 'very dilute concentrations', even when budget allowed acquisition. Early teaching and learning tools were often clunky, relatively slow, error-prone, and not always designed with a strong consideration of pedagogy or research into students' learning difficulties.

Sometimes these resources seemed to offer limited advances on existing approaches yet required teachers to commit time and effort to familiarisation, and even change aspects of how they worked. This relatively large 'activation energy' for minimal 'enthalpy' change did not support a high rate of uptake, nor promise an equilibrium position that would lead to the adoption ever 'going to completion'.

But in a relatively short period the speed of machines, the available memory, the quality of graphics, the sophistication of programming, and the increased focus on what the teachers want to support (rather than disrupt) their work have progressed considerably. The internet makes available through the World Wide Web an effectively infinite resource bank, encompassing professional scientific materials (but including dubious scientific reports in many low quality predatory open-access journals), professionally designed teaching and learning resources, and, increasingly, banks of materials shared by communities of, and individual, teachers. These extensive riches bring a need for new skills of quality assessment, syllabus matching, and curation.

One of the greatest advances of the digital age is the ability to copy, and potentially modify, materials without losing or destroying the original. One teacher's worksheet can in principle be shared with every other and customised as needed. Each student can annotate a digital textbook in their own way, and the teacher can read and respond to their notes without having to deny the students access. No more collecting-in, and handing back out, piles of books; nor excuses for why some seem to have gone missing during the process. Students can keep copies of ongoing work at different stages to monitor their own progress. A concept map for a new topic can be copied and updated from time to time.

Some of the 'new' technologies offer genuinely new opportunities.[1] There is technology offering haptic feedback that gives a feel for molecular structures. There are immersive simulations of virtual reality and augmented reality that superimposes the virtual on the real environment. New affordances, and new ways of working, can enrich teaching, but teachers need support in initial and continuing professional education to acquire the knowledge and skills to best make use of the technology. This became especially clear at the start of the COVID-19 global pandemic when schools and colleges closed down and shifted to on-line working in many parts of the world. This also highlighted the vast disparity in the level of resource that can be taken for granted both across and within different countries.

Whilst distance learning may never completely match *some* aspects of teaching undertaken face to face,[2] the pandemic showed just what is possible with the new technology. But it also showed how such transitions cannot be made effortlessly overnight. Good teaching, in whatever mode, requires extensive preparation. Even when there are endless resources accessible, it takes time to access, evaluate and incorporate them. Even when there are effective ways of group working on-line through chat rooms and break-out groups and the like, both teachers and students need familiarity with the systems available and support to learn how to make best use of them.

In other words, alongside investment in resources, research into their effective application is also needed. Tools need to meet teachers' needs, support their work, and make their professional lives easier and more productive. That requires research to develop solutions to educational challenges through resources that meet educational aims and enable effective pedagogies. It also needs the resulting knowledge and skills to be widely disseminated—by people who understand teaching and learning *as well as* new technology.

Teachers can only undertake evidence-based practice when the evidence has been collected and analysed; and conclusions have been drawn and validated; and the findings disseminated. There is an extensive developing research base on the effective use of digital tools in teaching, and a fair amount of this is discipline specific (and sometimes even topic specific). There is a 'state of the art' of our knowledge of *Digital Learning and Teaching in Chemistry* to be shared both amongst the research community, and with teacher educators and the teachers themselves.

This book offers a timely overview of the current state of the field of supporting chemistry teaching and learning with digital tools and includes diverse contributions from a wide range of experts from within the chemistry education research family across the world. I thank the editors for putting this book together, and all the contributors for collectively providing a resource (available in digital as well as physical form!) that I think will prove invaluable to the chemistry education community.

Keith S. Taber
Cambridge

References

1. K. S. Taber and X. Li, *Adv. Psychol. Res.*, 2021, **143**, 1–72.
2. K. S. Taber, *Cent. Educ. Policy Stud. J.*, 2021, **11**, 67–87.

Foreword by David F. Treagust

This book is well-placed to address the important issues of digital teaching and learning in chemistry that have been exacerbated by the recent Covid-19 pandemic. Indeed, even before the pandemic arrived, with the universal availability of high-speed internet, digital teaching and learning became more available as an option and a part of school and university course offerings. With the onset of the pandemic, digital teaching and learning became a necessity for education; teachers, lecturers and their students needed to rapidly adapt to a new style of teaching, learning and assessment online. These three aspects of education in chemistry are addressed in this book and demonstrate that teaching face-to-face and online have different dynamics that need to be learned and understood for their respective effectiveness. I discuss more on this point later from our own research.

In March 2019, before any concerns about Covid-19, The American Enterprise Institute[1] published 'The promises and limits of online higher education', based on a national survey of US college administrators, recognising the steady increase of higher education offerings of online course over the past two decades. While this review did not include science courses but focused on business, computing and statistics courses, some important points were that comparative learning improvements were made by students in single one-off online courses but for four-year relatively selective institutions, many studies identified negative effects of online instruction on course grade—in some cases much worse. There are messages here about the need for hybrid models of learning as demonstrated by several authors of this book.

Among the recommendations to enhance online learning and instruction that arise from this survey are that promoting interpersonal interactions is a key to successful learning by assigning students to peer groups, small group problem activities, and synchronous online discussions that are also

Advances in Chemistry Education Series No. 11
Digital Learning and Teaching in Chemistry
Edited by Yehudit Judy Dori, Courtney Ngai and Gabriela Szteinberg

Published by the Royal Society of Chemistry, www.rsc.org

advantageous for face-to-face classes. Later, I report similar findings from our own chemistry education research. An important consideration is the need to identify potential at-risk students online in a timely manner, which is likely to be more difficult than in face-to-face situations. Online students also need high levels of self-regulation and self-discipline, and this tends to favour adult learners. Another concern of online-only learning is the need to improve communication skills and making sense of concepts and communication through a different medium. Each of these issues most certainly apply to chemistry and have been noted by the authors in this book.

Returning to digital learning and teaching in chemistry—the focus of this book—the single biggest potential loss of online chemistry courses is the lack, or change in emphasis, of laboratory work. Fortunately, researchers[2] are addressing this issue by the creation of online resources such as *Labster*[3] and *Lab Sims*[4] that provide virtual laboratories across science for high school and university as well as laboratory techniques through *JoVE*,[5] a peer-reviewed and PubMed indexed video journal that publishes high quality demonstrations of methodology where students are presented with data and do the analysis. Using these *JoVE* videos, researchers[6] report investigations with university general chemistry students that show significant learning outcomes and reinforced conceptual understanding for important foundational concepts such as enthalpy, entropy, rate laws and Le Chatelier's principle. Likewise, it is possible for students to do practical science at home using online resources that are readily made available, as illustrated with first year undergraduates.[7] The chapters in this book address these issues.

Before the pandemic, with colleagues in chemistry and engineering at Curtin University, I recently completed an Australian Research Council (ARC) project to investigate the essential elements of laboratory-based learning for remote-access implementation of online science and engineering courses.[8] Working with first year university students in chemistry and engineering, our findings were that remote laboratories generated a feeling of undertaking a real experiment through a live video feed of the equipment. However, students reported that interactions with their peers were less important than in face-to-face laboratories. Similarly, students recognised the value of remote laboratories but noted that internet-mediated interactions could inhibit their acquisition of expected learning outcomes related to instrumentation, communication, experimentation, ethics, safety matters, and learning from failures.

Our research outcomes showed that students—under normal operation of universities—value remote-laboratory activities, not as a replacement for face-to-face laboratories but as a complement to the latter, traditional mode.[9] Our research has provided a much clearer understanding of the importance of instructor–student interactions and student–student interactions as well as interactions with the laboratory equipment in university science and engineering remote laboratories. For remote laboratories to be sustainably available in the future, we recommended that they need to be made more engaging and laboratory instructors should provide more autonomy for students learning in both remote and face-to-face laboratories. This ARC research project has provided both insights and tools to enable this to occur. This book brings different approaches and emphases to chemistry learning, teaching, and evaluation, both at higher education and school level. A solid research foundation runs along with the book's practical perspective, making it a valuable resource for educators and researchers in chemistry education, as well as those in other STEM fields, as the digitization of education systems continues to advance.

David F. Treagust,
Professor of Science Education, School of Education
Curtin University, Perth, Australia

References

1. D. Xu and Y. Xu, *The Promises and Limits of Online Higher Education; Understanding How Distance Education Affects Access, Cost and Quality*, American Enterprise Institute, Washington, DC, 2019.
2. R. LeBard, *Can Science Be Taught On-Line? Ideas for Learning and Teaching Biology without a Laboratory*, 2020, https://medium.com/age-of-awareness/can-science-be-taught-online-143603efbd83.
3. https://www.labster.com/.
4. https://www.learnsci.com/products/labsims.
5. https://www.jove.com/education/chem.
6. R. Ramachandran, E. M. Sparck and M. Levis-Fitzgerald, *J. Chem. Ed.*, 2019, **96**, 479–485.
7. D. J. Caruana, C. G. Salzmann and A. Sella, *Nat. Chem.*, 2020, **12**, 780–783.
8. J. Wei, D. F. Treagust, M. Mocerino, A. D. Lucey, M. G. Zadnik and E. Lindsay, *Discip. Interdiscip. Sci. Educ. Res.*, 2019, **1**(14), 1–16.
9. J. Wei, D. F. Treagust, M. Mocerino, A. D. Wallace and T. C. Brown, *J. Chem. Ed.*, 2022, **99**, 1201–1210.

Preface

During my childhood, in the very early morning hours, I used to help my mother, who was an exemplary elementary school educator, to prepare visualization aids—posters and cards—for her class. This is how I started developing an affection for visualizations and an understanding of their key role in education, especially in science and engineering education. I even embedded pictures and graphs in the *Chemistry for Nursing Students* book that I wrote as part of my Ph.D. research and in most of the papers I have published since.

Over time, technology has developed, and researchers and educators have increasingly recognized the importance and centrality of the various kinds of digital visualizations as a means to promote meaningful learning and deep understanding of natural and human-made systems and phenomena. Personally, I experienced this in the most profound way during the period in which I participated as the assessment leader in the Technology-Enabled Active Learning (TEAL) educational research project at the Massachusetts Institute of Technology. In the TEAL project, still and moving digital visualizations of magnetic fields constituted the basis for a transformative change in the way undergraduate physics is taught.

Nowadays, it is clearer than ever that beside the critical role of the teacher and educator to develop independent learners and evaluate them in multiple, non-traditional ways, digital visualizations and other technology-enhanced means of teaching are mandatory, indispensable components in any science or engineering course, including chemistry.

This book is a compilation of current research studies and perspectives from all over the world on how chemistry education benefits from, and can be advanced by, digital means to support students as they learn about chemical reactions and phenomena. This is a valuable resource for chemistry teachers

---

Advances in Chemistry Education Series No. 11
Digital Learning and Teaching in Chemistry
Edited by Yehudit Judy Dori, Courtney Ngai and Gabriela Szteinberg

Published by the Royal Society of Chemistry, www.rsc.org

and faculty, and for chemical education researchers who wish to keep abreast of educational technology innovations and familiarize themselves with the state-of-the-art in this fast-developing area.

I wish to thank Prof. Keith Taber and Prof. David Treagust for encouraging me to lead this endeavor and to Prof. Hannah Sevian, who introduced me to my co-editors Dr Courtney Ngai and Dr Gabriela Szteinberg. Together, the three of us collaborated with over 30 lead authors to complete this book. Thanks also to my Ph.D. student, Roee Peretz, who was most helpful in proof-reading the book. Last but not least, to my dear husband Prof. Dov Dori, thanks for accompanying me throughout our life.

Yehudit Judy Dori
Technion, Israel Institute of Technology and
Samuel Neaman Institute for National Policy Research
Haifa 320003, Israel

Biographies

Gyde Asmussen is currently a Ph.D. student at the IPN Leibniz-Institute for Science and Mathematics Education in Kiel, Germany. She studied chemistry and biology for teaching at the high school level and graduated with a Master's degree. Her research interest is in investigating and supporting students' problem solving in organic chemistry.

Shirly Avargil is an Assistant Professor at the Faculty of Education in Science and Technology, Technion, Israel Institute of Technology. She is the head of the chemistry education research group and the chemistry teaching diploma track. She received her Ph.D. in 2011 from the Technion and then spent two years as a postdoctoral researcher at University of Maine, USA. Her research interests include choosing chemistry as a major and career, students' higher-order thinking skills in chemistry and chemistry teachers' assessment knowledge. She is a member of the National Chemistry Education Committee in her country.

Miri Barak (Engr.) is an Associate Professor at the Faculty of Education in Science and Technology, Technion- Israel Institute of Technology. She investigates cognitive and sociocultural aspects of online learning and the promotion of 21st century skills. Dr Barak is a *Cum Laude* graduate of the Faculty of Food Engineering and Biotechnology, at the Technion, and worked in the past as an R&D Engineer. She received her M.Sc. and Ph.D. degrees at the Technion and was a postdoctoral fellow at MIT—Massachusetts Institute of Technology, USA. Currently, she is the Vice Dean for Graduate Studies and the Assistant to the SVPR for the Promotion of Learning and Teaching.

Catherine Barth studied graphic design at the Akademie U5 in Munich. Later on, she switched to teacher education studies and is currently studying

Advances in Chemistry Education Series No. 11
Digital Learning and Teaching in Chemistry
Edited by Yehudit Judy Dori, Courtney Ngai and Gabriela Szteinberg

Published by the Royal Society of Chemistry, www.rsc.org

chemistry and English teaching with a focus on lower secondary education at the Weingarten University of Education. In her bachelor thesis, she studied the creation of AR learning environments.

Nadja Belova is a PostDoc in the chemistry education group of the University of Bremen and a chemistry teacher at the Gymnasium Vegesack in Bremen. Her main research focus is media literacy in the science classroom. In this context, she explicitly addresses modern media types such as social media and has designed and published a number of curricular innovations involving different types of media using Action Research.

Sascha Bernholt works as a research scientist at the IPN Leibniz-Institute for Science and Mathematics Education in Kiel, Germany. His research interests lie on conceptual learning, competency modelling, as well as language, formulae, and terminology in chemistry education.

Ron Blonder is a Professor in the Department of Science Teaching at the Weizmann Institute of Science in Israel. She is the head of the chemistry group and the head of the Rothschild-Weizmann Master's Program for Excellence in Mathematics and Science Teaching. Her research focuses on chemistry teachers' professional development in the context of contemporary science using innovative technological tools and environments. She has published over 100 peer-reviewed and invited papers as well as book chapters in which she has explored chemistry teachers' self-efficacy beliefs and knowledge development when they learn contemporary research in chemistry (mainly nanochemistry) and when they incorporate technological tools in their chemistry teaching. She has also investigated differentiated instruction in heterogeneous chemistry classes with and without the aid of technology to promote personalization in chemistry teaching. Prof. Blonder has applied the results of her educational research to chemistry teachers' professional development and continuing education. Prof. Blonder is the recipient of the 2023 NARST Fellows Award.

Brittland DeKorver is an associate professor in the Chemistry Department at Grand Valley State University in Allendale, MI, USA. She earned her Ph.D. in chemistry education from Purdue University. Dr DeKorver has conducted research in informal science education, working with the Fusion Science Theater group to develop demonstration shows that teach science concepts. She has also conducted research about undergraduate students' experiences in chemistry laboratory coursework and digital badging to evaluate students' lab skills. Her current research interests explore issues of diversity and equity in chemistry coursework. She is also the administrator and moderator of Strategies for Teaching Chemistry Online, the virtual community that is the focus of this manuscript.

Dermot F. Donnelly-Hermosillo is an Associate Professor of Chemical Education at California State University Fresno, specializing in the impact of curriculum and learning technologies on student experimental practices. He has published in several leading peer-reviewed science and chemistry education journals and has reviewed over 100 journal submissions. He is a member of the American Chemical Society, the American Educational Research Association, and the California Association of Science Educators. Dermot earned his Ph.D. at the University of Limerick, Ireland, and completed postdoctoral work at Victoria University of Wellington, New Zealand, and at the University of California, Berkeley.

Yehudit Judy Dori is a Professor at the Faculty of Education in Science and Technology at the Technion, Israel Institute of Technology and was the Dean of the Faculty during 2015–2020. Professor Dori holds an MSc in Life Sciences and a PhD in Science Education from the Weizmann Institute of Science. She was intermittently a Visiting Professor or Scholar at Massachusetts Institute of Technology, MIT, between 2000 and 2020. Her research interests encompass learning in technology-rich environments, assessment, visualizations, 21st century STEM skills, and metacognition at the high school and university levels. She has mentored about 80 graduate and postdoctoral students, published more than 140 papers in peer-reviewed journals and book chapters, and co-edited two books on cognition and metacognition in STEM education, published by Springer in 2012 and 2018. Prof. Dori received the 2020 NARST Distinguished Contributions to Science Education through Research Award (DCRA) for her exceptional research contributions and strong national and international community engagements.

Julia Eckhard is currently a Ph.D. student at the Institute of Chemistry Education at the Justus-Liebig-University in Giessen, Germany. She studied chemistry, biology, and educational sciences for teaching at the high school level and graduated by the first *Staatsexamen*. Her research focuses on mechanistic reasoning, problem-solving strategies and designing meaningful instructions to foster students' learning in organic chemistry.

Rachel Eidelman is founder, manager, and a teacher for the 'chemistry online' project in the Davidson Institute in the Weizmann Institute of science. Dr Eidelman leads the design, development, and the research of the COBLE programme. She is an active high school chemistry teacher, and an academic lecturer.

Ingo Eilks FRSC is a Professor in chemistry education at the University of Bremen. His research interests encompass, among others, learning with and about digital media and education for sustainable development.

Ola Faris is a high school chemistry teacher in Peki'in, a Druze village in the Galilee, Israel. She received her B.Sc. and Ed. in 2010 with a specialization in Chemistry and Environmental Science, and her M.Sc. in science education, both from the Technion, Israel Institute of Technology. Her Master's thesis, supervised by Prof. Yehudit Judy Dori and Dr Orit Herscovitz, is entitled

"Adapting and implementing online assignments: Reflections of pre- and in-service chemistry teachers." Ola is leading a special project, Pre-ATIDIM, which encourages excellent students from underrepresented groups to acquire a high-quality matriculation certificate, with a focus on STEM disciplines.

Marcel Frailich holds a PhD in Science Education from the Weizmann Institute of Science. Since 1982, she has been teaching chemistry and at the same time has developed interactive online learning environments for middle and high school students, while guiding teachers and implementing the environments in the field. She was a partner in the development of the "General Chemistry and Industrial Chemistry at the Service of Mankind" website and the "We Have Chemistry!" National Projects Competition. She researched the impact of teaching and learning high school chemistry in an online environment through this website.

Fun Man Fung earned his postgraduate degrees from NUS and Technical University of Munich. As a passionate Singaporean educator, he has written on the Channel News Asia (CNA), *The Straits Times*, and *The Conversation*. He is the co-editor of two books: (1) *10 Things You Must Know About the International Chemistry Olympiad (IChO)* and (2) *Technology-Enabled Blended Learning Experiences for Chemistry Education and Outreach* (Elsevier). He served on the IChO steering committee for two terms (2014–2018). In 2022, Fun Man was elected council member of the Singapore National Institute of Chemistry (SNIC), and he represented SNIC on the IUPAC Committee on Chemistry Education since 2020. Since 2021, Fun Man served as a member of the Editorial Advisory Board at the *Journal of Chemical Education*, ACS Publications.

Libby Gerard is a Research Scientist and Lecturer in the University of California, Berkeley Graduate School of Education. Her research examines how innovative learning technologies can capture student ideas and help teachers use those ideas to make decisions about classroom instruction. Libby's recent projects explore the use of automated scoring of student written essays using natural language processing to provide immediate guidance to students as they progress through an inquiry project, and, to support teachers' customization of instruction both in real-time and between lessons. Libby's research is published in leading peer-reviewed journals including *Science* and the *Review of Educational Research*.

Nicole Graulich is an Associate Professor of chemistry education at the Justus-Liebig-University Giessen and director of the Institute of Chemistry Education. The current research of her group focuses on investigating instructional designs that influence meaningful learning in organic chemistry, by elucidating students' perception of organic representations with new technologies and the nature of mechanistic reasoning. In 2016, she received the prestigious Ars legendi award for teaching excellence. Since 2019, she is an Associate Editor for Chemistry Education Research and Practice.

Jia Yi Han graduated from the NUS with a B.Sc. in Chemistry in 2020 amidst the Covid-19 pandemic. He joined NUS as a Research Assistant through the SG United Traineeship scheme (a Singapore Government-funded scheme to train recent graduates during the employment uncertainty brought about by the pandemic) in the Department of Chemistry under the tutelage and supervision of his former environmental chemistry lecturer. During his traineeship, he assisted his supervisor in areas such as pedagogical research and videography.

Alina Heckenthaler is a full-time chemistry teacher at the Cato Bontjes van Beek-Gymnasium in Achim, Germany. Her master thesis focused on the inclusion of social media into the chemistry classroom.

Deborah Herrington is a Professor and Chair of the Chemistry Department at Grand Valley State University in Allendale, MI, USA. She obtained her Ph.D. in Chemistry Education, from Purdue University and her M.Sc. in Organic Chemistry from the University of Waterloo. Herrington is the co-developer of the award-winning Target Inquiry teacher professional development program and the director of the grant funded Retaining and Inspiring students in Science and Engineering program that provides support for academically talented, low-income STEM majors. Her current research focuses on supporting student learning and self-assessment using technology, three-dimensional instruction and assessment, and STEM faculty development.

Orit Herscovitz is a Senior Teaching Fellow, Senior Researcher, and the Head of the STEM Novice Teachers' Induction and Mentors Unit at the Faculty of Education in Science and Technology at the Technion, Israel Institute of Technology. Her research interests include developing and assessing higher-order thinking skills in chemistry education, promoting professional development (PD) of pre- and in-service science and technology teachers, and assessing the implementation of advanced educational technologies in education. She published over a dozen articles in peer-reviewed international and national journals, co-authored several book chapters and several learning units for high school chemistry students for Israeli students.

Felix Ho is Associate Professor at the Department of Chemistry—Ångström Laboratory at Uppsala University, Sweden, where he is also the subject coordinator for education programs in chemistry. He is currently a Titular Member and Secretary of the IUPAC Committee on Chemical Education. His research interests span both chemistry education and biophysical chemistry, including student reasoning and conceptual understanding in combining both chemistry and mathematics in problem solving, systems thinking in chemistry education, as well as mechanistic studies of water oxidation in Photosystem II during photosynthesis.

Avi Hofstein is an Emeritus Professor with a PhD in chemistry from the Weizmann Institute of Science. He served as head of the chemistry group, head of the National Center for Chemistry teachers, and head of the Department of Science Teaching at the Weizmann Institute. For more than 50 years, Prof. Hofstein has been involved in all facets of the curricular process in chemistry. In 2014 he was awarded the ACS-CEI, in 2016 the ACS award for his contribution to research in chemistry education, and in 2017 he was awarded the NARST DCRA (Distinguished Contribution for Research Award).

Muhamad Hugerat is a Professor of Science Education; Ph.D. in physical chemistry from The Hebrew University, Jerusalem. Was Vice-President of the Academic Arab College for Education (AACE) in Israel—Haifa, Dean of the science education department, Dean of Continuing Education Department. Editor of "DARUNA"—journal of AACE. Researcher in the centre of science and environmental education—The Galilee Society. He has conducted research in many areas of science education (*e.g.*, learning environment, laboratory work, science education at all age levels, microscale and nano laboratories, teaching science through research, argumentation, and sustainability in science education, history, and philosophy of science).

Johannes Huwer is a Professor in science education—with special emphasis on chemistry education—at the University of Konstanz. His research interests are digital learning in chemistry and science classes, education for sustainable development in formal and non-formal learning environments, individualization, and transformation processes of science education.

Krenare Ibraj is currently an undergraduate research assistant in the research group of Nicole Graulich at the Institute of Chemistry Education at the Justus-Liebig University in Giessen, Germany. She studies chemistry, English and educational sciences for teaching at the high school level.

Rhiannon Iha grew up in the Pacific Northwest and obtained a Bachelor's degree in Chemistry at the University of Puget Sound. She completed an M.Sc. in organic chemistry at Cornell University, and a Ph.D. in organic chemistry and polymer science at Washington University in St Louis. She teaches organic chemistry for the Post-Baccalaureate Pre-medical program, and serves as Chemistry Subject Librarian, focusing on increasing chemistry information literacy. Rhiannon has taught courses online and face-to-face, employing online resources for "flipping" the classroom and educating students on responsibly using resources, references, and literature in an accountable manner.

Maria Joselevich, Ph.D., is a Professor at the Institute of Engineering and Agronomy, Universidad Nacional Arturo Jauretche, Argentina where she teaches a general chemistry course for engineers. She also teaches a graduate-level course, Teaching Chemistry at Secondary and Superior Levels, at the School of Education, Universidad de San Andrés, Argentina. Dr Joselevich holds a Ph.D. and a Bachelor's degree in chemistry and a major in education and ICT. Her research interests include the study of how recent technologies such as remote laboratories and audiovisual resources may enhance learning at Secondary and University levels. She has worked in technology integration in science education both face-to-face and online settings.

Miri Kesner holds a PhD in Science Education from the Weizmann Institute of Science and served as Head of The Israeli Chemical Industry-Education Link Centre at The Weizmann Institute of Science for more than 20 years. She initiated and managed the development and implementation of the website "General Chemistry and Industrial Chemistry at the Service of Mankind" and the "We Have Chemistry!" National Projects Competition up to 2018. She has written textbooks on the chemical industry and was involved in their implementation. She also was involved in the professional development of science and mathematics teachers at the Davidson Institute of Science Education.

Joseph Krajcik serves as Director of the CREATE for STEM Institute and is a University Distinguished Professor and the Lappan-Phillips Professor of Science Education at Michigan State University. His work focuses on improving science education by engaging in innovation while working with science teachers to design and test learning environments to reform science teaching practices and research student learning. He received many professional awards and recognitions for his contributions to science education through the years. In 2010 Joe received the Distinguished Contributions to Science Education Through Research Award from NARST, and in 2020, he was elected to the National Academy of Education and received the prestigious McGraw Prize in Pre-K–12 Education in recognition of his many accomplishments and contributions to transforming education.

Joachim Kranz has been a teacher of chemistry and physics at the High School since 1981, and from 2009 he was a senior schoolteacher for natural sciences and economics, labour, and technology at the Senate Department for Education, Youth and Family in Berlin. He is a book author and further education trainer as well as co-author of the science standards for the middle school graduation in chemistry of the KMK Germany (2004) and currently a doctoral student at the Humboldt-Universität zu Berlin.

Axel Langner is currently a Ph.D. student at the Institute of Chemistry Education at the Justus-Liebig-University in Giessen, Germany. He studied biotechnology/biopharmaceutical technology and received a B.Sc. Furthermore, he studied chemistry and biology for teaching at the high school level and graduated by the first *Staatsexamen*. His research focuses on representational competence, eye-tracking and eye-gaze stimulated retrospectives in organic chemistry.

Rea Lavi, PhD, is a lecturer and curriculum designer with the Department of Aeronautics and Astronautics and with the New Engineering Education Transformation program, both in the School of Engineering, Massachusetts Institute of Technology (MIT), Cambridge, MA, USA. He is also expert-in-residence at the MIT Abdul Latif Jameel World Education Lab. Dr Lavi designs and teaches undergraduate courses on approaches to ill-defined problems. His research focus is on systems thinking, problem formulation, and creative ideation in engineering education. Dr Lavi received his PhD from the Faculty of Education in Science and Technology, Technion – Israel Institute of Technology, Haifa, Israel. Dr Lavi is the inventor of the SNAP Method® for structured creative problem-solving (US and UK trademarks).

Gwendolyn Lawrie is an Education-Focused Professor in the School of Chemistry and Molecular Biosciences (SCMB) at the University of Queensland (UQ). Her research considers how students learn through collaborative and self-regulated inquiry-based tasks in large classes to support diversity particularly in blended environments. Gwen has been the recipient of multiple institutional and national awards for her teaching and outstanding contributions to chemical education.

Sebastian Leach has recently completed an undergraduate science degree at The University of Sydney, culminating in the submission of an exemplary Honor's thesis in distance education and remote learning of chemistry. Deeply interested in education, Sebastian intends to complete a Master's degree in teaching to continue in this field and apply his enthusiasm and knowledge to teaching future students.

Marcia C. Linn is the Evelyn Lois Corey Professor of Instructional Science in the Graduate School of Education, University of California, Berkeley, specializing in science and technology. She is a member of the National Academy of Education and a Fellow of the American Association for the Advancement of Science, the American Psychological Association, the Association for Psychological Science, the American Educational Research Association, and the International Society of the Learning Sciences. Linn earned her Ph.D. at Stanford University where she worked with Lee Cronbach. She spent a year in Geneva with Jean Piaget and was a Fulbright Professor in Israel.

Jia Luo obtained her B.Sc. in chemistry at Xiamen University, China, and her Ph.D. in chemistry at Washington University. She is a Senior Lecturer, teaching Introductory General Chemistry, and the Director of the General Chemistry Transition Program. She oversees the Chemistry Peer Mentoring Program to provide support to first-year students. Dr Luo co-teaches Women in Science. She is part of the HHMI Inclusive Excellence Steering Committee, Undergraduate Work Committee of the Chemistry Department, Advisor for Washington University STEM Education Association, helps with the Catalysts for Change STEM outreach program, and is interested in education research.

Bryn Lutes is a Senior Lecturer and General Chemistry Lab Director at Washington University. After completing her Ph.D. in chemistry, she spent several years as an Assistant Director at The Teaching Center at Washington University in St Louis where she supported faculty in incorporating active-learning techniques and technology in their classrooms. Bryn also teaches General Chemistry for post-baccalaureate premedical students and Quantitative Reasoning.

Debora Marchak is the Head of the National Centre for Chemistry Teachers, at the Department of Science Teaching, Weizmann Institute of Science. She holds a PhD in Physical Chemistry from Tel Aviv University and completed her postdoctoral studies on Surface Chemistry at the Weizmann Institute of Science. Since 2013, Debora has been a chemistry educator. In parallel, she developed an arts-integrated approach to teaching chemistry, and she conducts research on the influence of implementing the approach. Her research interests are arts-integrated science education, neuro-pedagogy, and visualization in chemistry. Between 2018 and 2020, she was head of the "We Have Chemistry!" National Projects Competition.

Agustina Martínez, B.S., holds a Bachelor's degree in science teaching specialized in biology didactics, and majors in biology teaching for a scientific culture and science teaching. She is a Professor at the Facultad de Exactas, Universidad Nacional de La Plata, Argentina, in a university teaching initiation course, which provides training to new teachers entering the institution. She is an Assistant Professor at the Health Institute, Universidad Nacional Arturo Jauretche, Argentina, where she teaches biology. Her research interests include the didactic use of virtual learning objects.

Elizabeth Mavhunga is a Full Professor and a rated researcher in Science Education at Wits University, South Africa. Her research is focused on developing teachers' professional knowledge for teaching the difficult science content using a theoretical construct called Topic Specific Pedagogical Content knowledge (TSPCK). Her research contributions in this area have been acknowledged with two prestigious research awards from the two national professional bodies, SAARMSTE and SACI, respectively. Internationally, her work has contributed to the development of the famous model for describing teachers' professional knowledge called the Refined Consensus Model of PCK. Professor Mavhunga leads a research team pioneering the digitization of TSPCK. The digitized TSPCK empowers teachers to design and create engaging digital Science lessons in formats such as teaching videos and animations, the much needed skills for 4 IR participation. She has graduated many postgraduate students, delivered plenary talks at international conferences, and authored several book chapters and journal articles in accredited publications. She is regarded as one of the leading female professors of colour in science teacher education in South Africa.

Valery Milner is an Associate Professor of physics, at the University of British Columbia (UBC) Department of Physics and Astronomy in Vancouver, Canada. He conducts research in the field of chemical physics, specifically in the area of atomic molecular and optical physics. He has also been teaching secondary, undergraduate, and graduate physics courses, pioneering the implementation of data collection, visualization, and analysis tools in large and small classrooms.

Marina Milner-Bolotin, Ph.D. is a Professor of science, technology, engineering and mathematics (STEM) education at the Department of Curriculum and Pedagogy at the University of British Columbia (UBC), Vancouver, Canada. She studies how modern technologies can facilitate the development of teachers' capacity for implementing active learning environments.

Pablo D. Moro is a chemistry and physics high-school teacher. He is interested in teaching chemistry using NICTx for science learning. He has worked in the inclusion of ICT resources into Secondary school teaching and STEM learning materials. He has done research in the field of Organic Synthesis at the Department of Organic Chemistry, FCEyN, Universidad de Buenos Aires, Argentina, publishing articles and attending congresses of the specialty.

Alice Motion is the leader of the Science Communication, Outreach, Participation and Education Research Group at the School of Chemistry, University of Sydney. Alice is a research chemist and award-winning educator and science communicator who aims to connect people with science through her teaching, research, and practice.

Jacinta Mutambuki is an assistant professor and the Edward Bartlett Professor of Pedagogy in Chemistry, Department of Chemistry at Oklahoma State University. Dr Mutambuki's research involves investigating the effects of evidence-based classroom instructional approaches on students' cognitive and affective outcomes, and development of assessment instruments for characterizing instructional practices in college STEM classrooms. Dr Mutambuki teaches the General Chemistry courses, and a graduate-level course, Teaching and Research Innovations in Chemistry and STEM Education, aimed at preparing future STEM faculty on evidence-based instructional practices.

Caroline Muteti is a chemical education doctoral student in the chemistry department at Oklahoma State University. Caroline joined OSU in Fall 2019. Caroline's research interests focus on the infusion of metacognition instruction and Classroom Assessment Techniques (CATs) in the General Chemistry courses and assessing the effects of these approaches on performance and students' perceived learning experiences.

Courtney Ngai is the Associate Director of the Office of Undergraduate Research & Artistry at Colorado State University, where she manages a lab program supporting underrepresented students in STEMM. Her previous research focused on student thinking when solving problems using chemistry, and she has collaborated with K-12 teachers and university researchers on the Chemical Thinking Learning Progression project. Courtney's current research and facilitation work focuses on organizational culture and change in higher education.

Roee Peretz received his B.Sc. with plant sciences major (B.Sc. Agr.) from the Hebrew University of Jerusalem, and a master's degree in biology education at Oranim Academic College, *Summa Cum Laude*. He is currently pursuing a Ph.D. with the Faculty of Education in Science and Technology, Technion—Israel Institute of Technology, under the supervision of Prof. Yehudit Judy Dori. His research interests include systems thinking, conceptual modeling, and interdisciplinary learning, particularly in science and engineering online teaching and learning.

Reyne Pullen is currently employed as an education-focused lecturer in the School of Chemistry, University of Sydney. Reyne has experience in delivering online and blended learning experiences, and designing course-level blended learning models, both at the tertiary level. Reyne has also worked within the secondary education sector, teaching both science and mathematics.

Marc Rodemer is a research scientist at the Department of Chemistry Education at the University of Duisburg-Essen, Germany. His research interests lie on cognitive processes of conceptual learning with regard to models and representations from primary to tertiary science education.

Marissa Rollnick is Professor Emeritus in science education at Wits University. She holds a doctorate from Wits University and is a specialist in academic development and science education. Her professional career includes appointments as teacher, teacher educator, lecturer, and professor in chemistry education. She currently also has appointments at the Universities of Pretoria and Johannesburg. Her research areas include language in science education, student access to science in higher education and pedagogical content knowledge. She has published over 70 refereed publications. She has received several awards, most recently the distinguished contribution to research award from NARST in the USA.

Peter Rutledge is Professor of chemistry at the University of Sydney. He received his BSc and MSc from the University of Auckland and D.Phil. from the University of Oxford. Peter has held lectureships in Oxford and Dublin and visiting fellowships in Warwick and Cambridge. His current research interests include chemistry education, natural products chemistry, antibiotics discovery, molecular sensors, biosynthesis and biocatalysis. Peter is a Fellow of the Royal Society of Chemistry, a Fellow of the Royal Australian Chemical Institute, and a Senior Fellow of the Higher Education Academy.

Siegbert 'Siggi' Schmid graduated with a Ph.D. in inorganic chemistry from the University of Tübingen, Germany, and after post-doctoral research at the ANU, completed a Habilitation at his *alma mater*. He currently serves as Chair of the Chemistry Education & Communication Research theme at the School of Chemistry of the University of Sydney. His education research investigates changes to teaching practices that enhance the student experience and learning outcomes, with particular emphasis on inclusive teaching. He led an OLT-funded national project on Assessing the Assessments and was awarded the 2019 Fensham Medal of the Royal Australian Chemical Institute for sustained contributions to chemistry education in Australia.

Anat Shauly is a doctoral student at the Faculty of Education in Science and Technology at the Technion—Israel Institute of Technology, under the supervision of Dr Shirly Avargil. Her Ph.D. studies center on the professional development of chemistry teachers' communities. She holds B.Sc. Ed. in chemical education, a B.Sc., and M.Sc. degrees in chemical engineering from the Technion. Prior to joining the Faculty of Education, she had worked as a research assistant at the Rabin Desalination Laboratory at Technion, and as a characterization engineer at Semi-Conductor Devices (SCD).

Gabriella Shwartz was a postdoctoral researcher at the Faculty of Education in Science and Technology at the Technion—Israel Institute of Technology, under the supervision of Dr Shirly Avargil. During her postdoc her research dealt with chemistry teachers' assessment knowledge as reflected in educational escape rooms. She received her Ph.D. in 2019 from the Technion, focusing on choosing a teaching career, preparing novice STEM teachers, and their integration into the teaching profession. She is currently teaching academic courses in chemistry education and high school students who major in chemistry, and guiding novice chemistry teachers in their first years of teaching.

Yael Shwartz is a scientist at the Weizmann Institute of Science, incumbent of the Maurice Levy Staff Scientist Chair in the Science Teaching department. In 2014, she was awarded the ACS-CEI. She is the co-founder and researcher of the COBLE program. She is the head of the National Center for Science Teachers and Leads PeTeL—Personalized Teaching and Learning online environment for middle-school science.

Antje Siol is an analytical chemist. For many years, she works in outreach programs of the University of Bremen, especially in the design and operation of non-formal laboratory learning environments for secondary school students in the field of education for sustainable development.

Jaclyn J. Stewart is a Professor of teaching in the Department of Chemistry at the University of British Columbia in Vancouver, Canada. As an educational psychologist and chemistry educator, she is passionate about helping students learn and inspiring educators to transform curriculum and instructional approaches. Her teaching specialities are introductory and non-majors organic chemistry, as well as science communication. Dr Stewart also leads a Canada-wide research Consortium investigating equity in first-year science courses. The consortium brings educators and researchers together to implement and gather evidence regarding inclusive teaching practices with the goal of addressing systemic inequities.

Gabriela Szteinberg obtained her B.A. in chemistry at Lawrence University, and her Ph.D. in chemical education at Purdue University, with postdoctoral work at University of Massachusetts Boston. She has experience working with peer mentoring programs, such as Peer-Led Team Learning (PLTL). Dr Szteinberg is now an Assistant Dean for the College of Arts & Sciences at Washington University in St Louis, for which she serves as a four-year, Chemistry, and pre-health advisor, and progress counselor. Dr Szteinberg has a passion for helping students thrive and succeed, by promoting a growth mindset and resilience, through collaboration with peers and instructors.

Marina Tal is a Researcher at the Technion Research and Development Foundation, Haifa, Israel. She received her B.Sc. in chemistry in 2006, M.Sc. in physical chemistry in 2009, M.Sc. in chemistry education, *summa cum laude*, in 2018, and Ph.D. in water chemistry in 2014. Since October 2020 she has been a postdoctoral researcher at the Mofet Institute, Tel Aviv, Israel. Her research interests include water quality, water and wastewater treatment, teachers' knowledge types, and STEM teachers' sustainability and systems thinking.

Vicente Talanquer is a University Distinguished Professor in the Department of Chemistry and Biochemistry at the University of Arizona. His current research focuses on undergraduate chemistry education. He has published over 125 peer-reviewed and invited papers where he has explored the conceptual difficulties that students face when learning chemistry and the effect of different teaching strategies on student understanding. He has also investigated prospective science teachers' reasoning and practices. Dr Talanquer has applied the results of his educational research to the development of innovative curricula for undergraduate chemistry education. He has received various awards during his academic career, including the 2021 ACS Award for Achievement in Research for the Teaching & Learning of Chemistry by the American Chemical Society.

Rüdiger Tiemann studied chemistry and physics for high-school teaching. Since 2005 he has been the head of the chemistry education group at the Humboldt-Universität zu Berlin. Prof. Tiemann conducts research on problem solving and mental modeling as well as on digital teaching and assessment methods (www.tiemannlab.de).

Jan Mathis Tietjen is a full-time chemistry teacher at the Wilhelm-Kaisen-Oberschule in Bremen, Germany. His master thesis dealt with teachers' views on social media in the chemistry classroom.

Marcy H. Towns is the Bodner-Honig Professor of Chemistry and Director of General Chemistry at Purdue University. She is a member of the National Academies of Science, Engineering and Medicine Board on Science

Education. She received the 2017 American Chemical Society Award for Achievement in Research for the Teaching and Learning of Chemistry, the 2017 James Flack Norris Award for Outstanding Achievement in the Teaching of Chemistry from the Northeast Section of the American Chemical Society, and the 2019 Nyholm Prize in Education from the Royal Society of Chemistry, in addition to numerous teaching awards from Purdue University. She is a fellow of the Royal Society of Chemistry, the American Chemical Society, and the American Association for the Advancement of Science.

Maya Usher is a Research Associate at the Faculty of Education in Science and Technology, Technion – Israel Institute of Technology. She received her PhD in Learning Technologies from Technion in 2019. During 2020–2021, she has been a postdoctoral researcher at the School of Education at Tel Aviv University, Israel. Her research interests include designing and evaluating learning processes in diverse online learning environments.

Denise van der Merwe is currently completing her Ph.D. in science education at Wits University, South Africa. She works as an acting Principal at a secondary school in Gauteng, South Africa. Her studies aim to empower teachers to create pedagogically reasoned teaching content on digital platforms using animations. She holds a Master's degree in science education, as with her Honours degree, *cum laude* despite being a part-time student from the same University. She began her teaching career in 1999, teaching mathematics, life sciences and physical sciences, which she continued at the time of studies. She has co-authored two articles focusing on topic-specific pedagogical content knowledge in science education.

Merav Versano received her B.Sc. in chemical engineering in 1995 and her M.Sc. as a MABATIM II–Views graduate, in 2020, both from Technion, Israel Institute of Technology. Her Master's supervisors were Prof. Yehudit Judy Dori and Dr Orit Herscovitz. She is the head of a team of programmers at a high-tech company, and also works at the Davidson Center, Weizmann Institute of Science, Rehovot, Israel. At this center, she teaches chemistry virtually, prepares Arab high-school students for matriculation exams in machine learning, and participates in an Intel-led pilot project to prepare middle school students to be young scientists.

Shane Wilkinson is a lecturer in the School of Chemistry, University of Sydney. He has a keen interest in the adoption of technology to enhance how we teach and deliver chemistry in both classroom and laboratory environments. An example is augmented reality (AR) and how it can be used as a tool in education or as means of inclusivity for disadvantaged students. Shane is adept in the use of "teaching analytics" software to deliver personalised student experiences in large scale teaching environments. He also has a passion for laboratory pedagogy with a focus on authentic and competency-based laboratory assessments.

Stephen George-Williams is an education-focused lecturer based at the School of Chemistry, University of Sydney. His research has explored several areas including the purpose of the chemistry teaching laboratory, the relationships built between students and teaching assistants in the teaching laboratory, and the use of virtual reality (VR) to support teaching chemistry in the classroom.

Kim Yongbeom is a computer science student at the National University of Singapore. While he studies computing, his keen interest in chemical education and the learning sciences has led him to write in Channel News Asia (CNA), the Journal of Chemical Education and the IEEE International Conference of Teaching, Assessment and Learning for Engineering (TALE). Lastly, as an international student in Singapore, he hopes to be part of education systems that are more diverse, accessible, and inclusive.

Christian Zowada is an associate member of the chemistry education group of the University of Bremen and a full-time chemistry teacher at the Gymnasium Osterholz-Scharmbeck, Germany. His research focuses on curricular innovations in the context of education for sustainable development.

Dana Zreke is an M.Teach Senior in education of chemistry at The Arab Academic College for Education in Israel, Haifa. She received her B.Sc. degree in chemistry from the Ben-Gurion University of the Negev in 2020. Her research interests include sustainability and environmental education.

Contents

Advances in Chemistry Education Series No. 11
Digital Learning and Teaching in Chemistry
Edited by Yehudit Judy Dori, Courtney Ngai and Gabriela Szteinberg

Published by the Royal Society of Chemistry, www.rsc.org

CHAPTER 1

Digital Learning and Teaching in Chemistry—What We Know and What We Wish to Investigate Further

Y. J. DORI*[a,b], R. PERETZ[a], C. NGAI[c] AND G. SZTEINBERG[d]

[a]Faculty of Education in Science and Technology, Technion, Haifa, Israel; [b]Samuel Neaman Institute for National Policy, Haifa, Israel; [c]Office of Undergraduate Research and Artistry, Colorado State University, Fort Collins, CO, USA; [d]College Office, College of Arts & Sciences, Washington University in St. Louis, St. Louis, MO, USA
*E-mail: yjdori@technion.ac.il

1.1 Introduction

With education rapidly moving into the digital age, the role of the teacher and lecturer is significantly evolving in an imperative attempt to align with technological advances. In chemistry education, educators face many challenges due to changes in the learning environment, including administering and monitoring the learning process. Teachers and researchers around the world are looking for creative and efficient ways to move from teacher-centred to student-centred classrooms, and from summative, assessment *of* learning to formative, assessment *for* learning methods,[1] especially in online and digital learning environments.[2] Therefore, gathering the most up-to-date

Advances in Chemistry Education Series No. 11
Digital Learning and Teaching in Chemistry
Edited by Yehudit Judy Dori, Courtney Ngai and Gabriela Szteinberg

Published by the Royal Society of Chemistry, www.rsc.org

techniques for teaching and evaluating chemistry learning in digital environments is essential.[3]

In this book, we provide high-school teachers and university and college lecturers with both theory and practice to support their pedagogical decisions regarding online teaching, learning, and assessment methods for chemistry courses. There is great potential in deploying digital technologies in purposeful ways, giving chemistry teachers and lecturers a variety of options for a comprehensive examination of learners' progress.[4] Digital technologies can also influence how new assessment criteria are set to match online learning and teaching methods and their unique characteristics for teaching science in general and chemistry in particular.

Technology has been integrated into classrooms with varying degrees of success even before the COVID-19 pandemic, and the pandemic has forced educators to turn to digital learning to support their students. This book is motivated by a desire to collect and share effective and cutting-edge practices in teaching chemistry digitally, especially those that have been recently developed due to the pandemic. Furthermore, teaching chemistry digitally has the potential to bring greater equity to the field of chemistry education and foster access to high quality learning. This book will contribute to that goal. With more than 70 authors from nine countries, our book presents the ways digital learning and teaching of chemistry are being implemented around the world. At the same time, because of the universal nature of technology, the topics shared in this book will apply to chemistry teachers, researchers, and learners globally.

1.2 Word Cloud Analysis of the Book Chapters

A world cloud was generated to summarize the main themes of the chapters in this book. The most frequent words in the full-text chapters (excluding abstracts) were *students* (more than 2000 times) and *learning* (more than 1100 times) followed by these words in descending order, from more than 850 times to fewer than 270 times: *teachers*, *chemistry*, *teaching*, *groups*, *online*, *knowledge*, *digital*, *feedback*, *support*, *content*, *environmental*, and *social* (see Figure 1.1). *Knowledge* was included in the top-ten list of both full-text and abstract analyses and the word *content* was in the top-fifteen of the full text. Surprisingly, given the prevailing educational discourse surrounding students' *skills*, this word was not included in the top 15, emerging with less frequency. The word *practices* was included toward the end of the list of the 55 most common words in the abstracts of the book (see Figure 1.2). Accordingly, a significant part of this book deals with skills, competencies, and practices, in addition to content knowledge. Concepts, understanding, ideas, experiences, hands-on activities, modelling, and information and media literacy are all shown in the full-text word cloud illustration (see Figure 1.1).

The word cloud analysis of the abstracts (Figure 1.2), which appear online, is similar to that of the full-text analysis for the top 10 words, in a slightly different order. However, it is interesting to note that in both

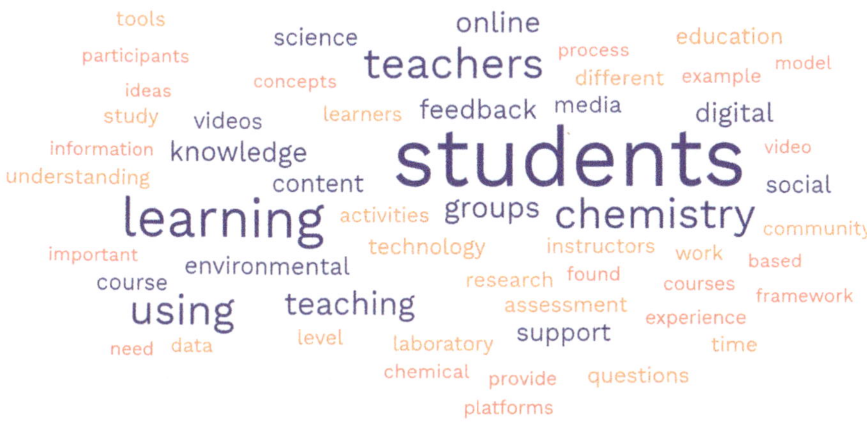

Figure 1.1 Word cloud analysis of the full text.

Figure 1.2 Word cloud analysis of the abstracts.

clouds *students* and *learning* were the most widely mentioned words. When comparing the 20 most frequently used words, the word *pandemic* is relatively frequent in the abstracts, evenly distributed among all book themes, whereas in the full-text analysis this word does not even make the top 100 (which is also the trend for *COVID* and *COVID-19*). The COVID-19 pandemic appears to have more of an impact on the transition to digital online learning than is explicitly indicated in the study's background, motivation, and structure. The fact that the abstracts were written at the first stages of the pandemic and the chapters only about a year later when the transition to digital media had already begun and was practically a *fait accompli*, might explain that gap.

1.3 Contribution

This book aims to benefit a wide range of relevant populations—school and higher education educators, laboratory instructors, educational content developers, education researchers, and policymakers. This book has several key contributions that can be beneficial for any of the above populations: research-oriented, theoretical, practical, and methodological contributions.

1.3.1 Research-oriented and Theoretical Contributions

This book has notable contributions to educational research and theory in a variety of aspects and levels. Highlighting various educational issues that have arisen in recent years, certainly since March 2020 when the pandemic began to erupt, may give other researchers opportunities to explore new directions, particularly in the aspect of digital chemistry learning in the post-pandemic era. These new directions are evidence-based, so continuing the research avenues shared in this book may yield more important insights that can generate more practical contributions.

In the first theme of this book, *Best Practices of Teaching and Learning Digitally*,[5] several chapters focus on expanding existing educational theories regarding digital resources for the learning and teaching of chemistry. They add fresh perspectives and characteristics related to this kind of learning, both in a student-centred and a teacher-centred approach. Other chapters included in the second theme, *Digital platforms for increasing inclusion in chemistry education*,[6] add new knowledge to what we already know about the various aspects of inclusion in online chemistry learning, whether through YouTube, social media, or massive open online courses (MOOCs). The third theme, *Using Visualization and Laboratory to Promote Learning in Science*,[7] is mostly focused on practical and methodological aspects in the application of visualization and technologies in chemistry education. However, a couple of chapters clearly illustrate the importance of relying on existing theoretical frameworks (such as knowledge integration), which can add significant value to the development of chemistry learning processes that integrate visualizations and technology. The following theme of *Digital Assessment*[8] includes a number of chapters that add new layers to existing theoretical frameworks dealing with different aspects and approaches to assessment. These chapters include topics such as formative feedback in laboratory activities, teachers' assessment knowledge, the need for extensive training in this regard, and the need to provide multi-modal supportive feedback to increase student engagement in learning. In the final theme, *Building Communities of Learners and Educators*,[9] all the chapters discuss community formation, bringing new theoretical viewpoints to the field of chemical education. This contribution is particularly significant considering the major disturbance caused by the COVID-19 pandemic.

1.3.2 Practical and Methodological Contributions

Several chapters in this book present studies of a distinctly practical and methodological nature, either by elaborating on different existing methodologies and technologies or by presenting results and insights that can and should be applied in educational practice. The third theme, which focuses on visualization and laboratory in chemistry education,[7] includes studies that are more application-based. In this theme, digital technologies that can be used in both remote and in-person chemistry learning are presented, such as virtual and augmented reality, instructional videos, and smartphone applications. In the chapters that make up this theme, specific guidelines for using similar digital platforms for other educational purposes (*e.g.*, for other chemistry topics or other disciplines) are described in detail and with full transparency.

The opening theme, *Best Practices of Teaching and Learning Digitally*,[5] illustrates how technical and theoretical principles are applied in converting in-person chemistry courses into a remote learning platform, and in creating an informal learning environment enriched with augmented and virtual reality in the context of recycling electronic devices. Other notable practical and methodological contributions found in the book include identifying the most crucial elements for successful blended learning for better utilization in education systems; adapting an open-access chemistry learning website to the new needs and limitations that emerge over the years; designing an adaptive digital environment for personalized chemistry learning; introducing a modelling tool and methodology that has been proven to facilitate the development of systems thinking; and laying down guidelines for an innovative online assessment that includes a two-stage test to encourage group interaction as a component of inclusive collaborative assessment. Most of these technological practices and methods can be used by any chemistry instructor, since the theoretical background, the generalizations and their validity, and the insights arising from these generalizations are all visible and detailed as required by evidence-based education.[10]

1.4 Recommendations for Future Studies

As mentioned in section 1.2, the word *skills* was among this book's 55 most frequently used words (see Figures 1.1 and 1.2) in the full-text. The frequency of the word *skills* was about one third of the word *knowledge* and about half of the word *content*. The relatively medium-level frequency of *skills*, also referred to as *competencies* or *practices*, throughout the book may indicate the importance of future studies that will focus more on these terms and find out how to connect them better to content knowledge in chemistry.

Another aspect where further research may be helpful to establish the necessary knowledge and tools in digitally assisted learning and teaching is teacher preparation and professional development (PD) programs. To support pre- and in-service teachers in using learning technologies to help students develop a deeper understanding of the importance of scientific ideas, more needs to be done in science teacher PD and pre-service training programs to smoothly incorporate technology into the curriculum. Research-based programs based on scientific theories are preferable to programs that lack proper research-theoretical foundations, which mandates more pertinent teacher-centred studies. Considering the findings presented in the *Digital Assessment* theme,[8] such training is particularly important regarding teachers' assessment tools and knowledge in online environments. The connection between assessment and online learning environments in the context of teacher training should be subjected to more research.

Regarding the inclusion of learners from different societies, needs, and backgrounds, additional studies are required to generate insights and knowledge adapted globally as well as to each country and each different target population. Adaptation is required to produce reliable research evidence for different target populations, especially in issues related to equity and inclusion—societal aspects of education. Some of the studies presented in the book can and should form the basis for similar studies among other populations, in order to expand existing theories or even establish new ones.

Acknowledgements

The editors wish to express gratitude to the chapter authors for their contribution to this book, and to Roee Peretz, a co-author in this chapter, for providing much needed editorial support and helping to bring this book to completion.

References

1. M. Birenbaum, K. Breuer, E. Cascallar, F. Dochy, Y. Dori, J. Ridgway, R. Wiesemes and G. Nickmans, *Educ. Res. Rev.*, 2006, **1**, 61–67.
2. A. Littlejohn, N. Hood, C. Milligan and P. Mustain, *Internet High. Educ.*, 2016, **29**, 40–48.
3. G. Shwartz, S. Avargil, O. Herscovitz and Y. J. Dori, *Chem. Educ. Res. Pract.*, 2017, **18**, 214–232.
4. L. D. Lapitan, C. E. Tiangco, D. A. G. Sumalinog, N. S. Sabarillo and J. M. Diaz, *Educ. Chem. Eng.*, 2021, **35**, 116–131.
5. See Chapter 2: V. Talanquer, in *Digital Learning and Teaching in Chemistry*, ed. Y. J. Dori, C. Ngai and G. Szteinberg, Royal Society of Chemistry, United Kingdom, 2023.
6. See Chapter 9: R. Blonder, in *Digital Learning and Teaching in Chemistry*, ed. Y. J. Dori, C. Ngai and G. Szteinberg, Royal Society of Chemistry, United Kingdom, 2023.
7. See Chapter 16: J. Krajcik, in *Digital Learning and Teaching in Chemistry*, ed. Y. J. Dori, C. Ngai and G. Szteinberg, Royal Society of Chemistry, United Kingdom, 2023.

8. See Chapter 22: C. Ngai, in *Digital Learning and Teaching in Chemistry*, ed. Y. J. Dori, C. Ngai and G. Szteinberg, Royal Society of Chemistry, United Kingdom, 2023.
9. See Chapter 28: M. H. Towns, in *Digital Learning and Teaching in Chemistry*, ed. Y. J. Dori, C. Ngai and G. Szteinberg, Royal Society of Chemistry, United Kingdom, 2023.
10. P. Davies, *Br. J. Educ. Stud.*, 1999, **47**, 108–121.

CHAPTER 2

Theme Introduction: Best Practices of Teaching and Learning Digitally

V. TALANQUER*

Department of Chemistry and Biochemistry, University of Arizona, Tucson AZ, 85721 USA
*E-mail: vicente@u.arizona.edu

2.1 The Promise of Digital Learning

Chemistry educators around the world have used diverse digital technologies to support and foster student learning for many years now. For example, we have taken advantage of animations[1] and interactive simulations[2] to help students build connections between the macroscopic world and the sub-microscopic models of matter used to explain, predict, and control the properties of chemical substances and processes. We have also developed a variety of video resources to guide student learning in flipped classrooms and other blended learning environments.[3] We now have access to a variety of applications for smartphones and tablets that facilitate the creation of multiple chemical representations,[4] and different groups of chemistry educators have developed virtual labs where students get to perform different experiments by manipulating virtual chemical reagents and equipment.[5] In recent years, chemistry educators have also begun to explore the educational potential of augmented and virtual reality applications.[6]

Advances in Chemistry Education Series No. 11
Digital Learning and Teaching in Chemistry
Edited by Yehudit Judy Dori, Courtney Ngai and Gabriela Szteinberg

Published by the Royal Society of Chemistry, www.rsc.org

Despite the availability of multiple and diverse digital resources for the teaching of chemistry, there has been limited discussion, reflection, and research on how to best take advantage of these educational tools to create well-integrated digital learning environments that actively and productively engage students in meaningful learning. Many questions remain unanswered on how to use available technological resources to facilitate chemistry learning by giving students more control over the time and place, as well as the path and pace of their studies. We would also benefit from having more and better insights into how to guide and support the work of chemistry teachers and instructors as they plan and implement lessons and embark on the assessment of student learning in the digital world.

Digital teaching and learning can be quite enriching but can also pose diverse challenges to both students and educators. In digital environments learning does not need to be restricted to the school day or the school year. The use of technological devices with internet access gives students the opportunity to learn at anytime, anywhere, and everywhere. Nevertheless, ensuring equal access for all students to digital technologies and Internet services is not easy and failure in this area may be the source of major educational disparities.[7] In digital environments students may gain agency and ownership in their learning as they get to make decisions on what, how, and when to learn. Nevertheless, making productive choices and actively engaging in the educational process often demand high levels of motivation, self-regulation, and self-efficacy that some learners may need to develop.[8]

Interactive and adaptive software opens the door for engaging and individualized learning that can be tailored to the needs and interests of every student. It creates the opportunity for students to progress at their own pace, following a learning path that suits their knowledge and skills. But designing and developing effective adaptive educational resources is costly and time-demanding, and it requires a level of technological expertise that most teachers do not possess. Interactive and adaptive technologies can provide teachers with valuable formative assessment data that they can use to adjust and adapt instruction to meet the needs of each student. Unfortunately, most teachers do not have the time and institutional support required to engage in the analysis of such data.

Teaching in digital environments creates opportunities for teachers to design and implement creative lessons that not only engage students in the analysis of central ideas and concepts in chemistry, but also connect those understandings with questions and problems of relevance in their surroundings. Learning in the digital world can be contextualized by focusing on current social, political, economic, and environmental issues being shared and discussed in social media that students access on a daily basis. Nevertheless, the design and development of these instructional tasks demand specialized pedagogical content knowledge that many instructors may need to acquire.

The educational challenges and opportunities associated with digital teaching and learning have been made all the more apparent in the wake of the COVID-19 pandemic which forced chemistry educators at all educational levels to transform curricula, instructional strategies, and assessment practices to work in digital environments. The challenges that chemistry instructors and students faced adapting to online teaching and learning were diverse, spanning from lack of access to the technological resources needed to meaningfully participate in educational activities to difficulties to translate practical experimental work into remote or virtual tasks. The pandemic exposed major inequities in access to resources and social capital that students need to advance in their studies,[7] as well as serious educational issues linked to conventional chemistry curricula, instruction, and assessment.[9] On the other hand, educational work under the pandemic has spurred many innovative approaches to the teaching of chemistry that have enriched the educational practices of many instructors and diversified the approaches to learning of multiple students.[10]

2.2 The Chapters within the Theme "Best Practices of Teaching and Learning Digitally"

Given our limited knowledge about chemistry teaching and learning in digital environments, this book is an invaluable contribution to chemistry education as it brings together the knowledge and experiences of diverse chemistry educators across the world. In particular, the first part of the book focuses on "Best Practices of Teaching and Learning Digitally" and the authors of this initial set of chapters describe, discuss, and reflect upon many of the educational opportunities and challenges mentioned in the previous section. These initial chapters address issues and present examples in three broad areas. Chapters 3 and 4 focus on the perspectives and lessons learned about digital teaching and learning by both chemistry students and instructors during the COVID-19 pandemic. Chapters 5 and 6 describe and discuss major characteristics of students and instructors that support learning and teaching in the digital world. Lastly, Chapters 7 and 8 describe innovative approaches to chemistry teaching and learning in digital environments.

Chapter 3 by Ho, Ibraj, and Graulich[11] summarizes the results of a research study focused on characterizing the perspectives of chemistry students and instructors about factors and conditions that support online teaching and learning. The authors characterized challenges that students and instructors faced as they transitioned from in-person to online teaching and identified actions and characteristics of learning environments that helped address such challenges. Student learning was affected by lack of social contact, difficulties in adapting to online learning, and lack of direct hands-on experiences in the laboratory, while instructors struggled to rapidly adapt their teaching to online environments with limited support. Core strategies to address these issues relied on the creation of conditions and structures that enhanced personal contact, community building, and collaboration.

In Chapter 4, Iha, Luo, Lutes, and Szteinberg[12] describe in detail how four large general chemistry and organic chemistry courses were adapted for fully remote and hybrid instruction due to the COVID-19 pandemic. The authors identify instructional strategies that students found particularly useful in their learning or improved student performance in the courses, such as lecture review questions targeting specific concepts introduced in videos, the integration of asynchronous and synchronous activities, and the use of pathway questions in combination with short videos in asynchronous online teaching. The educational knowledge gained through the evaluation of these four instructional adaptations, together with the insights presented in Chapter 4 about needed supports for effective online teaching and learning, should be of great value to chemistry instructors teaching both in-person and online, and to online instructional designers.

The ideas and results presented in Chapters 5 and 6 help us identify core characteristics of successful digital learners and instructors. In particular, Eidelman and Shwartz[13] in Chapter 5 identify predictors of student success in a chemistry online blended learning environment (COBLE). Students in this three-year online program participate in weekly synchronous sessions, complete asynchronous tasks, and engage in laboratory activities in various formats (home lab kit, virtual, and face-to-face two or three times a year). Researchers in this study built self-regulated learning (SRL) profiles of participating students characterizing, among other things, their attitudes and interest, their concentration and attention to academic tasks, and their motivation, diligence, self-discipline, and willingness to work hard. The authors show that students' SRL profile and their level of participation in synchronous activities are significant predictors of success in the blended learning environment as measured by student performance in a matriculation exam.

In Chapter 6, Mavhunga, Rollnick, and van der Merwe[14] direct our attention to the competencies that teachers need to develop, design, implement, and assess effective chemistry lessons in digital environments. In particular, the authors present a framework for digital topic-specific pedagogical content knowledge (d-TSPCK) that enables teachers to create and teach pedagogically transformed science content presented through online teaching and learning platforms. In this framework, teachers' knowledge is conceptualized as a resource for critical thinking and exploration beyond context, in contrast to knowledge as an end in itself. d-TSPCK is knowing connected to teaching activity that integrates digital teaching competence and considerations for learning through multimedia; it allows teachers to transform disciplinary content knowledge into comprehensive versions for digital learners. The authors illustrate the application of this type of knowledge through the analysis of an instructional activity on chemical equilibrium.

The two closing chapters within the theme "Best Practices of Teaching and Learning Digitally" present and describe two concrete examples of innovative approaches to the teaching of chemistry in digital environments. In Chapter 7, Huwer, Barth, Siol, and Eilks[15] summarize the core components of an activity designed to help students understand the chemistry behind the recycling of digital hardware. This task was designed for a non-formal learning environment, and it engages students in the analysis of practical information using augmented reality and virtual reality components. The activity fosters self-regulated and autonomous learning by allowing students to dismantle the components of an iPad modelled with augmented reality and then use interactive 3D animations and virtual labs to explore a variety of chemical processes. This digital resource has been built using a modular approach that facilitates enrichment, revision, and adaptation to different learning environments.

In Chapter 8, the last one in this set, Belova, Heckenthaler, Tietjen, and Zowada[16] illustrate how social media can be effectively used to engage students in meaningful chemistry learning. The researchers initially present and discuss the results of a study on teachers' perceptions and beliefs about the use of social media as an instructional tool. Their findings suggest that a significant number of educators are sceptical about and reluctant to the use of modern media tools in the science classroom. As a counterpoint to this position, the authors present and discuss an instructional activity in which social media is used to contextualize the learning of a chemistry topic (use of parabens as preservatives in commercial products) and create a need to know. Through this task, students learn not only chemistry with social media but also about social media itself and critically reflect on the nature of the information presented in those environments. Students' evaluations of the task were quite positive both in terms of interest in the chemistry topic and the use of social media to motivate and contextualize the learning of chemistry.

2.3 Final Comments

The collection of chapters under the theme "Best practices of teaching and learning digitally" provides critical insights for chemistry education in digital environments. They highlight the importance of identifying and characterizing major factors that affect the learning and teaching of the discipline in online settings, as well as the conditions and structures that support students' and teachers' activity in these environments. In particular, several of the studies point to the critical role that human interactions, a sense of community, and collaboration play in motivating, supporting, and sustaining productive engagement in the virtual world. They also outline the knowledge, skills, and dispositions that students and teachers at all educational levels need to have, or develop, to effectively work in and benefit from digital education.

In terms of student learning, several of the contributions in the next six chapters make explicit the importance of properly scaffolding student work and interactions in digital environments.[11,12] Successful students are aware and make use of the various resources available to them in online settings, know how to manage them, and have the ability to self-regulate their learning by being proactive, diligent, and metacognitive.[13] Thus, most students are likely to benefit from structured learning environments that facilitate task organization and time management, and that foster the development of active, interactive, and supportive communities of learners with common goals.[11] Student performance can also be enhanced by careful design of academic tasks that guide and direct students' attention to the analysis and discussion of the central ideas and concepts to be learned.[12,15,16]

In terms of teacher preparation, several of the studies within the first theme in this book provide insights into conditions, knowledge, and dispositions that chemistry instructors should have to plan, implement, and assess instructional activities that foster meaningful learning. Most instructors are likely to benefit from participation in communities of practice that are a source of knowledge, resources, and support for teaching online.[11] Teaching in digital environments demands specialized pedagogical content knowledge that integrates understanding of central ideas in the discipline, topic-specific pedagogy, understanding of multimedia learning, media literacy, and digital teaching competence.[14,16] It also requires the disposition to embrace diverse, and sometimes unconventional digital technologies and environments to develop curricular and instructional innovations that effectively exploit the educational affordances of these tools.[15,16] Teachers need time and institutional support to enhance their knowledge, reflect on their beliefs, and develop the skills needed to create productive learning environments and experiences in the digital world.

References

1. R. Tasker, Using multimedia to visualize the molecular world: educational theory into practice, in *A Chemist's Guide to Effective Teaching*, ed. N. J. Pienta, T. J. Greenbowe and M. M. Cooper, Prentice Hall, Upper Saddle River, NJ, 2004, ch. 16, pp. 256–272.
2. E. B. Moore, J. M. Chamberlain, R. Parson and K. Perkins, PhET interactive simulations: transformative tools for teaching, *J. Chem. Educ.*, 2014, **91**, 1191–1197.
3. M. K. Seery, Flipped learning in higher education chemistry: emerging trends and potential directions, *Chem. Educ. Res. Pract.*, 2015, **16**, 758–768.
4. D. Libman and L. Huang, Chemistry on the go: review of chemistry apps on smartphones, *J. Chem. Educ.*, 2013, **90**, 320–332.
5. N. Ali and S. Ullah, Review to analyze and compare virtual chemistry laboratories for their use in education, *J. Chem. Educ.*, 2020, **97**, 3563–3574.
6. S. Bernholt, K. Borman, S. Siebert and I. Parchmann, Digitising teaching and learning – Additional perspectives for chemistry education, *Isr. J. Chem.*, 2019, **59**, 554–564.
7. T. Dreesen, S. Akseer, M. Brossard, P. Dewann, J. P. Giraldo, A. Karme, S. Mizunoya and J. S. Ortiz, Promising practices for equitable remote learning: Emerging lessons from COVID-19 education responses in 127 countries, *Innocenti Research Brief*, Unicef 2020-10.
8. F. I. Winters, J. A. Greene and C. M. Costich, Self-regulation of learning within computer-based learning environments: A critical analysis, *Educ. Psychol. Rev.*, 2008, **20**, 429–444.

9. V. Talanquer, R. Bucat, R. Tasker and P. Mahaffy, Lessons from a pandemic: educating for complexity, change, uncertainty, vulnerability, and resilience, *J. Chem. Educ.*, 2020, **97**, 2696–2700.
10. See papers published in the "Journal of Chemical Education Special Issue on Insights Gained While Teaching Chemistry in the Time of COVID-19," 2020, **97**(9).
11. See Chapter 3: F. Ho, K. Ibraj and N. Graulich, in *Digital Learning and Teaching in Chemistry*, ed. Y. J. Dori, C. Ngai and G. Szteinberg, Royal Society of Chemistry, United Kingdom, 2023.
12. See Chapter 4: R. Iha, J. Luo. B. Lutes, and G. Szteinberg, in *Digital Learning and Teaching in Chemistry*, ed. Y. J. Dori, C. Ngai and G. Szteinberg, Royal Society of Chemistry, United Kingdom, 2023.
13. See Chapter 5: R. Eidelman and Y. Shwartz, in *Digital Learning and Teaching in Chemistry*, ed. Y. J. Dori, C. Ngai and G. Szteinberg, Royal Society of Chemistry, United Kingdom, 2023.
14. See Chapter 6: E. Mavhunga, M. Rollnick and D. van der Merwe, in *Digital Learning and Teaching in Chemistry*, ed. Y. J. Dori, C. Ngai and G. Szteinberg, Royal Society of Chemistry, United Kingdom, 2023.
15. See Chapter 7: J. Huwer, C. Barth, A. Siol, and I. Eilks, in *Digital Learning and Teaching in Chemistry*, ed. Y. J. Dori, C. Ngai and G. Szteinberg, Royal Society of Chemistry, United Kingdom, 2023, ch. 7.
16. See Chapter 8: N. Belova, A. Heckenthaler, J. M. Tietjen and C. Zowada, in *Digital Learning and Teaching in Chemistry*, ed. Y. J. Dori, C. Ngai and G. Szteinberg, Royal Society of Chemistry, United Kingdom, 2023.

CHAPTER 3

Supportive Aspects of Online Learning and Teaching—What Does a Rapid Transition Teach Us?

FELIX M. HO*[a], KRENARE IBRAJ[b] AND NICOLE GRAULICH[b]

[a]Department of Chemistry—Ångström Laboratory, Uppsala University, Box 523, 75120 Uppsala, Sweden; [b]Justus-Liebig-University Giessen, Institute of Chemistry Education, Heinrich-Buff Ring, 35392 Giessen, Germany
*E-mail: felix.ho@kemi.uu.se

3.1 Introduction—Digital Learning and Teaching

Digital learning allows more flexibility, accessibility, and convenience in organizing and structuring the learning process for learners and instructors. Students have more opportunities to control their pace, time, and how and what they learn. This opportunity, however, can represent a challenge as well if students are not equipped with the ability to self-regulate their learning in this more open space or if the learning environments provided do not match the needs of the students. For students, the challenges, and supportive aspects they encounter while switching to primarily online learning influence how they can profit from digital learning scenarios. Faculty members were required to act quickly in this situation to provide students with appropriate material for learning chemistry digitally. More than an IT help desk, a fit-for-purpose support infrastructure that aids faculty in integrating both

Advances in Chemistry Education Series No. 11
Digital Learning and Teaching in Chemistry
Edited by Yehudit Judy Dori, Courtney Ngai and Gabriela Szteinberg

Published by the Royal Society of Chemistry, www.rsc.org

digital technology and digital pedagogy is crucial (see Chapter 31 in this book by DeKorver and Herrington,[1] and Chapter 30 by Shauly *et al.*[2]).

The unique COVID-19 pandemic situation in 2020 accelerated the transition into complete online learning and teaching. This allowed us to observe and characterize students' and instructors' supportive factors. This exceptional circumstance served as a magnifying glass to showcase issues of digital learning and teaching and outline supportive factors that possibly could be adopted in future online learning (see also Pullen *et al.*[3]).

This chapter investigates factors that have facilitated successful digital learning and teaching in response to the abrupt transition to remote education. Two cases are presented herein. First, we characterize chemistry students' experiences from across four countries, and how they adjusted to and resolved challenges during this rapid transition to online chemistry learning. Second, we highlight an example of a multi-tiered network model for supporting faculty in digital teaching that goes beyond generic technical tips and tricks, instead catering for discipline-specific support at the local level that also connects to and informs broader initiatives at a higher institutional level. The results provide interesting insights and points of discussion on creating favorable pre-conditions for digital chemistry learning and teaching.

3.2 Case 1: Supportive Factors for Chemistry Students Transitioning to Online Learning

3.2.1 A Theoretical Perspective

The transition into online learning and teaching in chemistry classes happened suddenly and caused various challenges for students. Laboratory classes were cancelled and students needed to adapt to online learning and a changing way of organizing their chemistry learning.[4] Collaborative learning in classes or self-organized learning groups, which allowed students to check-in with each other, to help each other to be engaged and persist, required new ways of communication. This transition was experienced differently by students, with 80% of students describing it as a loss and a challenge, as documented by Jeffery and Bauer.[4]

It seems to be more than a question of access to technology. One student is possibly equipped to participate in digital learning from an infrastructure point of view (*e.g.*, hardware and Wi-Fi), but probably struggles to feel comfortable in a collaborative online learning environment. Some students have no compensatory chemistry classes online, whereas others benefitted from the implementation of virtual labs or digital resources for learning chemistry.[5,6] Although these virtual or online chemistry activities were initially intended as additions to typical laboratory courses, during the pandemic, these tools can offer alternative digital laboratory activities at home.[7]

While digital tools and online learning environments allow more opportunities to engage with the course content, they can place a high demand on students and rely heavily on students' self-regulated learning strategies.[8] The literature reveals that students with a high competence in self-regulated learning strategies performed better in online learning environments than those with lower self-regulated learning strategies.[9,10] Besides the necessity to self-regulate learning during the transition to an online environment, collaborative learning processes also change due to the shift towards online learning. Collaborating with peers, exchanging ideas, and learning together in a seminar or laboratory becomes much more challenging when it is transferred to online meetings. New ways of communication and social interaction need to be established online and used by students as well as instructors. How students experience this transition and what they report as supportive with regard to their challenges gives us insights into future online teaching strategies. This case study seeks to explore chemistry students' learning experiences with regard to their perceived challenges in terms of online learning and social distancing and support actions that emerged to overcome their problems, either actively or passively experienced. This case presented here is, thus, guided by the following questions:

1. What types of challenges emerged for students with regard to transitioning into online chemistry learning and changing social interactions?
2. What factors positively supported students?
3. To what extent are these supportive factors provided by the institutions or sought by the students?

3.2.2 Study Design and Methods

In this case study, chemistry major students from around the world shared their own story of how they experienced the transition to online learning in their chemistry classes and what they perceived as helpful. Based on students' self-reports, and in accordance with the research questions, we tried to trace: (1) the challenges that emerged with regard to transitioning into online chemistry learning and changing social interactions; (2) the factors that positively supported students; and (3) the extent to which these supportive factors are provided by the institutions or sought by the student.

Students' self-reports were collected *via* short online interviews over Zoom (approximately 40 min each) with 12 international second year chemistry students—three males and nine females, with an average age of 23 years, and from four different countries: Germany (three female), Costa Rica (two female, one male), Sweden (two female, one male), and the United States (two female, one male) to get a broad multi-national perspective on this topic. None of the participants self-reported to have any type of learning disability, or visual or hearing impairment. The interviews, conducted online *via* Zoom between October 2020 and March 2021, followed a semi-structured interview protocol containing open-ended questions focusing on students' description of their individual experiences. The study followed ethical guidelines as the participants were informed about their rights and handlings of their data before the interview started. Furthermore, the participants were asked to provide their consent for the recording of the interview beforehand. The interview typically began with the question: “Can you describe how you experienced the situation in spring 2020?” This prompt allowed the participant to warm up and frame the situation. Afterwards the focus was put on the transition to online learning by asking “How did your chemistry courses change due to the lockdown?” Then the focus of the interview shifted to the encountered challenges and support factors by asking questions such as “How did you experience the first weeks of transitioning into online teaching, what would you say was the most challenging for you and why?”, “What was supportive with regard to this challenge?”, and “What did you do to overcome this?”.

3.2.2.1 *Data Analysis*

Students’ self-reports in terms of their interview transcripts were transcribed *verbatim* and analyzed through thematic content analysis. We chose to use three broad categories that have been discussed in the literature as having major impacts on chemistry students learning during the pandemic.[4,11] We broadly categorized the challenges students reported deductively into: (1) lack of social contacts (participants’ reports about their experience of social distancing); (2) adapting to online learning (participants’ reports about struggling with learning online); and (3) laboratory access (participants’ reports about the cancellation of labs) (see Figure 3.1). We further coded supportive factors inductively, such as actions or environments either actively sought by the students or provided by the university. Students, for instance, reported about virtual study groups that were either implemented

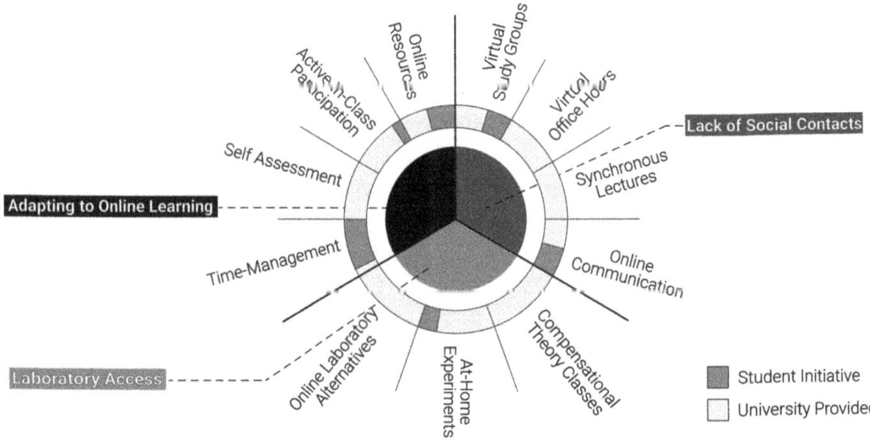

Figure 3.1 Overview of the three challenges and the respective support factors (outer ring).

as part of their courses (code: 'university provided') or organized by the students themselves (code: 'student initiative').

3.2.3 Results: Challenges and Supportive Factors Reported by the Chemistry Students

Figure 3.1 illustrates the challenges and the respective supportive factors and indicates the relative proportion of factors initiated by students or provided by the university.

Unsurprisingly, a common theme that emerged primarily during the interviews was the lack of social contacts, as illustrated by the quote from Charly (see Figure 3.2).

The common or familiar communication between peers in normal chemistry classes and the social contacts has been interrupted and is often not yet fully established during online settings. The quote from Polly (see Figure 3.2) further exemplifies the small occasional conversations that seem to help her in normal lectures, *i.e.*, having peers to ask for clarification. The quick transition into online learning often did not prepare the students to maintain the same amount of communication with their peers. The use of group work in breakout rooms or virtual study groups were slowly established by the instructors and by the students. Furthermore, students in our cohort seemed to value online classes that used breakout rooms, or when online study groups were organized as a convenient means of communicating with peers or with the instructor.

The second challenge most of the students reported was adapting to online learning and changing ways of regulating one's own learning, *i.e.*, scheduling and being organized, as well as self-discipline to attend lectures, were reported as being more difficult than before. All students described going through a period of adapting and coping with the changing situation. Adapting to online learning was mainly facilitated by students' own efforts to stay organized, by getting up early, maintaining a schedule, and by adopting strategies to benefit from the online resources provided, such as learning

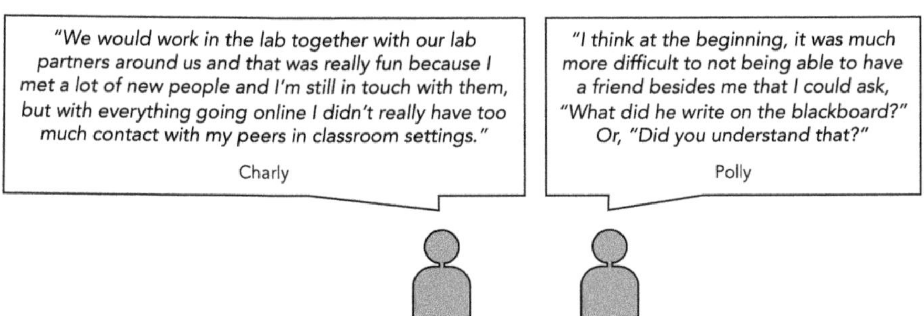

Figure 3.2 Quotes from Charly and Polly about how they perceived the lack of social contacts.

how to follow online lectures by rewinding or taking notes. Surprisingly, most of the aspects related to time-management that supported students to organize their days and learning online were initiated by their own motivation to keep up with the content. Rarely did professors give advice on how to regulate learning online or how to engage with the material while learning alone at home. Students additionally reported that they used and shared online resources for chemistry learning, either provided by the university as optional learning material for the chemistry courses or through external online resources such as Khan Academy.

Most of the self-reports were devoted to students' experiences of transitioning to online learning and keeping contacts to peers. Students also mentioned the cancellation of laboratory classes as being detrimental to their learning and well-being. Laboratory classes are a place to meet peers and do experiments and are spaces that make students feel like a chemist. Students in our cohort missed this inherent activity of being in the lab and the hand-on experiences that were hard to replicate at home or online (Lizzy's quote, see Figure 3.3). Similar findings have been described in the study by Jeffery and Bauer.[4]

Students reported that in their laboratory classes, when switching to online, the instructors usually provided videos of experiments to watch at home or covered the content in compensational theoretical classes. Two students of the cohort valued the opportunity to analyze at-home experimental data they received online. (Arthur's quote, see Figure 3.3). This provided an alternative to practice certain lab-related skills online, such as data evaluation and documentation, instead of passively watching content in terms of videos. Various sources for comparable digital lab activities that mimic in part the real laboratory environments (*e.g.*, analyzing given data), are now available to allow lab-related experiences when teaching online.[5,12,13]

Other students in our sample could opt for doing at-home experiments. Although this is a typical alternative reported in various studies,[14,15] the students in our cohort appreciated the opportunity but felt that this is not the same as being in the lab and handling lab equipment.

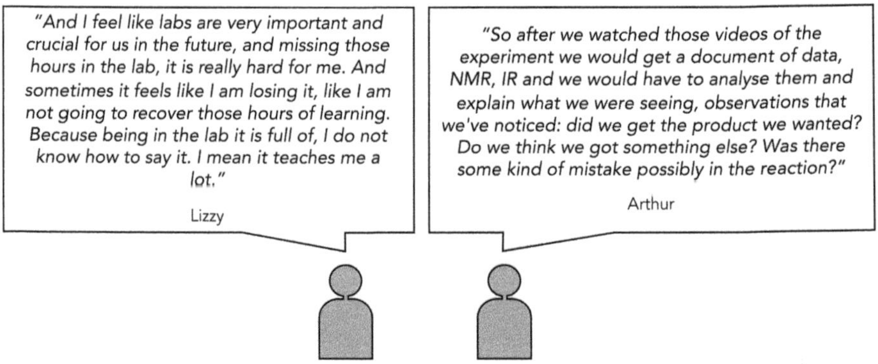

Figure 3.3 Quotes from Lizzy and Arthur about their lack of laboratory classes.

3.2.4 Implications for Future Online Learning

Students in our sample came from across the world and still did not differ largely from each other in how they experienced the transition. All students had access to computers and did not report any socio-economic issues. Our analysis, as summarized in Figure 3.1, elicited various supportive factors that instructors can incorporate in their planning and organization of online teaching or use it as a diagnostic tool to check what they are already providing, such as initiating virtual study groups, easy-to-access office hours, or providing online resources. The collection of supportive aspects derived in this case study is consistent with those already described in the literature.[16,17]

The analysis of students' self-reports clearly showed that there is a huge need for social contacts with peers when it comes to online learning, especially ways to stay in touch. Designing meaningful collaborative learning scenarios and hybrid formats will be the next challenge to incorporate this aspect in future chemistry classes.[18] To further provide access for students to find peers for study groups or exchange, universities should do more to provide platforms for meeting peers online and setting up interest groups. This could be a valuable add-on, besides facilitating online collaboration in Zoom breakout rooms. In terms of facilitating online learning, some aspects students mentioned in their self-reports are quite easy to implement but had a huge effect on students' perceived learning. Foremost, providing recordings of synchronous lectures online allowed students to rewind and replay sections of the lectures. As opposed to regular lectures where students typically relied on their course notes, recordings now provide access to course content for those students who struggle with certain topics or who want to process the content or take notes at their own pace.

With regard to laboratory classes, it was evident from students' self-reports that conducting experiments can hardly be replaced completely. However, during this time of transition into online learning in laboratories, various alternatives have been explored by instructors, such as providing students with experimental data to evaluate them at home, at-home experiments, or using simulations. The hands-on aspect of these alternative lab-activities should be maintained when incorporating them in future classes. In pre-pandemic times, being in the lab was a unique learning situation, often only accessible for those who could attend the lab. Students who missed days or were not allowed in the lab due to other reasons, *e.g.*, pregnancy, had clear disadvantages. The huge potential for future chemistry learning is in incorporating these alternatives in our future laboratory classes to create a more inclusive learning.[6,7]

Although students were focusing mainly on their own struggles and difficulties, they were often very attentive towards the other side, the instructors, professors, and their challenges of switching to online classes. This feeling of understanding is expressed by Andrea who closed her interview by formulating what she would say to her younger self: "*First I would say to be patient, to have a lot of patience with yourself, but also with others, like with your lecturers, with the university. Because it is going to take time for you and for them to adapt to this situation. And then I would also say like yes, maybe try to not be so ambitious. Maybe just accepting the situation, and I think it is more important to understand and tell yourself that it is a difficult situation for everyone."*

3.3 Case 2: Establishing a Support Network for Instructors for Digital Teaching

Not only do students need support for learning digitally, but instructors also need support for teaching digitally. In this section we provide a description of a model for supporting faculty in transitioning to online learning and teaching, which in providing support to instructors can also lead to their being able to cater to student needs. This case describes the design and establishment of a network for supporting instructors' digital teaching at the Faculty of Science and Technology (TekNat) at Uppsala University, and we provide specific illustrative examples from the chemistry departments.

The sudden transition to digital teaching quickly highlighted a number of challenges in instructor support. The IT support departments at the faculty received as expected a very high volume of requests for assistance. Many seemingly technical questions were in fact (also) enquiries concerning pedagogical issues, which the IT support staff were not well placed to advise on. The level of support an individual instructor could receive was very dependent on the context in which individual instructors happened to find themselves in, and there was no structure in place to coordinate these efforts across departments and faculty. Those with colleagues who were experienced and/or enthusiastic about digital teaching and learning were well placed to adapt to the digital transition, while others were left to fend for themselves to a much higher degree.

3.3.1 Design Principles and Description of the Support Network

When considering provision of support for effective digital teaching and learning, one can turn to the Technological Pedagogical Content Knowledge framework (TPACK) as a framework highlighting the need to look beyond instructors purely technological knowledge, and specifically account for its overlaps and confluences with their content knowledge, pedagogical knowledge, and pedagogical content knowledge. The framework has since been discussed widely.[19–21] Much more than mere technical competence in using digital and teaching technology *per se* is required to support and enhance student learning.

A need for a more coordinated support structure was therefore identified, with specific principles in mind:

1. Successful transition to digital teaching requires more than infrastructure with hardware and software. It needs to be combined with pedagogical expertise of both digital learning environments (in general) and

the specific teaching contexts (including both content and pedagogy) in question.

2. A distributed organizational structure is needed to provide rapid, fit-for-purpose support and solutions at the local level. A broader competence base for handling routine enquiries at the local level could avoid overloading central supporting units, supplemented by more specialized support and training as the need arises.
3. A platform across departmental boundaries for sharing ideas, experiences and expertise can provide mutual support, capture, and build on successful ideas, and avoid unnecessary duplication of effort.
4. As a positive legacy of the digital transition, the support structure should be sustainable and continued in the future, taking advantage of the digital competences and solutions rapidly accumulated during this time, as well as the communities of practitioners formed.

These guiding principles can be compared with the literature discussions of a Community of Practice (CoP), which, as described by Wenger and colleagues,[22] involve "groups of people who share a concern, a set of problems, a passion about a topic, and who deepen their knowledge and expertise in this area by interacting on an ongoing basis" (p. 4), see also Cox.[23] CoP can take many forms in terms of size, lifetime, distribution, diversity of members, and the manner and form of their establishment, but have common features of a shared *domain* of knowledge (here: digital teaching and learning), a *community* of people (satellites distributed across the faculty: see below), and their sharing *practice* to become more effective in the domain (in providing support to teachers and sharing their experiences in doing so).[22] CoP is a framework that has often been employed for designing and researching training and continuing professional development of teachers, including chemistry teachers.[24] Furthermore, the focus on local, customized support and solutions can be compared with calls in the literature for shifting the paradigm for institutional change from dissemination to propagation.[25] Propagation aims at a higher level of interaction with adopters of potential innovations to interactively support them in adapting and implementing innovations, having regard to the local context, including structural factors affecting whether sustained adoption of change can be achieved,[26] rather than simply providing "best practice" examples to raise awareness through webpages, videos or publications.

A schematic diagram of the digital teaching network that was initiated at the TekNat based on these design principles is shown in Figure 3.4. The network is described in more detail below.

To create a distributed organizational structure, each head of department nominated one member of staff with experience in both teaching and digital instructional technology as a "departmental satellite", with a focus on identifying individuals who could combine pedagogical expertise and digital competences, with close knowledge of the local institutional and education contexts. Departments were encouraged to inform their staff members to seek assistance from their local satellite as the first port of call, who would

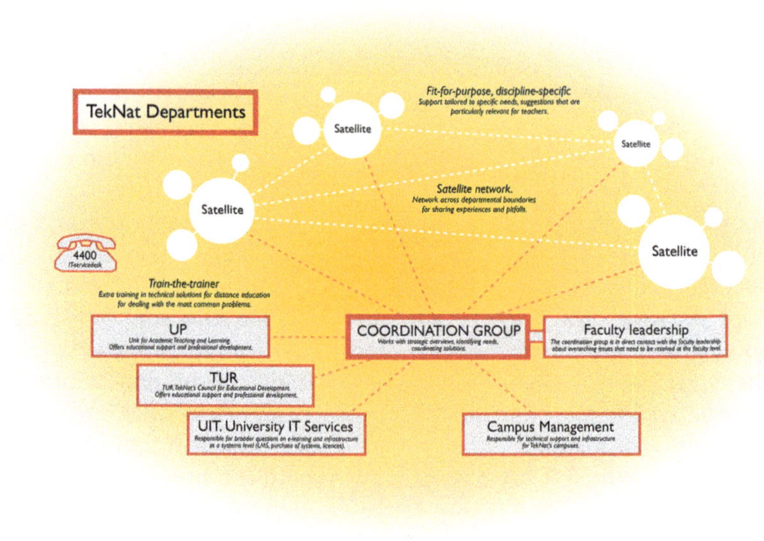

Figure 3.4 Organization diagram of the digital teaching network at the Faculty of Science and Technology (TekNat), Uppsala University. Image credit: Mats Kamsten.

provide support as well as perform a "triage" function, referring more challenging issues to specialized assistance by other units as necessary. Through a "train-the-trainer" model, satellites had access to additional training and support from other support units (Figure 3.4). The departmental satellites were also encouraged to share and discuss with each other, as a way to propagate ideas and experiences across departmental boundaries.

A small coordination group was also set up, consisting of experienced university instructors with substantial digital experience and IT support managers. Apart from coordinating the satellite network, the coordination group also liaised with faculty leadership to ensure a clear vertical channel of communication on matters requiring more substantial resources or policy changes, and a way for the faculty leadership to gain insights into the needs that have arisen from the field. The coordination group was also responsible for liaising with campus management, for issues such as physical installations of equipment or alterations of teaching facilities as necessary. Again, it was designed to provide a way to coordinate the workflow and requests, and clearer channels for communication and discussions.

3.3.2 Examples from Chemistry

A first example is the conversion of a number of second-year quantum chemistry laboratories to an online format that required specific licensed software and computing power beyond what students had available at home. With the contacts and knowledge of the departmental satellite concerning university supercomputer facilities, it was possible to arrange access to these servers for the required calculations. The familiarity of the satellite with the pedagogical requirements of the laboratory allowed appropriate testing and drafting of detailed instructions for remote access, in addition to student consultation sessions *via* Zoom. Security and other technical issues of remote access *via* a virtual private network (VPN) or opening of specific tunnels could also be addressed for smooth running of the laboratories. In a specific instance, significant connection and lag time issues prevented the use of the standard protocol, but an alternative solution was developed together with the course instructors, involving an interactive demonstration format where students "directed" instructors on campus performing the calculations *via* Zoom, then discussing the results online. While the students did not perform the calculations "hands-on", this was a pedagogically sound solution that provided an interactive and valuable learning experience while overcoming technical issues, and demonstrated the value of the collaboration between the instructors and the departmental satellite in combining both technical know-how with content and pedagogical expertise. These experiences and protocols for setting up remote computer laboratories were also summarized and communicated to other chemistry teachers.

A second example involved weekly online drop-in support and discussion meetings for teachers at the department. At these meetings hosted by the department satellite, teachers were invited to ask questions about digital teaching, to both satellite and also other teachers present. Questions ranged from technical issues of the use of handwriting real-time in online teaching and audiovisual aspects of dealing with simultaneous in-person and online teaching in hybrid format classes, to issues with design, submission and marking of assignments and lab reports online. These meetings allowed teachers to discuss and exchange their experiences, concerns, tips, and ideas on a regular basis, both receiving and providing each other with support. As such, this can again be compared with the model of building a CoP, now at a local departmental level.

3.3.3 Experiences from Departmental Satellite Support for Chemistry Instructors

While it is still too early in the development and implementation of this network to conduct a full research-oriented evaluation, responses to a small-scale written survey administered to both of the satellites at two departments (both researchers/instructors in chemistry) and instructors (five professors of various ranks) at the two departments of chemistry at TekNat illustrate their support experiences. The satellites were asked about what support requests they received, what challenges did they consider instructors faced, whether and in what way they felt their role was mostly technical, pedagogical or both, whether they felt they could support instructors in meaningful ways, what aspect of their assistance did they feel was most appreciated by instructors, as well as general evaluations of the initiative. Instructors were asked what challenges they faced, what support did they seek from the satellites and the results, whether and in what way they felt the satellites offered mostly technical, pedagogical support or both, whether they felt the satellites could support them in meaningful ways, as well as general evaluations of the initiative. Analysis of the responses (performed in NVivo, QSR International) revealed some interesting emerging themes.

When asked whether they felt the support was mostly technical or pedagogical in nature, or both, the instructors mainly self-reported receiving technical support. This involved the choice, installation, and use of hardware and software. They generally did not regard the support as pedagogical in nature. By contrast, it was interesting to note that the two departmental satellites responded that they regarded the support they provided as also being pedagogical in nature (see Figure 3.5).

The discrepancy between the instructors' and departmental satellites' views of whether pedagogical support was provided is interesting. Further examples of questions were given by a satellite: "Does my usual teaching really work *via* Zoom? How can I replace the blackboard in distance teaching? How do I make a meaningful electronic lab based on my usual lab? What types of questions can I ask in a take-home exam?". These demonstrated very well that the line between what is technical and what is pedagogical is not a hard one. For questions of "how best" to teach in a digital environment, the technical and pedagogical aspects in fact overlap significantly. The pedagogical support was noticed explicitly by departmental satellites but was not something that seemed to strike the instructors themselves, even if the requests for support implicitly raised such issues. As noted above, even when questions may initially be couched as purely technical in nature, the real question could also be pedagogical in figuring out how to best support student learning. There may also be different phases in the required support: whereas the immediate need might be technical support, more educational and pedagogical support may be appropriate in subsequent phases (see Figure 3.6).

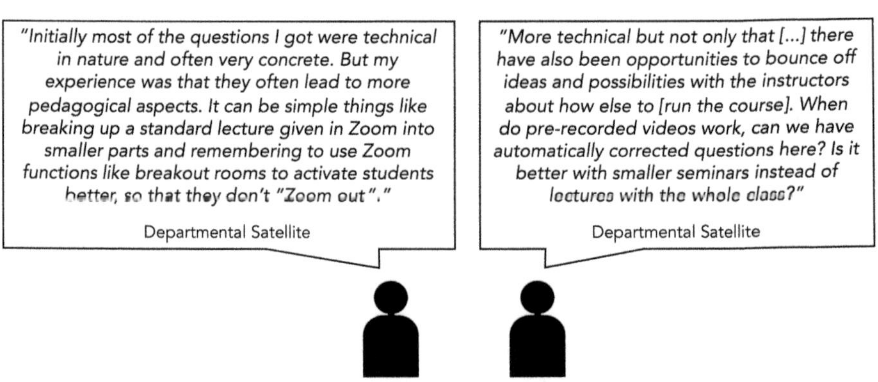

Figure 3.5 Impressions given by two departmental satellites about the type of support required.

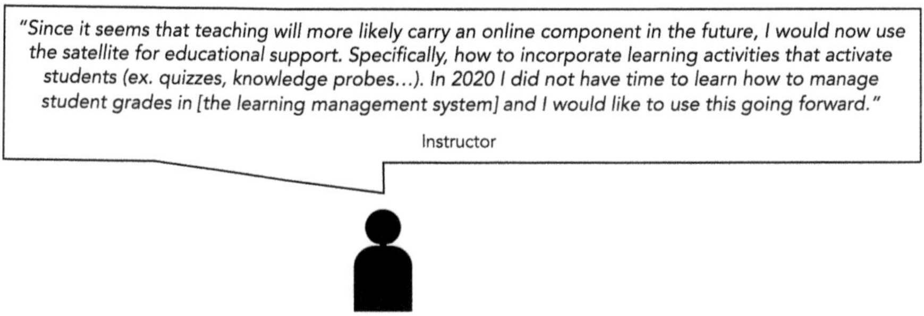

Figure 3.6 Quote from an instructor about possible phases in the kinds of support required.

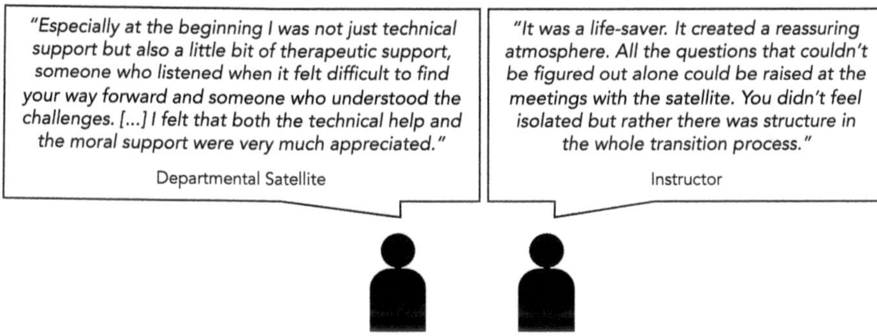

Figure 3.7 Quotes regarding moral support, as expressed by a departmental satellite and an instructor.

Having the departmental satellite at the local level allowed them to even provide moral support, and also create forums (such as weekly drop-in meetings) that facilitated discussion and exchange of ideas and experiences between instructors (see Figure 3.7).

Rather than becoming obsolete with time, the departmental satellites have the potential to continue to support digital teaching and learning and build on the experiences from this intense transition period. Their role can evolve to suit the changing needs of the instructors, thanks to their close proximity both in terms of technical, pedagogical, and subject expertise to those needing support. Facilitating and supporting a CoP would also increase the distribution of expertise beyond a limited number of people and sustain the sharing of ideas and experiences (see Figure 3.8).

A clear challenge with the digital teaching network is how to coordinate and incentivize the continuation of the initiative. While the heads of department were very forthcoming with nominating departmental satellites, there was variation in whether these became formal duties with allocated work time. Furthermore, while it was a core idea that the departmental satellites would decide for themselves what the needs at the local level were, some felt that their task was rather undefined and that it was challenging to reach

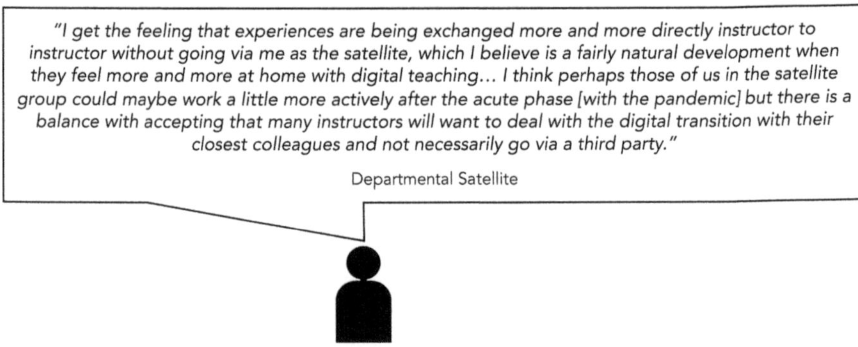

Figure 3.8 Further support and community-building aspects, as expressed by a departmental satellite.

sufficient awareness amongst instructors. At the departments of chemistry, the more formalized appointments of satellites, the initiation of specific activities such as drop-in sessions and discussion meetings, and information dissemination *via* e-mail seemed to be a useful model, based on instructor experiences. At the faculty level, more formal allocation of time and resources would also be needed for the members of the coordination group to better facilitate the mutual support and sharing of experiences envisioned in the design of the network, and to increase awareness of its existence and potentials to the instructors at the faculty.

3.4 What's Next? Outlook into Future Digital Learning

Compared to prior studies (see *e.g.* the COVID-19 Special Issue in the *Journal of Chemical Education*[11]), our case descriptions of students' and instructors' experiences presented herein show that the basics of learning and teaching, such as access to learning material and ensuring effective communication, dominated the discussion. Early on in this time of transitioning, virtual labs and remote laboratory settings only slowly started to emerge and were not yet implemented. Thus, the instructors and students in our cases were mainly occupied by ensuring teaching and keeping up with the learning content. Although aspects of pedagogical knowledge and implementing meaningful learning experiences were evident in instructors' search for support, aspects of how to achieve meaningful chemistry learning online are seldom brought up by the students. This points towards the challenge of providing meaningful chemistry learning online, such as evidence-based remote laboratories and virtual labs, when the basis aspects of access, communication and organization of teaching and learning online are not yet fully established. Instructors need support in organizational aspects such as how to provide online lectures and organize breakout rooms, but pedagogical advice how to teach lab skills in online courses with available educational resources as well.

A post-COVID learning and teaching landscape with enhanced digitalization of chemistry education should ideally learn from the experience that we all gained in this transition. There is a plethora of digital teaching and learning methods and a high level of digital ingenuity among both instructors and students. A significant potential here is that a broader range of means for accessing and learning course content, and for providing/receiving support, can be made available. But the digital tools themselves are no panacea and instructors and students alike need support in learning how to select and make use of appropriate tools—a systematic support system for instructors and students addressing both technical and pedagogical questions is needed. The model of having experienced chemists and chemistry educators that offer both technical and pedagogical support with in-depth understanding of both the chemistry disciplines and the needs of the instructors in the specific local context is something that should be further developed.

Acknowledgements

We would like to thank Ginger Shultz and Carlos Antonio Arias Álvarez for facilitating the recruitment of the participants. We would further like to thank all the students who voluntarily participated in the interviews and the departmental satellites and instructors who provided their impressions and experiences.

References

1. See Chapter 31: B. DeKorver and D. Herrington, in *Digital Learning and Teaching in Chemistry*, ed. Y. J. Dori, C. Ngai and G. Szteinberg, Royal Society of Chemistry, United Kingdom, 2023.
2. See Chapter 30: A. Shauly, G. Shwartz and S. Avargil, in *Digital Learning and Teaching in Chemistry*, ed. Y. J. Dori, C. Ngai and G. Szteinberg, Royal Society of Chemistry, United Kingdom, 2023.
3. See Chapter 20: P. Pullen, A. Motion, S. Schmid, S. George-Williams, S. Wilkinson and S. Leach, in *Digital Learning and Teaching in Chemistry*, ed. Y. J. Dori, C. Ngai and G. Szteinberg, Royal Society of Chemistry, United Kingdom, 2023.
4. K. A. Jeffery and C. F. Bauer, *J. Chem. Educ.*, 2020, **97**, 2472–2485.
5. J. L. Davenport, A. N. Rafferty and D. J. Yaron, *J. Chem. Educ.*, 2018, **95**, 1250–1259.
6. L. Groos, K. Maass and N. Graulich, *J. Chem. Educ.*, 2021, **98**, 1919–1927.
7. C. L. Dunnagan, D. A. Dannenberg, M. P. Cuales, A. D. Earnest, R. M. Gurnsey and M. T. Gallardo-Williams, *J. Chem. Educ.*, 2020, **97**, 258–262.
8. F. Zimmermann, I. Melle and J. Huwer, *J. Chem. Educ.*, 2021, **98**, 1863–1874.

9. M. Barak, R. Hussein-Farraj and Y. J. Dori, *Int. J. Educ. Technol. High. Educ.*, 2016, **13**, 35.
10. R. Hussein-Farraj, M. Barak and Y. Dori, *Interdiscip. J. E-Learn. Learn. Objects*, 2012, **8**, 115–135.
11. T. A. Holme, *J. Chem. Educ.*, 2020, **97**, 2375–3470.
12. W. J. Howitz, T. A. Thane, T. L. Frey, X. S. Wang, J. C. Gonzales, C. A. Tretbar, D. D. Seith, S. J. Saluga, S. Lam, M. M. Nguyen, P. Tieu, R. D. Link and K. D. Edwards, *J. Chem. Educ.*, 2020, **97**, 2624–2634.
13. D. A. Guarracino, *J. Chem. Educ.*, 2020, **97**, 2742–2748.
14. J. I. Selco, *J. Chem. Educ.*, 2020, **97**, 2617–2623.
15. M. Schultz, D. L. Callahan and A. Miltiadous, *J. Chem. Educ.*, 2020, **97**, 2678–2684.
16. E. A. Perets, D. Chabeda, A. Z. Gong, X. Huang, T. S. Fung, K. Y. Ng, M. Bathgate and E. C. Y. Yan, *J. Chem. Educ.*, 2020, **97**, 2439–2447.
17. O. Villanueva, D. A. Behmke, J. D. Morris, R. Simmons, C. Anfuso, C. M. Woodbridge and Y. Guo, *J. Chem. Educ.*, 2020, **97**, 2458–2465.
18. B. M. McCollum, in *Active Learning in College Science: The Case for Evidence-Based Practice*, ed. J. J. Mintzes and E. M. Walter, Springer International Publishing, Cham, 2020, pp. 621–637.
19. L. M. Archambault and J. H. Barnett, *Comp. Educ.*, 2010, **55**, 1656–1662.
20. L. Brantley-Dias and P. A. Ertmer, *J. Res. Technol. Educ.*, 2013, **46**, 103–128.
21. M. Koehler and P. Mishra, *Contemp. Issues Technol. Teach. Educ.*, 2009, **9**, 60–70.
22. E. Wenger, R. McDermott and W. M. Snyder, *Cultivating Communities of Practice*, Harvard Business School Press, Boston MA, 2002.
23. A. Cox, *J. Inf. Sci.*, 2005, **31**, 527–540.
24. G. Szteinberg, S. Balicki, G. Banks, M. Clinchot, S. Cullipher, R. Huie, J. Lambertz, R. Lewis, C. Ngai, M. Weinrich, V. Talanquer and H. Sevian, *J. Chem. Educ.*, 2014, **91**, 1401–1408.
25. J. E. Froyd, C. Henderson, R. S. Cole, D. Friedrichsen, R. Khatri and C. Stanford, *Change: The Magazine of Higher Learning*, 2017, vol. 49, pp. 35–42.
26. C. Henderson, A. Beach and N. Finkelstein, *J. Res. Sci. Teach.*, 2011, **48**, 952–984.

CHAPTER 4

Adapting Large Intro-level Chemistry Courses to Fully Remote or Hybrid Instruction

RHIANNON IHA[a,b], JIA LUO[a], BRYN LUTES[a] AND GABRIELA SZTEINBERG*[c]

[a]Department of Chemistry, Washington University in St. Louis, MO, USA; [b]Washington University Libraries, Washington University in St. Louis, MO, USA; [c]College Office, College of Arts & Sciences Washington University in St. Louis, MO, USA
*E-mail: gsztein@wustl.edu

4.1 Introduction

The Department of Chemistry at Washington University in St. Louis offers introductory chemistry courses to a wide range of science, technology, engineering, and mathematics (STEM) and non-STEM undergraduate students every year. Because of our dedicated mission to undergraduate education, we aim to create an inclusive and positive learning environment for all students through best teaching practices and effective pedagogical approaches or strategies. During Spring 2020, due to the interruption caused by COVID-19, Washington University shifted the undergraduate education instruction from in-person to fully remote mid-semester. Many instructors had to quickly adjust their courses to this new modality, while trying to maintain the same quality of education before the pandemic. These hybrid or fully remote formats were continued during the 2020–2021 academic year.

Advances in Chemistry Education Series No. 11
Digital Learning and Teaching in Chemistry
Edited by Yehudit Judy Dori, Courtney Ngai and Gabriela Szteinberg

Published by the Royal Society of Chemistry, www.rsc.org

In this chapter, we describe full-semester adaptations for several introductory chemistry courses: Introductory General Chemistry I and II (Chem 105/106, summer, fall, and spring semesters), General Chemistry I (Chem 111A, fall semester), and Organic Chemistry I and II (Chem 251/252, summer, fall, and spring semesters). All courses adopted a version of the flipped classroom approach, in which students learned the material by watching lectures before attending class remotely. During classes, all instructors employed active learning strategies to maintain student engagement and solidify understanding of core concepts,[1,2] while employing variations of this approach for each set of students and instructors. For each course, instructors and/or instructional teams prepared video lectures (using recorded lectures from the previous year or creating new lectures) for students to watch asynchronously in preparation for the synchronous (yet still virtual) component of the class where students had the opportunity to work on problems and ask questions. Based upon knowledge acquired first-hand and from colleagues at other institutions,[3] course structures were designed to create inclusive learning experiences while ensuring complete delivery of all content relative to previous iterations of the course. In each section below, we describe the specific adaptations that we found to be the most successful and conducive to students' learning. Table 4.1 shows a summary of the comparison of all of our courses.

4.2 Course Descriptions and Summary of Adaptations

Table 4.1 summarizes the courses and adaptations described in this chapter. The sections below describe specific aspects of the adaptations utilized in each course.

4.2.1 Use of Asynchronous Video Lectures

Lectures pertaining to both general and organic chemistry were provided asynchronously to students in all courses. This was done in order to accommodate students taking these courses completely remote; an additional benefit of this approach is that the use of video lectures in combination with other strategies can help improve student learning and retention.[4] For these courses, two major teaching approaches were adopted when implementing asynchronous lectures. One approach included providing three one-hour lectures per week to create a comprehensive remote learning environment. These asynchronous digital lectures fully reflected the in-person lecture course. The second approach included creating shorter lectures that focused on a single topic or concept and assigning these shorter lectures such that the time spent watching the videos was comparable to the time students spent in lecture during previous in-person semesters. Both approaches were employed in a manner to try to alleviate

Table 4.1 Summary of the comparison of all our courses.[a]

Courses	Chem 105/106 (Fall, Spring)	Chem 105/106 (Summer)	Chem 111A (Fall)	Chem 251/252 (Fall, Spring, Summer)
Student body/ Course goals	STEM majors, and/or completing pre-health requirements that either taking chemistry for the first time or struggled with chemistry in the past. Aim: provide introduction to chemistry concepts and prepare students to think and solve problems like chemists, while developing study skills that will serve them for the rest of their time in college and beyond.	STEM majors, and/or completing pre-health requirements that either taking chemistry for the first time or struggled with chemistry in the past. Aim: provide introduction to chemistry concepts and prepare students to think and solve problems like chemists, while developing study skills that will serve them for the rest of their time in college and beyond.	STEM majors, and/or completing pre-health requirements. Aim: to prepare students to think and solve problems like chemists, while developing study skills that will serve them for the rest of their time in college and beyond.	Summer predominately rising sophomores that are completing the course for pre-health requirements or STEM major requirements; fall and spring post-baccalaureate premedical students. Aim: provide a survey of organic chemistry to prepare students for pre-health requirements like the MCAT and biochemistry courses. It also emphasizes skills for problems solving and using references and resources responsibly to help inform that process.
Number of students	250–400	20	About 500	Summer 70 Fall and Spring 30
Instructor team/ AIs/TAs *etc.*	Four instructors, one project coordinator for supplemental programs, two or three graduate students who served as Assistants in Instruction (TA position for graduate students), two or three undergraduate TAs, and ~30 undergraduate peer mentors/leaders.	One instructor, two graders.	Two professors who are leading lecturers; one lecturer helping out during lectures; one project coordinator for supplemental programs; 15 undergraduate TAs and five graduate student Assistants in Instruction. About 40–50 peer leaders.	One instructor, three graders.

(*continued*)

Table 4.1 (*continued*)

Courses	Chem 105/106 (Fall, Spring)	Chem 105/106 (Summer)	Chem 111A (Fall)	Chem 251/252 (Fall, Spring, Summer)
Lecture format	Asynchronous; 45 min pre-recorded lecture video per lecture; posted at 9 am Mondays, Wednesdays, and Fridays onto Canvas.	Asynchronous pre-recorded lecture videos; synchronous help sessions.	Asynchronous pre-recorded lecture videos, some which were only available after students answers pathway questions; synchronous review and problem-solving lectures on Zoom.	Asynchronous pre-recorded lecture videos; synchronous activities/help sessions.
Synchronous component	Required recitation (once per week for 50 or 80 min). Optional instructors' online help sessions/office hours (four times per week for 1.5 or 2 h). Optional Peer-Led Team-Learning (PLTL) study groups (once per week for 2 h). Optional "walk-in" Residential Peer Mentoring (RPM) sessions (multiple times per week for 2–4 h).	All scheduled course time was help session/office hours with the instructor.	Review lecture with announcements and problem-solving; voting on polls; recitations on Zoom; help sessions; PLTL; RPM.	Scheduled course time was divided between small group work, student presentation of work, and question and answer sessions with the instructor (*e.g.*, like office hours).

Asynchronous component	Lecture review questions; graded online homework.	Lecture videos designed, structured, and recorded for summer 2020. Videos were split by topic with the goal of each video being between 7 and 20 min long.	Lecture videos from 2019 and pathway questions.	Lecture videos designed, structured, and recorded for summer 2020. Videos were organized by topic or reaction. Video lectures were between 5 and 30 min long.
Assessment strategies	Closed book, closed note 25 min quizzes; closed book, closed note 120 min exams.	Open note, topic-specific quizzes. Closed note exams—written as "hour-long" exams, but students were given 3 h to complete and upload their work.	Closed-notes quizzes and exams, proctored *via* Zoom (except for those with accommodations to take exam at other times).	Closed note quizzes on fundamental topics and skills. Open note exams with application problems not directly from lecture; written as "hour-long" exams, but students were given 3 h to complete the and upload their work.
Supplemental programs/Course support	PLTL, RPM, transition program.	—	PLTL and RPM; help sessions; recitation.	—

[a]Key: TA: Undergraduate students who serve as Teaching Assistants; AI: Graduate students who serve as Assistants in Instruction; PLTL: Peer-Led Team Learning; RPM: Residential Peer Mentoring.

Zoom fatigue and also incorporated other practices like asking questions in the middle of the lecture to help students engage the material as it was taught.

The first approach was employed for lectures in Chem 105/106 during the 2020–2021 academic year, when the course was offered entirely online. The major course components are summarized in the Table 4.1. When comparing to 2019–2020, while the asynchronous video lectures were a key adaptation to the course, the most promising adaptations were low-stake assignments, including lecture review questions and graded online homework, which is discussed in depth in a subsequent section. For Chem 105 and 106, each semester included 41 lectures of approximately 50 min, accompanied by 41 sets of lecture review questions. (Note: in Spring 2021, the University schedule allowed only 40 lectures in Chem 106.) This timeline corresponded to the typical number of lectures that students would attend in-person during a typical semester.

The second approach was employed in Chem 111A, the summer session of Chem 105/106, and in the summer organic chemistry series of Chem 251/252. For all courses, most lectures were incorporated digitally as short (10–15 min) videos to render the material more digestible and easier to watch, and to provide a topical focus. During previous in-person lectures, students were often rushed or were unable to write everything down, and lectures often ended mid topic or thought due to time constraints. By making the lectures asynchronous and employing a flipped classroom approach, students could watch lectures at their own pace within the schedule of the course and could also review and re-watch lectures as part of their learning process. Students expressed their appreciation for having more than one chance to learn and review fundamental lecture material.

4.2.1.1 Chem 111A

The Chem 111A course in Fall 2020 had a total of 485 students. To provide a comparable experience for remote and in-person students, the instructors decided to use a flipped classroom approach, utilizing pre-recorded lecture videos from Fall 2019 (taught by the same lecturers as in Fall 2020) that were edited to exclude any announcements or other parts not directly related to the presentation of course material. The videos were also reorganized to focus one topic at a time. All lecture videos were posted to Canvas, the Learning Management System (https://www.instructure.com/canvas), about 48 or 72 h before the synchronous lecture sessions that were held on Zoom. Students were asked to watch the recorded videos in sequence. While each video focused on one single topic, students had to demonstrate some mastery through completion of a short quiz (a pathway question) to proceed to the next video in the sequence. Even though students were required to complete the asynchronous components prior to the synchronous sessions, instructors did not check for completion. This means that students had the freedom to complete the asynchronous steps when they were able to do so.

4.2.1.2 Summer Courses Chem 251 and Chem 105/106

Both general and organic chemistry were taught as accelerated online courses during the summer term. The summer 2020 offerings of Chem 105 and 106, Introductory General Chemistry I and II, respectively, were taught online asynchronously with optional synchronous components. The same was true for Chem 251 and 252, organic chemistry 1 and 2, respectively. Lecture material was presented *via* video in a variety of ways including PowerPoint presentations with instructor voice-over, videos of hand-written work with instructor voice-over, and/or chalk talk videos of the instructor presenting lecture material. Additionally, in organic chemistry asynchronous lectures questions were assigned for the students to try out, which were directly related to lecture material. These problems were presented mid-lecture for students to test their knowledge and also provided a jumping off point for synchronous sessions so that students could have discussions related to problem solving techniques.

Regardless of the style of the video used, students generally showed a favorable attitude with respect to these lectures. While student evaluations should not be the determining factor in pedagogy, it has been documented that video lectures have been shown to have a positive effect on student learning when combined with other pedagogies.[4] Students' apparent willingness to engage with remote lectures and their generally favourable attitude related to this teaching method may suggest that this method can be used effectively.

We present student comments from the course evaluation related to lectures (for Chem 111A):

- "The lectures are concise and easy to understand and can be watched in nice 20–30 min chunks."
- "I loved how the notes appeared on the screen before they were written, this was very helpful."
- "I liked how I could watch the lecture recordings on my own time."
- "I LOVED how the lectures were broken into 10–30 min videos and organized into modules. These worked well for my attention span and broke material into digestible chunks in which it was easier to determine the important core concepts. This worked so much better than hour and a half long live lectures for me. Also seeing the concepts written out in front of me rather than in a PowerPoint or just spoken out-loud enhanced my ability to really wrap my head around concepts and concrete problem-solving skills."
- "The lecture videos were much more helpful for me than an in-person lecture. I was able to stop them and rewind them and rewatch them. I could take notes at my own pace, and this was helpful."

The following are student comments on aspects of the course that they liked related to asynchronous lectures (for organic chemistry, Chem 251/252):

- "I actually liked having the video lectures because it allowed me to be able to rewatch videos to concepts or problems I did not understand. In a normal class setting I would have to rely on the notes I took or have to ask around for help and this is all within a structure time such as office hours *etc.* Having the videos allows me to revisit at any time and catch details that I might have missed the first time watching."

- "Video lectures allowed me to rewatch material if needed."
- "The recorded lectures also allowed us to repeat concepts that were harder to grasp than the others and review materials better."

4.2.2 Assessments and Assignments for Engagement

With any course, assessments are a key component to determining the degree of mastery of topics. In all three courses a variety of low-stake assessments or assignments were used to try to ensure that students were watching the lecture material in a timely manner, because these were automatically graded or checked on Canvas, and to enhance students' mastery of the fundamental material and concepts presented in the asynchronous lectures.

4.2.2.1 Chem 105/106

Since Chem 105/106 were offered fully online during the 2020–2021 academic year, all lectures were presented asynchronously and the instructors incorporated low-stake assignments, including lecture review questions, and graded online homework; these are adaptations to the online instructional modality that have not been used by Chem 105/106 previously. The implementation of the lecture review questions is one of the highlights that is unique in Chem 105/106. It contributes 6% to students' overall course grade (30 out of 500 points), and the instructors perceive it as helping the students' engagement with the course content and possibly as leading to positive learning outcomes from students.

Each set of lecture review questions is worth 1 point, with two attempts. It is composed of five short-answer questions (*i.e.*, multiple choice, multiple answer, true/false, fill in the blanks, *etc.*). They are directly relevant to the associated lecture video content, and work at the "Remember and Understand" levels or occasionally at the "Apply" level, according to Bloom's taxonomy.[5] Each question asks about one concept that is introduced in a roughly 10 min segment of the lecture video. For example, Question 1 is for a concept introduced between 0 and 10 min, Question 2 is for another concept introduced between 10 and 20 min, *etc.* This is designed to deter students from playing the video at a fast speed or skipping content, which could lead to missed details. However, these questions are open book/note, meaning students can complete them while watching the video rather than after watching the video. On average, students should be able to complete these questions in no more than 10 min.

The time window given for completion (more than 48 h) is to encourage students to watch a lecture video before the next one is posted because of the cumulative nature of lecture content. However, this window also maintains a flexible due time for international students in significantly different time zones. In addition, the asynchronous style gives students plenty of time to digest a newly learned concept before putting it into application. Based on our previous in-person synchronous lecture experience, we noticed that not all students are comfortable solving at the "Apply" level of questions after just hearing the concept in the lecture. By providing a time window separated from the lecture, we eliminate the anxiety for this cohort of students, which enhanced the low-stakes nature of this assignment.

Since unexpected interruptions can occur in the midst of the pandemic for students, we also incorporated a practical drop policy regarding lecture review questions. Each semester has three mid-term exams and one final exam. Each mid-term exam includes content from 11 to 13 lectures. We allow students to drop 11 sets of lecture review questions in total: three sets relating to material from each mid-term exam and two sets from post-exam 3 material. This is to prevent students from dropping 11 consecutive sets resulting in them missing some major topics necessary for their learning. (Note: a similar drop policy is also applied to the graded online homework.)

Based on the data collected from 2020–2021, among 299 students in Chem 105 Fall 2020 and 259 students in Chem 106 Spring 2021, only one student did not attempt any lecture review questions. Over 88% of students received at least 25 points (out of 30 points) for this assignment (88.3% for Chem 105 and 91.5% for Chem 106). While more studies should be done in this area, it is promising that a positive correlation between the lecture review question score to the total course score was observed for both Chem 105 and 106 (Figure 4.1).

From the end-of-semester course evaluations, which are provided by the university to all students enrolled in courses and then made available to instructors the following semester, students left comments indicating a certain degree of appreciation of the lecture review questions. For both semesters, we asked students to comment on how each of the instructional approaches or strategies was helpful in learning the course content in a voluntary, anonymous end-of-semester course evaluation questionnaire. In Chem 105, 64 students answered this question. Among them, 27 students specifically commented on the usefulness of the lecture review question: 16 of them were positive, 7 were negative, and 4 were neutral. While in Chem 106, 30 students answered this question. Among them, 13 students specifically commented on the usefulness of the lecture review questions: 9 were positive and 4 were negative. Selected representative quotes from these course evaluation responses are shown below.

- "The lecture review questions every other day were really great to keep me motivated to stay on task during lectures and keep me from falling behind."
- "Lecture review questions were very beneficial for making sure I understood the topics presented in the lectures."
- "Lecture review questions—incredible! [They] really helped me to succeed in the class! I don't think I would have kept pace without them."
- "The lecture review questions allow me to pay more attention to the videos and take more notes."

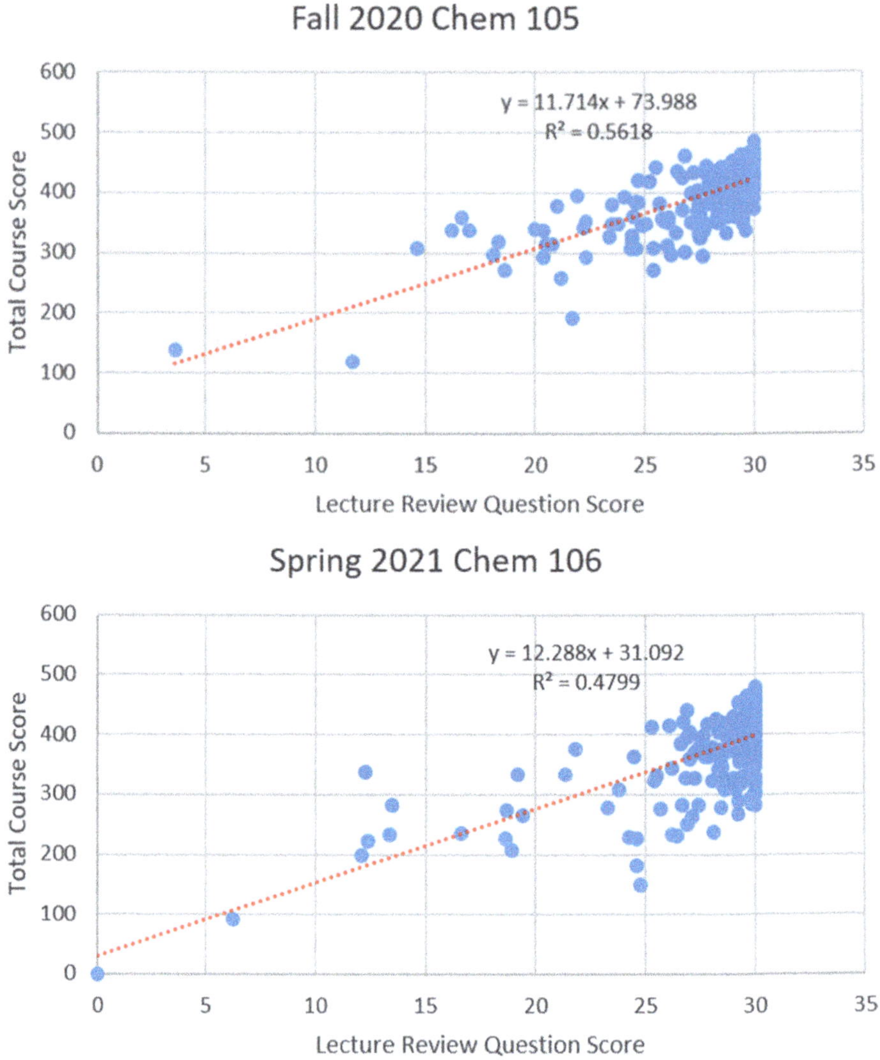

Figure 4.1 Correlation between the lecture review question score and the total course score (Chem 105 – top; Chem 106 – bottom). Each blue dot represents one student. Red lines are the trend lines to show the positive correlation.

- "Lecture review questions were helpful as a quick review on the content we just learned so I didn't just write it down and forget about it."
- "Lecture review questions were not helpful. It just moved our grade down if they asked a weirded question."
- "The lecture review questions were a bit more stressful than helpful in my opinion, because of all the other opportunities we have that don't have point values assigned to them."

- "The lecture review questions were pretty hard, but I wish they had been more for participation than graded so that we wouldn't be so scared when submitting and could focus on the problem solving."
- "While the lecture review questions were not super helpful in regard to the material, they kept me accountable for watching lectures on time."
- "I liked the change from having one chance to two. That really helped me realize what misconceptions I have coming out of the particular lecture."

In addition, when being asked if there were any features of this remote course they would encourage the instructor to retain when the course returns to an in-person format, five out of 44 responses mentioned lecture review questions and commented on keeping them in Chem 105, while eight out of 22 expressed the same desire in Chem 106. Considering this assignment only contributes six percent to a student's overall course grade, we believe this assignment may have a positive impact on students' learning. We acknowledge that this evidence is not robust, therefore at this point we are inferring from anecdotal evidence.

4.2.2.2 Chem 111A

As part of watching the asynchronous lectures for this course, students were asked to watch the recorded videos in sequence, each focusing on one single topic. After the first video presenting one topic of each daily series, subsequent videos were only accessible when students would answer pathway questions which were presented in the form of a Canvas quiz. Figure 4.2

Figure 4.2 Sample of how the videos were presented to students on Canvas. Students had to watch videos in sequence, and answer pathway questions, before obtaining access to the next video. The parts that are inaccessible appear in faint font.

shows a sample of how the videos were presented to students on Canvas. This technique allowed students to be actively engaged with the material presented in the videos and sought to avoid fatigue, both through the use of shorter videos (10–15 min long), and through opportunities for students to put the new knowledge into practice.

4.2.2.3 Summer Courses

For both the summer general chemistry and organic chemistry series, low-stakes and frequent student quizzes were given to ensure that students were completing lectures at a pace appropriate for an accelerated course. For each course, the quizzes consisted of a combination of auto-graded questions such as multiple choice and fill-in-the-blank questions, and manually graded questions, which required students to upload images of their hand-written work.

The summer general chemistry (Chemistry 105/106) quizzes were open-note and open-book to help develop students' skills on fundamental chemical concepts and chemical calculations. Students were then expected to apply these skills on closed-note, closed-book exams as a final test of mastery with application to new situations—namely, to examine the transfer of learning.[6] This was done to ensure that students were proficient in fundamental chemical concepts needed for future chemistry courses.

In organic chemistry (Chem 251/252), quizzes were closed-note and closed-book to make sure that students had a solid understanding of basic reactions and mechanisms before being asked to apply and extend these skills on exams. In general, students seemed to appreciate these small assessments along the way, and when asked in course evaluations what they liked about the course one student commented "Frequent quizzes" and another said, "Frequent quizzes also helped me keep up with the material."

While quizzes were closed-note and closed-book in organic chemistry (Chem 251/252), the five exams were open-note/open-book. Scientific areas are collaborative in nature, areas where researchers look for what they do not yet know. It is therefore important to document and cite the references that one uses when doing scientific work. To help improve information literacy in chemistry and teach students how to use information ethically and responsibly, as part of the course they were taught to evaluate the authority, reliability, and reputability of different kinds of sources, and also how to use these sources as citations. On open-note/open-book exams, they were required to provide citations for any references or resources they used on their exams. In general, this improved the academic integrity surrounding course exams since using books was not considered dishonest. Anecdotally, there were no discovered postings of exam questions to outside websites from closed-note/closed-book course exams, whereas some of the closed-note quizzes from the course had been shared to outside sites to obtain answers. As part of the course, students had to determine which references were going to be appropriate and useful for answering a question. By engaging with references, they often learned more than they would from their notes alone and gained skills with respect to evaluating sources. Since they were asked to cite their sources, the importance of ethical use of information as part of the scientific process may be more emphasized for them.

4.2.3 Active Learning During Synchronous Meetings

4.2.3.1 Chem 111A

In Chem 111A, after watching a sequence of asynchronous lectures, students were required to attend synchronous Zoom sessions taught by the same instructors in the asynchronous lecture videos. The synchronous sessions focused on problem-solving, served as the space in which students heard announcements about logistics and related course events (such as help sessions), and acted as a forum where students could ask questions about the recorded lectures. In these sessions, students were presented with problems related to the topics of the asynchronous videos. The instructors then opened breakout rooms for students to work together on these problems, with the option to move to a breakout room for students who preferred to work by themselves. During regular in-person classes, students are encouraged to work together during questions in the class. The Zoom breakout rooms allowed for more varied opportunities for students to work with other people they would not have met otherwise. Once the breakout rooms were closed, students had to answer the question *via* a Zoom poll (shown with multiple-choice answers). These sessions occurred three times a week, for 45 min each time. Instructors used the data obtained from the Zoom poll as a way check attendance and to monitor student participation in each session.

The course also featured other synchronous components:

- 45 min problem-solving recitation sessions
- instructor Zoom help sessions four times a week, for about 1.5 h each
- weekly optional Residential Peer Mentoring (RPM) sessions *via* Zoom
- weekly Peer-Led Team Learning (PLTL) that took place mostly over the weekend or Monday nights.

Entry to the PLTL program was optional at the beginning of the semester, and attendance was strictly enforced and followed throughout the semester. Attendance was recorded by Peer Leaders, which was shared with the Program Coordinator who checked for absences. This information was sent to the Director of the program, who contacted students to warn them about dismissal or being close to program dismissal.

Another resource for students was Piazza (https://piazza.com/), an online forum in which students can ask questions about the course material to each other and instructors. This platform was embedded in the Canvas course, and students could post questions anonymously. Questions were typically answered by someone from the instructional team and/or other students within 24 h.

To accommodate students in different time zones, instructors offered synchronous lecture meetings, recitation sessions, and help sessions at times that would accommodate these time zone differences at 8 p.m. Central Time, allowing all students to engage with the instructors in a synchronous manner.

We had students who liked this course format and mentioned this in course evaluations:

- "Synchronous lectures were very engaging with the professors with the use of polling questions."
- "I'm not really sure what in-person classes would look like or if these are already normally included, but I liked the pathway questions because they were simple ways to check my understanding and they would let me know if I needed to spend more time on that topic or not."
- "I really liked that the lecture videos were a series of shorter videos that had flexible timing. I liked that they weren't overwhelmingly long, and I enjoyed completing the pathway questions to test how much I'd learned through the video."
- "I think the flipped classroom worked well for the most part, and I would encourage continuing that. It's easy to look at the CCV's (course content videos) or recordings of Zoom if I feel there's a gap in my notes or if I have a question but can't remember what was said in lecture."
- "I would keep the flipped classroom approach because it really helped me take the time and think about the ideas before I needed to ask questions and apply the material. It also forced me to spend more time on work for the class that I might not have spent in a normal semester."
- "A feature of the remote course I would encourage the instructors to retain when the course returns to in-person format would be the availability of the professors and other members of the teaching team, as I feel the extra help played a large factor in my success in the class."

Even though the instructors provided several opportunities for engagement, and although we were trying to prevent Zoom fatigue, for some students this fatigue was unavoidable, and they also felt disengaged from the class:

- "I have had to watch course content videos many times after being mentally exhausted from an exam. Maybe after major quizzes or exams, or random days sprinkled throughout, there should be breaks from these videos."
- "I think that having more opportunities to meet classmates during class would make it easier to find people to work and study with. Being able to choose breakout rooms was nice but I often found myself going into the individual workspace room because it felt weird to join a room with people who decided to join together."
- "The course content videos were very long and often left me confused because I would take notes on them but I didn't fully understand them until days later when it was talked about in class."

In Fall 2021, this course was held in-person. The instructors continued to offer several opportunities for students to connect with them and each other, as well as maintained the active learning components in class with questions throughout lectures that make students reflect on the newly learned material and apply the new knowledge.

4.2.3.2 Summer Courses

Due to the rapid pace of a summer course, for students to learn effectively—especially in science—they need to practice and apply learned knowledge.[7,8] In both summer general chemistry (Chem 105/106) and organic chemistry (Chem 251/252), synchronous office hours and problem-solving sessions were provided to students. The advantages of this structure include the benefits of a traditional flipped classroom.[2] The time students spend outside of "class time" is more productive—they watch lecture videos as many times as they need, take notes, and start working on practice problems. Then, if they get stuck on the practice problems, they do not need to struggle alone. Class time is dedicated to working through problems. It also leaves enough time to discuss problem-solving strategies and sets students up to be more successful the next time they work on the problems. All contact with the instructor is interactive and there is very little passive information transfer. For general chemistry a specific module was designated for each synchronous session. This was to help keep students on track with content. Synchronous sessions were more like office hours in that students were required to come with questions prepared on the related module.

For organic chemistry (Chem 251/252), synchronous sessions were more about problem solving and building application skills. Asynchronous lectures would assign questions for students to work on their own during mid-lecture or at the end of a lecture, providing a point to begin discussions related to problem solving techniques. Some synchronous sessions were devoted to POGIL (Process Oriented Guided Inquiry Learning) activities where students would work in small groups to extend their understanding on a specific topic (such as conformational analysis, spectroscopy, and substitution reactions). The students were divided into breakout rooms and urged to explore and make conclusions about topics or reactions that they would then present to the class as part of a discussion. Another use of these sessions involved dividing students into small groups using Zoom breakout rooms to work on class problem sets or a challenge problem that required extending and combining major concepts from lecture. Upon return to the large groups, students would engage with each other and the instructor to establish methods for approaching problems rather than focusing on finding the answer.

Here are some student comments on aspects of the course that they liked related to synchronous sessions (organic chemistry):

- "The problem sets and you–do material were generally very helpful and arguably the activities that helped me learn the most from this class."
- "Practicing the problem sets allowed me to actually learn what the video lectures said."

- "Synchronous sessions that included break out rooms for either scheduled group activities or just time to discuss problems from the Problem Sets or You Dos from the lectures were the most helpful."
- "The group sessions working through problems with other students are extremely helpful."

The comments on these evaluations were only positive about the lectures and their benefits. In passing, students did comment to the instructor that it was occasionally hard to manage time to watch the asynchronous lecture before synchronous sessions. Students commented to the instructor that they did feel that given the pace of the class during the summer there were too many assessments (in the form of quizzes). However, students also expressed liking having several assessments because this helped them stay on track with their studying of the material. Because these comments were communicated verbally to the instructor, we do not have direct quotes that we can share.

4.3 Discussion and Conclusions

All courses described in this chapter had strengths as well as points of improvement. Our primary goal in designing courses for online delivery was to have students understand the material and prepare them for their next courses while maintaining the educational quality of an in-person semester. All of the instructors had the goal of having integrations of technology that would accommodate to and be inclusive of all students, which helps with educational continuity.[9] As we transition back to mostly in-person instruction, we now bring with us lessons on inclusivity and accessibility along with an incorporation of technology and active learning that provided a positive contribution to our courses.

The flipped classroom approach provides students with opportunities to set their own pace to watch lecture content, while leaving in-class time to interact with instructors and peers to have more meaningful and in-depth discussion of the content. The asynchronous component with a sufficient window for completion provides flexibility to students in different time zones, especially international students, to be equally engaged with the lecture content.

Additionally, the use of short lecture videos (10–15 min each) in Chem 111A and summer courses provided insight on lecture design. It is likely that after every 10–15 min of lecture, a different activity could be implemented to re-engage students and redirect their attention to course material. These activities could be a Clicker question (or a pathway question), a demonstration, or a problem-solving activity.

While video lectures can improve learning and grades, it is important to remember that synchronous activities and a flipped classroom approach whether remote or in-person can increase engagement and learning in both general and organic chemistry courses.[10,11]

In a similar vein, the low-stake lecture review questions implemented in Chem 105/106 (Fall/Spring) served a purpose similar to short lectures and pathway questions. It was found to be beneficial to student learning by ensuring that they watched the lecture videos on a regular basis rather than binge watching the night before quizzes and exams. The short-answer questions served as a quick self-assessment for students to check their understanding of the basic knowledge, in receiving instant feedback on the correctness of their answers through the learning management system. Moreover, providing a second attempt allows student to learn from their own mistakes, and as a low-stakes assignment it provides an alternate way for students to demonstrate their understanding of the course content, in addition to traditional high-stakes assessments (*i.e.*, quizzes and exams). The low-stakes assignments are designed to keep students on track with the lecture content and help all students complete course work and stay engaged with the course content in a guided manner. Such formative feedback is beneficial for students.[12] We realize that students who have high motivation and tend to complete any type of class work frequently also get better grades, but this approach makes sure that the success of students does not solely rely on students' own motivations and discipline but is supported by instructors to give all students an equal foundation. The fact that more than 88% of the students completed the majority of these assignments suggests the high engagement of students to this course, which is the main goal we would like to achieve through this assignment.

Considering there was positive feedback from students, we hope to continue implementing these practices in introductory chemistry courses even when the instructional modality returns to in-person instruction. We believe that the two approaches implemented during the pandemic can provide a benefit to all students regardless of modality. We learned from work published in the past on how to properly design a flipped classroom and learned important lessons that have shaped our courses for the future. The strength of our data was limited by what information we could gather and present. However, these anecdotal data helped us determine that we were going in the right direction. Connecting with the students was and is an integral aspect of our courses, and we have observed that students and instructors believe that these methods will help in establishing connections and creating inclusive and equitable environments that lead to meaningful chemistry learning.

Acknowledgements

The authors would like to thank the rest of the instructor team for letting us write this chapter. We also want to thank the students and everyone at Washington University in St. Louis who had to adapt to the difficult academic year during the pandemic.

References

1. J. W. Baker, in *Selected Papers from the 11th International Conference on College Teaching and Learning*, ed. J. Chambers, Florida Community College at Jacksonville, Jacksonville, FL, 2000.
2. B. Tucker, The Flipped Classroom Online instruction at home frees class time for learning, https://www.educationnext.org/the-flipped-classroom/.
3. N. Dietrich, K. Kentheswaran, A. Ahmadi, J. Teychene, Y. Bessiere, S. Alfenore, S. Laborie, D. Bastoul, K. Loubiere, C. Guigui, M. Sperandio, L. Barna, E. Paul, C. Cabassud, A. Line and G. Hebrard, *J. Chem. Educ.*, 2020, **97**, 2448–2457.
4. M. Noetel, S. Griffith, O. Delaney, T. Sanders, P. Parker, B. del Pozo Cruz and C. Lonsdale, *Video Improves Learning in Higher Education: A Systematic Review*, 2021, vol. 91.
5. B. Bloom, M. Englehart, E. Furst, W. Hill and D. Krathwohl, in *Handbook I: Cognitive Domain*, David McKay., New York, 1956.
6. D. N. Perkins and G. Salomon, in *International Encyclopedia of Education*, Pergamon Press, Oxford, England, 2nd edn, 1992, pp. 6452–6457.
7. B. Moran, NYTimes.com, 2013, 11–14, https://www.nytimes.com/2013/11/03/education/edlife/how-to-get-an-a-in-organic-chemistry.html.
8. R. L. Stowe and M. M. Cooper, *J. Chem. Educ.*, 2017, **94**, 1852–1859.
9. See Chapter 17: J. Y. Han and F. M. Fung, in *Digital Learning and Teaching in Chemistry*, ed. Y. J. Dori, C. Ngai and G. Szteinberg, Royal Society of Chemistry, United Kingdom, 2023.
10. S. R. Mooring, C. E. Mitchell and N. L. Burrows, *J. Chem. Educ.*, 2016, **93**, 1972–1983.
11. J. F. Eichler and J. Peeples, *Chem. Educ. Res. Pract.*, 2016, **17**, 197–208.
12. See Chapter 27: G. A. Lawrie, in *Digital Teaching and Learning in Chemistry*, ed. Y. J. Dori, C. Ngai and G. Szteinberg, Royal Society of Chemistry, United Kingdom, 2023.

CHAPTER 5

Personalized Support for Students Learning Chemistry Online—The Development of a Prediction Model

RACHEL EIDELMAN[a] AND YAEL SHWARTZ*[b]

[a]Davidson Institute, Weizmann Institute of Science, Rehovot, Israel; [b]Department of Science Teaching, Weizmann Institute of Science, Rehovot, Israel
*E-mail: yael.shwartz@weizmann.ac.il

5.1 Introduction

5.1.1 COBLE

In Israel, chemistry studies are an optional science choice for students who are interested in deepening their knowledge of the subject, and students usually study chemistry for three years. The Chemistry Online Blended Learning Environment (COBLE) is a three-year distance learning programme (10–12th grades), aimed at learning chemistry as a major in high school, offering a new, and hopefully an upgraded and appealing way to study chemistry while attaining the matriculation exam. Students learning through COBLE are those who do not have the option to study chemistry in their schools, because it is not offered (*e.g.*, students from small schools in the periphery, ultra-orthodox schools restricted by social rules, *etc.*).

Advances in Chemistry Education Series No. 11
Digital Learning and Teaching in Chemistry
Edited by Yehudit Judy Dori, Courtney Ngai and Gabriela Szteinberg

Published by the Royal Society of Chemistry, www.rsc.org

COBLE contains regular weekly synchronous lessons and asynchronous tasks, as well as weekly tutorial lessons. Chemistry laboratories are held in three different ways: (1) home-lab kits; (2) virtual labs; and (3) face-to-face labs held 2–3 times a year, mostly during holiday breaks. The course development is based on the following foundations: (1) cognitive—the active and interactive learning theory;[1] (2) design—according to lesson plans and virtual environments;[2–4] and (3) content—according to the Israeli curriculum.[5]

5.1.2 E-learning and Blended Learning

COBLE design and implementation leans heavily on the research and development regarding virtual and blended learning. The International Association for K-12 Online Learning (iNACOL) defined online learning as education in which instruction and content are delivered primarily *via* the internet and blended learning as learning that combines two modes of instruction, *e.g.*, online and face-to-face, but at potentially different points in time.[6] Blended courses combine e-learning and other components, as in our case, where students participated in face-to-face chemistry inquiry laboratory activities. The blended form attempts to deliver "the best of both worlds" (p. 3),[6] attempting to combine the advantages of online learning with traditional classroom benefits. E-learning has the potential to influence the way learners all over the world are educated, providing access to the best teachers, the best international resources, and subjects that are not otherwise offered, provided that the culture, policy, and support structures are congruent.[7]

Several researchers have tackled the issues that emerge when students encounter e-learning. Tallent-Runnels *et al.*[8] reviewed 76 papers and found that e-learners were more successful when the course was well-designed and emphasized the importance of the designer's role in determining the educational design theory used, and in the overall success of the students taking the course. They also addressed four main categories that emerged when students encountered e-learning: (1) course environment (classroom culture, structural assistance, success factors, interaction online, and evaluation); (2) learners' outcomes (understanding teaching and learning processes in the virtual environment); (3) learners' characteristics (motivation to take a virtual course, learner's goals and needs); and (4) institutional and administrative factors (clear policies for virtual courses, such as a support system). Boelens *et al.*[9] reviewed 20 studies on the design of blended learning environments and discovered four key challenges: incorporating flexibility; stimulating interaction; facilitating students' learning process; and fostering an effective learning climate. They concluded their review by stating that social interaction deserves more attention in future designs because it enhances the students' sense of belonging (to a school or a class). E-learning was once regarded by many as a negative form of learning, since both students and teachers believed that technology had a negative effect on their studies.[10,11] Today, there seems to be a more positive attitude toward e-learning, and there is an understanding that this form of learning is beneficial in the long term.[12]

5.1.3 Self-regulated Learning (SRL)

Self-regulated learning (SRL) skills are important for learning and may pose difficulties when flawed. They are especially important in a virtual learning environment, since students lack the immediate ability to seek help from teachers.[13] Self-regulated learners are defined as proactive seekers of information, and as those who take the necessary steps to master what they learn.[14] These learners are conceptualized as being metacognitive, since they plan, set goals, organize, self-monitor, and continuously self-evaluate themselves during learning processes. They are also motivated, hold high self-efficacy, engage in self-attributions, and have intrinsic task interest.[14,15] Teaching students to self-regulate their academic learning can be successful only when students experience the benefits of SRL.[16] Four design principles were suggested in order to promote SRL skills: (1) SRL activities are to be explicitly presented to the students; (2) learning events should include opportunities to use SRL strategies; (3) SRL skill interventions are mandatory; and (4) students must experience success while using SRL skills in order to continue to use them regularly.[16]

In the following paragraphs, several SRL representations will be introduced. Each one of them provides useful data on student's learning and SRL characteristic, however—identification of influential indicators from all representations allowed the building of a prediction model. The model can be used to predict student's ability to learn in an online environment and enable future planning of personalized support based on students' SRL skills.

The SRL representations are: self-regulated learning on-going profiles of COBLE students and dropouts; involvement score, which characterized students' synchronous and asynchronous behavior (also referred to as 'effort mark'); and activity pattern, which integrated SRL components with students' achievements over time.

5.2 Following Students Learning

5.2.1 Self-regulated Learning Profiles

During the first cohort of COBLE, an SRL study was conducted. We created SRL profiles and followed the changes in student's profiles during 10–12th grades. We explored the correlation between student's SRL profiles and changes in these profiles, tracking their level of success during and at the end of the course.

To do so, we used a modified and translated (from English to Hebrew) form of the Learning and Study Strategies Inventory (LASSI) which is a Likert-type questionnaire, containing 80 items attributed to ten different categories.[17] The modified version contained 48 items related to six out of ten categories, taken as they are from the original LASSI questionnaire: attitude and interest (ATT), concentration and attention to academic tasks (CON), motivation, diligence, self-discipline, and willingness to work hard (MOT), use of support

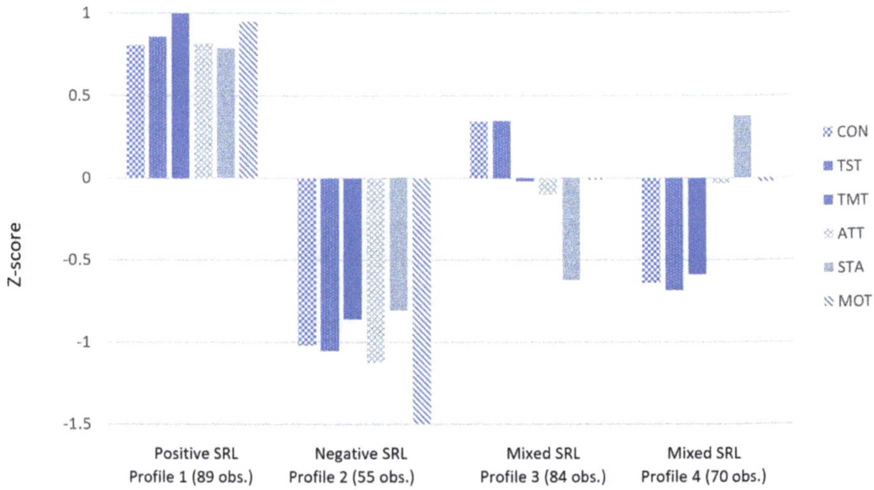

Figure 5.1 The four SRL profiles and their distribution ($N = 298$).

techniques and materials (STA), use of time management principles for academic tasks (TMT), and finally, test strategies and preparing for tests (TST). For establishing a SRL reference score, the questionnaire was administered to 9th-grade students. Their scores were considered the level of self-regulated learning that is typical before starting 10th-grade studies. Four profiles emerged as result of this study: all positive SRL profile—Profile 1, all negative SRL profile—Profile 2, mixed SRL profiles—Profiles 3 and 4. SRL profiles and their distribution are presented in Figure 5.1. For detailed analysis see Eidelman & Shwartz.[18]

5.2.2 Involvement Score

Interaction and involvement of all sorts are important since the lack of interaction can result in disengagement and be the cause of student dropout.[19] Involvement in the course was imperative for students' progress. An involvement score was defined and calculated using big-data methodologies. The involvement score was composed of synchronous behavior and asynchronous behavior (see Figure 5.2).

Synchronous behavior contained all synchronous non-obligatory actions in the course and was divided into two components: (a) participation in the synchronous lesson—despite being obligatory, participation in the synchronous lesson was also considered as synchronous behavior since it is related to the synchronous lesson after all; (b) synchronous effort mark (SEM)—three synchronous effort components were counted during each lesson and were scaled together to form a unified synchronous effort mark (SEM): chat volume (placing messages of various types in the synchronous lesson chat

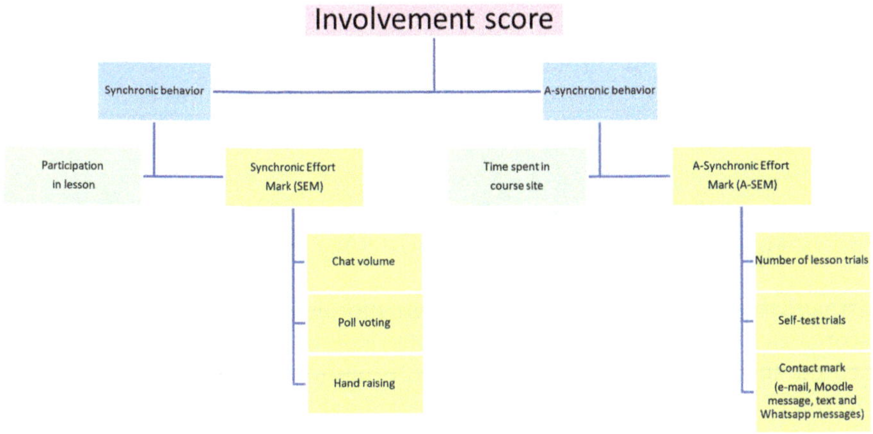

Figure 5.2 Involvement score components.

area);[†] hand-raising (to ask a question, participate, or as a response to the teacher's request); and poll-voting (as poll-like questions were used during the lessons, students were expected to cast their vote when asked to do so). A personal code was assigned to each student, linking directly between them to any action they initiated. All students' actions during the synchronous lessons were recorded and stored and the data were counted and scaled by using the R statistical software and programming language.[20]

Asynchronous behavior, which contained all asynchronous optional assignments and tasks: (a) time spent at the site—the existence of "time spent at the site" in the A-SEM is not acknowledged, since it is impossible to differentiate the actual time of student engagement or idle connectivity to the site; (b) 'asynchronous effort mark (A-SEM)'—students' non-obligatory actions and effort during asynchronous activities were measured by recording all relevant activities such as: (i) percentage of student attempts in homework task assignments—each assignment had up to two optional attempts (to improve scores, maximum = 108 possible attempts were available during the whole course); (ii) percentage of self-tests used by the student—students could self-assess their knowledge by answering self-tests (maximum = 15 self-tests were provided during the course); and (iii) the calculated percentage of contact attempts, *via* any contact means—public course forum, privately or directly to the teacher, e-mailing to the course mailing address, text or WhatsApp message *via* personal phone. These actions were counted and scaled for each student to form a unified asynchronous effort mark (A-SEM).

[†]Chatting was the major aspect of SEM and it is hereby elaborated: the majority of chat messages were chemistry content-related messages (63%). This indicates that students were engaged in the lesson and asked about, and/or widely used the scientific terminology.

5.2.3 Achievements

All score assignments throughout the course were recorded and analyzed by frequency means comparison. A semester score was given to each student, and their matriculation (MAT) exam score was also collected.[18]

5.3 Activity Patterns: the Construction of a Prediction Model

An activity pattern for each student was plotted, using the SEM semester scores and last achieved SRL profiles. Matriculation score (MAT) is also presented above the student's plot. The information includes three interconnected main parameters of students' learning: (1) student's SRL profile and changes in this profile over time; (2) student's synchronous activity mark (SEM) and changes in this mark over time; and (3) student's achievements including their score in the final matriculation examination. The division formatted four groups (one for each SRL profile) presenting clear relationship and logical explanation for all variables.

5.3.1 General Interpretation of the Structure of Activity Patterns

Figure 5.3 demonstrates a typical activity pattern graph for each student in the course. It is important to emphasize that the A-SEM is not represented in the activity patterns directly, but as a part of the student's semester score.

G.A. is the name code of a student who joined the class in the second semester of the 11th grade. Components of G.A.'s activity pattern graphs

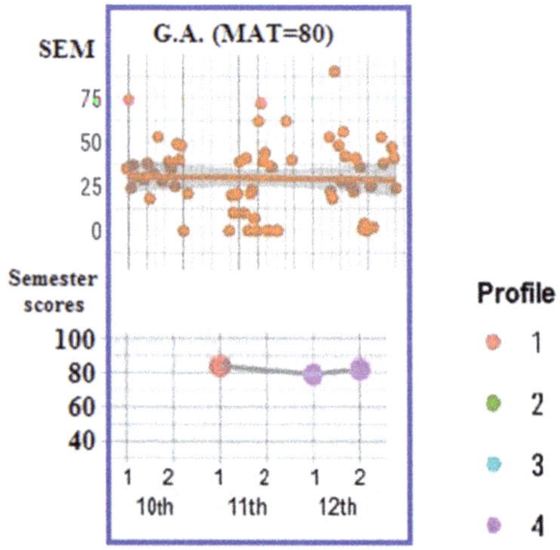

Figure 5.3 A typical activity pattern graph.

are: (1) the student's name code (G.A.); (2) the student's matriculation score (MAT = 80); (3) year and semester (scale on x-axis: 10th [1–2], 11th [1–2], 12th [1–2]); (4) SEM score (scale on y-axis: 0–100%, represented by the small orange dots throughout the three-year course, during the summer breaks there are no dots since there was no SEM activity); and (5) semester scores, ranging from 30–100 on the y-axis. The color of the large dot represents the profile type of the student at the given point in time. In this example, the student was profiled as all-positive for the first semester of the 11th year, and as the fourth mixed type the year after.

Figures 5.4–5.10 show the activity patterns of various groups of students.

5.3.1.1 *Very Active Students*

High-achieving active students, who were also active during the synchronous lessons. Interestingly, very active students hold various SRL profiles, which for students G.A. and U.S. also changed over time. The matriculation scores for the students in this group did not fall below 80 (Figure 5.4).

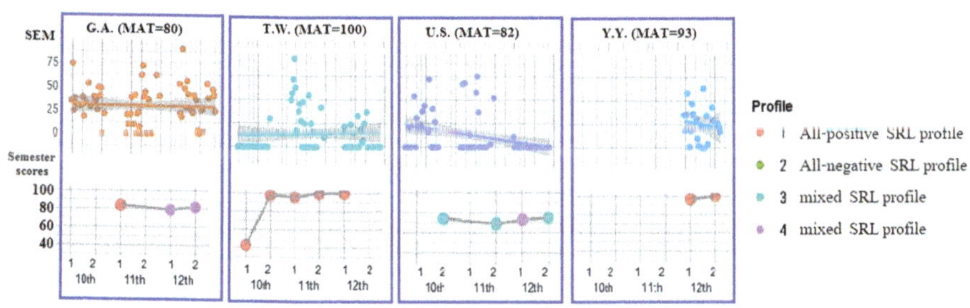

Figure 5.4 Activity patterns for very active students.

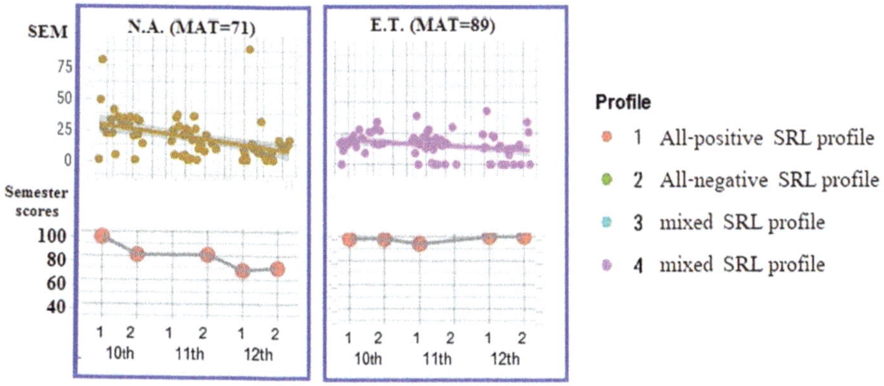

Figure 5.5 Activity patterns for mildly active students—all-positive SRL profile.

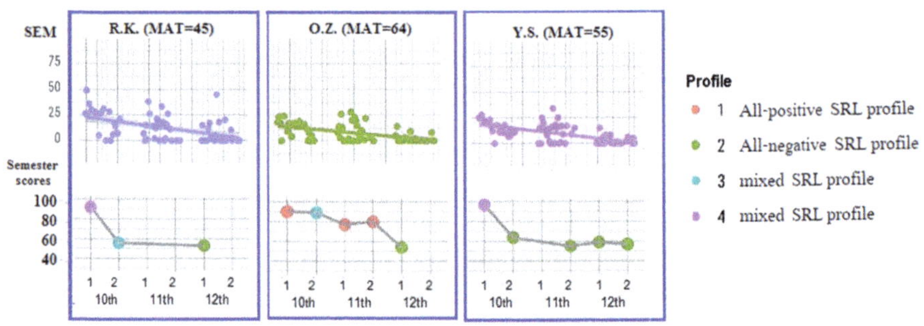

Figure 5.6 Activity patterns for mildly active students—all-negative SRL profile.

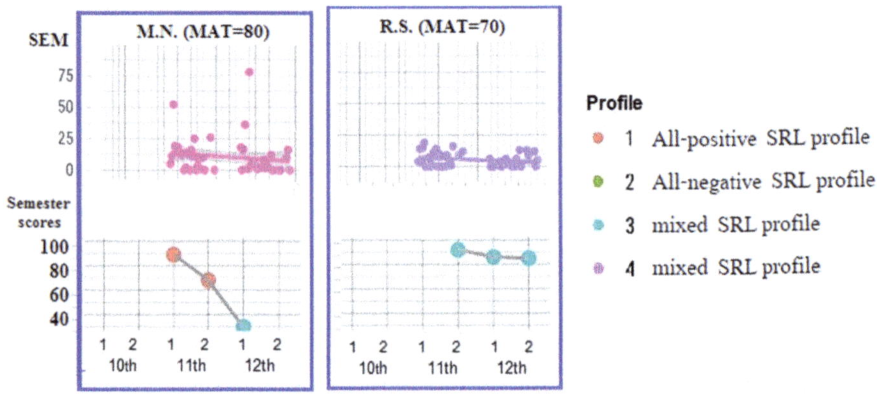

Figure 5.7 Activity patterns for mildly active students—last type 3 mixed SRL profile.

5.3.1.2 Mildly Active Students

The matriculation scores of students with this pattern were widely ranged. We also found sub-trends within this group, according to the last exhibited SRL profile.

(a) Last SRL profile: all-positive SRL (Figure 5.5).
(b) Last SRL profile—all-negative SRL (Figure 5.6): students in this group were very low-functioning students who finally deteriorated to the all-negative SRL profile. All students in this group achieved low MAT scores.
(c) Last SRL profile 3 (a mixed SRL profile, CON, TST > 0; STA, ATT < 0; TMT, MOT ≈ 0) (Figure 5.7).
(d) Last SRL profile 4 (a mixed SRL profile, STA > 0; TMT, CON, TST < 0; ATT, MOT ≈ 0) (Figure 5.8).

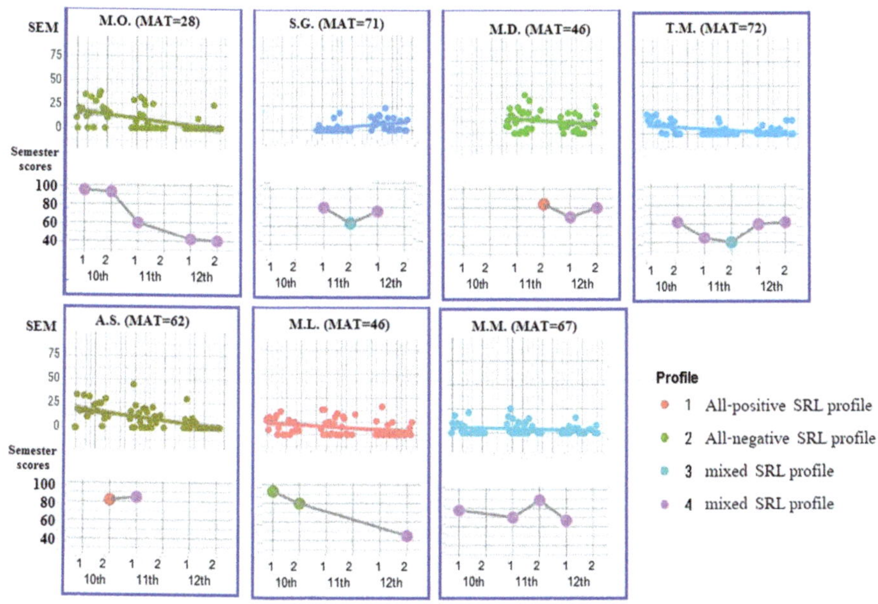

Figure 5.8 Activity patterns for mildly active students—last type 4 mixed SRL profile.

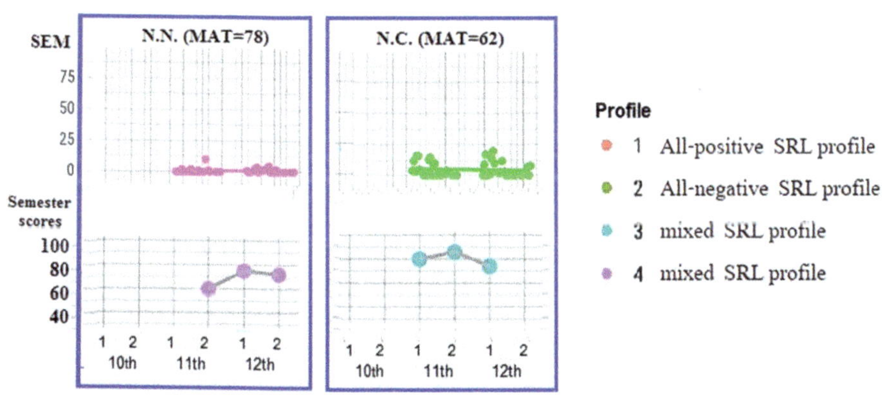

Figure 5.9 Activity patterns for non-active students.

5.3.1.3 *Non-active Students*

In this group, we can see a very low SEM activity and medium to low MAT scores. All students in this group had mixed SRL profiles with no changes along their three years of studies (Figure 5.9).

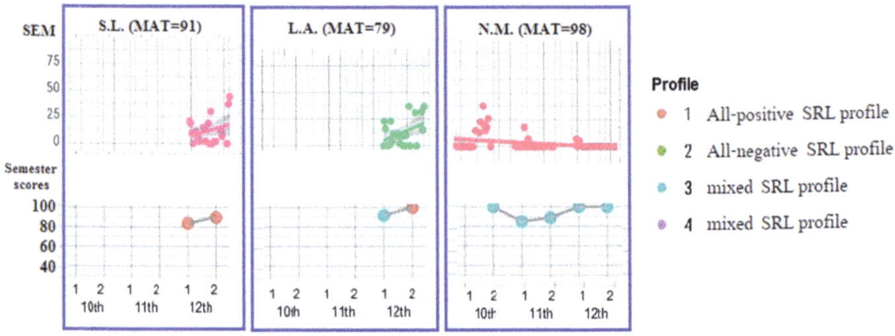

Figure 5.10 Activity patterns for diversely active students.

5.3.1.4 Diversely Active Students

In this group, we had students who became more active over time (positive trend) and students that decreased their activity over time (negative trend). Most students in this group became less active over time until they dropped out (Figure 5.10).

What have we learned from creating a personal activity pattern? The combination of students' activity, SRL profile and semester achievements allows teachers to better interpret and understand student's remote learning process, and address students' personal needs for support.

However, can it be used to create a model that will allow prediction of a student's success? Identifying such predictors and following them may increase success and lower the number of dropouts by allowing teachers to develop support strategies according to changes occurring in the student's activity patterns, thus minimizing dropout.

We identified a clear relationship between the last-observed SRL profiles of the students and all other components: achievement group, MAT score and activity pattern. All-positive last-observed SRL profile were high achievers that were highly active during the lessons. Students with all-negative last-observed SRL profile, were low-achievers (except for one medium achiever) and low MAT scoring students, all were mildly active students. Students with mixed last-observed SRL profiles (type 3 and 4) were mainly medium and high achievers with medium MAT score students.

5.3.2 Success Predictors

Beyond the value of creating a methodology for tracking students' activity profiles in a virtual environment, our main goal was to locate the most influential factors that may serve as predictors, promoting student's success and highlighting difficulties and needs, to support and minimize dropouts.

Different predictors can be taken into consideration in order to locate the most influential ones that may promote student's success.[21] As discussed, there were quantitative measures that were related to success, although one can argue that there are many types of success: for example, engagement, completion of the three-year course, leaving the course with a positive perception of chemistry, *etc.* In this study, students' matriculation scores served as the outcome of the prediction. A success model was created, using multiple linear regression to predict future MAT scores, based on five independent variables: SRL profiles; achievement group; SEM; A-SEM; and activity patterns. Two promising success predictors were identified: involvement (expressed by SEM) and SRL profile type. The multiple linear regression results regarding the two influential predictors mean that a student possessing an all positive SRL profile, can achieve a score of up to 38.53 points higher than a student possessing other SRL profiles. Additionally, student's MAT scores may be increased by 24.25 points, for students who possessed a 'mildly active' activity pattern (over other activity patterns).

5.4 Discussion

COBLE is a dynamic environment undergoing constant changes and redevelopments that has succeeded to cope with rapid growth in student numbers, making the necessary adjustments to improve students' achievements. COBLE is now a leading virtual programme for high-school students in Israel. Its main contribution is that it allows more students to study chemistry throughout the country.

The programme's success is expressed by: exponential growth in participating students; enabling schools from the periphery to offer chemistry courses in a special urban model programme; and creating and maintaining an all-girl ultra-orthodox class (in schools where core subjects such as mathematics and English are not always taught, with limited and controlled access to the internet).

In addition, the COVID-19 pandemic caught the programme in a mature form, which allowed us to continue our online routine learning, while all other schools were closed. We were able to support other chemistry teachers in the country by allowing open access to our teaching materials. We also allowed all chemistry students in the country to participate in our test-preparation sessions before the matriculation examination.

From a research point of view, the aim of this study was to investigate the characteristics and self-regulated learning skills of high school students in a blended learning environment. During the three-year study, there were individual and group changes in the intensities of some of the categories in the SRL profiles. Not surprisingly, graduating students showed a more positive SRL profile than dropouts did. In a different study we found that COBLE students showed a more positive SRL profile than chemistry face-to-face students did. This suggests that COBLE actually helps promote the development of self-regulated learning skills over time.[18] The need to explicitly promote SRL skills amongst high-school students learning online, is supported by Ho, Ibraj & Graulich,[22] (see Chapter 3) who found that during the COVID-19 outbreak, challenges reported by university students were difficulties adapting to online learning and changing ways of regulating one's own learning. Furthermore, they reported that professors rarely gave advice on how to regulate while learning online or how to engage with the learning materials while learning alone at home.

The data that have been collected led to the composition of a holistic activity pattern and enabled the assumption of a basic success prediction model for students who study in a chemistry-blended learning environment. This prediction model was based on the relationship between students' self-reported SRL skills and success in their studies *via* a blended learning environment, linking between student's involvement and their success in the course (expressed by their matriculation scores). An additional link between students' SRL profile (expressed by modified LASSI questionnaires) and their success in the course (expressed by their matriculation scores) was found. Prediction models in educational learning environments are not new. Agudo-Peregrina *et al.*[23] examined two different learning modalities: (1) a virtual learning environment (VLE) supported by face-to-face lessons; and (2) online learning. They defined three system-independent classifications of interactions and evaluated the relation of their components through academic performances. They found that different types of interactions within each classification were related to student academic performance, but only in courses delivered online and not in VLE-supported face-to-face courses. Similarly, Calvo-Flores *et al.*[24] created a prediction model that was based on the interactions between the students and the course learning management system (Moodle), free of demographic values or marks in other subjects in order to predict the students' performance. Their proposed model showed that it was possible to detect students who may have problems in successfully completing a course. These findings line up with the current study in the sense that the suggested prediction model does not depend on outside sources, but the data within the LMS (in this study's case: Moodle), and because they highlight interaction as a predictor for success, as involvement in this case study.

Some researchers indicated that students learning in a blended learning environment achieve higher average scores than their non-blended learning environment peers, as well as reporting an increase in course completion rates, improved retention and increased overall student satisfaction in blended learning environments.[25,26] In this study, there is evidence for this to be true; the intervention students' average MOE exam score was almost 6% higher than the national Israeli students' average MOE exam score for the same year.

Some limitations of this study and any prediction model lay in the fact that the outcomes of learning may be affected by many factors, some of which have no connection with the model. In this study, the focus was set on the influence of student's characteristics on the student's success in terms of scores. Nevertheless, since the study was relatively long (three years), many causes could have influenced the adolescent students throughout and have been the cause of difficulties and dropouts. Another limitation of this study is the need to test the prediction model on a larger sample of students.

5.4.1 Design Implications

There are also some important design implications. Although the course was initially designed as a virtual course, we have realized the importance of blended learning, and the need for more opportunities for face-to-face interaction (science camps, laboratory days and face-to-face exercise lessons). The design should support and maximize students' abilities, by challenging strong students and supporting weak students (either by virtual or by face-to-face means). Enrichment activities and linkage to daily life and personal interest areas are imperative to promote students' satisfaction, interest, and positive attitudes. Encouraging students' active involvement and participation during the synchronous or asynchronous lessons is crucial for success in terms of achievements and therefore, must be carefully considered and incorporated in the design as well. Students should receive feedback regarding their knowledge in order to make progress in their studies.

An interesting SRL category that needs future study and focus is motivation (MOT). Motivation is absent from the mixed SRL profiles generated from the LASSI questionnaires. Dropouts in this study reported the lack of intrinsic motivation and sufficient interest in the subject, as well as the need for more face-to-face contact as the main reasons for not continuing their chemistry studies. Other researchers also highlight motivation as one of the factors that may have an impact on success and retention in online courses; 62% of the satisfied online participants and 82% of the unsatisfied participants reported the impact of motivation on their success in online learning.[27] COBLE students have dealt with many logistical problems which face-to-face school students do not normally encounter that may have influenced their motivation to study, such as overlapping lessons, finding a quiet place with a computer connected to the internet at school, or getting home in time to connect to the synchronous lesson. In fact, one might expect a more prominent expression of this feature in this study since it is believed that motivation is a crucial factor for success in a blended environment.[8] Motivation is also linked to involvement in learning processes,[28] and was found as a predictor for success in this study's prediction model.

One of the major research implications of this study is that the success prediction model can serve as a productive framework for studying SRL skills in a blended learning environment in more detail, moving the research area in the direction of a larger research scale. The success prediction model developed in this study is based upon the SRL characteristics of the chemistry students, but these SRL characteristics may serve as predictors for any blended environment, in any given subject. Personalizing the learning processes may also benefit from the prediction model and may result in the automatic production of an individual adaptive support system for students willing to study this way, promoting self-regulated learning that they may or may not exhibit at the beginning of their studies.

5.4.2 Teaching Implications

Teachers who want to incorporate blended learning in their classrooms should be aware that there is a need for face-to-face support, and that it is imperative to develop the bond between the teacher and the students.[29] This can be achieved by easy access to teachers or other students and available contacting methods, as in the science camps planned for this study.[30] In online learning, learner satisfaction may be a factor influencing learning outcomes and dropout rates. Two factors may influence students' satisfaction: (1) learner–instructor interaction;[31] (2) the quality of technology use[31–33] and (3) understanding the learning management system and its tools.[34] There is therefore a need for student instruction prior to starting the course, in order for students to later succeed, save time, prevent frustration and enhance satisfaction.

Some of the teachers' goals worldwide are to increase learners' levels of knowledge construction, analytical skills, and the ability to assess and to critically evaluate knowledge sources. Blended learning can be a means to achieve these goals by utilizing the independent and flexible virtual space created for the construction of knowledge while monitoring the learning processes in the classroom, face-to-face, by the teacher. Such an example can be found in Chapter 13 by Joselevich, Moro and Martínez.[35] They found that students actively and autonomously use YouTube chemistry videos instead of textbooks. Learning from videos was found to be familiar and attractive for students with diverse learning styles, contributing among other things to the increase of self-efficacy and relevance.

Scaffolded assignments should slowly guide students through the learning process, and also gradually transfer the responsibility over to the students. Personalization of the learning processes is another means of support that teachers can benefit from and utilize for enhancing learning for all levels of student knowledge in each class. This assistance can be offered whether a teacher teaches in a virtual, face-to-face, or blended learning environment since all students can gain from this.

Students' SRL skills should be mapped at the beginning of their studies and monitored by their teachers in order to prevent any decrease in their SRL skills throughout their studies; maintaining interest and attitude is crucial for the creation of motivation, and for students' ability to invest time and use the learning materials available.

5.5 Recommendations for Education

Education systems are complex entities with limited resources. Often, innovative teachers are stopped in their tracks by the hindrance of the education system bureaucracy. One of the main problems that need to be resolved is the lack of conditions available in schools in order to enhance computer literacy and computer skills amongst teachers and students since this could be the cause of failure leading to frustration and eventually to students dropping out in e-learning and blended learning courses. Blended learning requires a high level of computer competence[36] and confidence that is gained by practice, which eventually becomes capability to participate in the blended learning environment.[37] Schools must direct attention and resources to the development and maintenance of the hardware in the schools, so there can be an effective and reliable support system for the teachers and students, enabling them to participate in external e-learning courses. There is an added bonus for students who are exposed to innovative means of learning in the education system. Employers value technologically trained employees who are creative and innovative and are not afraid of trying new paths of action in order to solve problems that might emerge. These future workers can add value to their companies and will be potentially beneficial to them. Furthermore, Kintu *et al.*[29] recommend that higher education institutions should also encourage the integration of blended learning in high schools so that students will not experience a technology shock when encountering the higher education system, and in order to enable effective learning and research through technology in a developing world.

References

1. M. T. Chi, Active-constructive-interactive: A conceptual framework for differentiating learning activities, *Top. Cogn. Sci.*, 2009 1, **1**, 73–105.
2. G. Collison, B. Elbaum, S. Haavind and R. Tinker, *Facilitating Online Learning: Effective Strategies for Moderators*, Atwood Publishing, Madison, WI, 2000.
3. B. Elbaum, C. Mclntyre and A. Smith. *Essential Elements: Prepare, Design, and Teach Your Online Course*, Atwood Publishing, Madison, WI, 2002.
4. R. Palloff and K. Pratt, *Lessons from the Cyberspace Classroom*, Jossey-Bass Publishers, 2001.
5. Ministry-Of-Education, http://cms.education.gov.il/EducationCMS/Units/Mazkirut_Pedagogit/chimya, July 2022 (in Hebrew).
6. M. B. Horn and H. Staker, *The Rise of K-12 Blended Learning*, 2011, Innosight Institute, http://www.christenseninstitute.org/wp-content/uploads/2013/04/The-rise-of-K-12-blended-learning.pdf July, 2022.
7. C. McNaught, R. Phillips, D. Rossiter and J. Winn, *Developing a Framework for a Useable and Useful Inventory of Computer-facilitated Learning and Support Materials in Australian Universities*. AusInfo, Canberra, ACT, 2000.
8. M. K. Tallent-Runnels, J. A. Thomas, W. Y. Lan and S. Cooper, S. Teaching Courses Online: A Review of the Research, *Rev. Educ. Res.*, 2006, **2006**(1), 93–135.
9. R. Boelens, B. De Wever and M. Voet, Four key challenges to the design of blended learning: A systematic literature review, *Educ. Res. Rev.*, 2017, **22**, 1–18.
10. Oxford Group, Blended learning-current use, challenges and best practices, 2013, https://pdf4pro.com/cdn/blended-learning-current-use-challenges-and-2bdc4b.pdf.

11. Q. Li, Student and Teacher Views about Technology: A Tale of Two Cities? *J. Res. Technol. Educ.*, 2007, **39**(4), 377–397.
12. E. Allen and J. Seaman, *Grade Change: Tracking Online Education in the United States*, Babson College Survey Research Group, Needham, MA, 2014.
13. M. H. Cho, *The Effects of Design Strategies for Promoting Students' Self-Regulated Learning Skills on Students' Self-Regulation and Achievements in Online Learning Environments*, Association for Educational Communications and Technology, 2004.
14. B. J. Zimmerman, Self-regulating learning and academic achievement: an overview, *Educ. Psychol.*, 1990, **25**(1), 3–17.
15. B. J. Zimmerman, Investigating Self-Regulation and Motivation: Historical Background, Methodological Developments, and Future Prospects, *Am. Educ. Res. J.*, 2008, **45**(1), 166–183.
16. K. Ley and D. B. Young, Instructional principles for self-regulation, *Educ. Technol. Res. Dev.*, 2001, **49**(2), 93–103.
17. C. E. Weinstein, D. R. Palmer and A. C. Shulte, *LASSI: Learning and Study Strategies Inventory*, H&H Publishing Company, Inc, Clearwater, FL, USA, 2nd edn, 2002.
18. R. R. Eidelman and Y. Shwartz, in *Fifty Years of Curriculum Development, Implementation, and Research in the Department of Science Teaching, the Weizmann Institute of Science, Israel*, ed. A. Hofstein, A. Arcavi, B. S. Eylon, and A. Yarden, Springer, 2021, vol. 3, pp. 44–70.
19. A. R. Artino, Think, feel, act: Motivational and emotional influences on military students' online academic success, *J. Comput. High. Educ.*, 2009, **21**(2), 146–166.
20. R Development Core Team, R: A language and environment for statistical computing. R Foundation for Statistical Computing, Vienna, Austria, 2017, http://www.R-project.org.
21. R. S. Baker and K. Yacef, The state of educational data mining in 2009: A review and future visions, *J. Educ. Data Min.*, 2009, **1**(1), 3–17.
22. See Chapter 3: F. Ho, K. Ibraj and N. Graulich, in *Digital Learning and Teaching in Chemistry*, ed. Y. J. Dori, C. Ngai and G. Szteinberg, Royal Society of Chemistry, United Kingdom, 2023.
23. Á. F. Agudo-Peregrina, S. Iglesias-Pradas, M. Á. Conde-González and A. Hernández-García, Can we predict success from log data in VLEs? Classification of interactions for learning analytics and their relation with performance. In: VLE-supported F2F and online learning, *Comput. Hum. Behav.*, 2014, **31**, 542–550.
24. M. D. Calvo-Flores, E. G. Galindo, M. P. Jiménez and O. P. Piñeiro, Predicting students' marks from Moodle logs using neural network models, *Curr. Dev. Technol.-Assisted. Educ.*, 2006, **1**(2), 586–590, https://pdfs.semanticscholar.org/5b8c/9dc63e4df7dcbd7316c4b445036cd7781f63.pdf.
25. E. Stacey and P. Gerbic, Teaching for blended learning: research perspectives from on-campus and distance students, *Educ. Inf. Technol.*, 2007, **12**, 165–174.

26. J. Kenney and E. Newcombe, Adopting a blended learning approach: Challenges, encountered and lessons learned in an action research study, *J. Asynchronous Learn. Networks*, 2011, **15**(1), 45–57.
27. L. Song, E. S. Singleton, J. R. Hill and M. H. Koh, Improving online learning: Student perceptions of useful and challenging characteristics, *Internet High. Educ.*, 2004, **7**(1), 59–70.
28. P. R. Pintrich and D. H. Schunk, *Motivation in Education: Theory. Research, and Applications*, Merrill Prentice Hall, Columbus, Ohio, 2nd edn, 2002.
29. M. J. Kintu, C. Zhu and E. Kagambe, Blended learning effectiveness: the relationship between student characteristics, design features and outcomes, *Int. J. Educ. Technol. High. Educ.*, 2017, **14**, 7, 10.1186/s41239-017-0043-4, https://educationaltechnologyjournal.springeropen.com/articles/10.1186/s41239-017-0043-4.
30. P. A. Willging and S. D. Johnson, Factors that influence students' decision to drop-out of online courses, *J. Asynchronous Learn. Networks*, 2009, **13**(3), 115–127.
31. K. Swan, Virtual interactivity: design factors affecting student satisfaction and perceived learning in asynchronous online courses, *Distance Educ.*, 2001, **22**(2), 306–331.
32. M. A. Naaj, M. Nachouki and A. Ankit, Evaluating student satisfaction with blended learning in a gender-segregated environment, *J. Inf. Technol. Educ. Res.*, 2012, **11**, 185–200.
33. G. Piccoli, R. Ahmad and B. Ives, B. Web-based virtual learning environments: A research framework and a preliminary assessment of effectiveness in basic IT skills training, *Manag. Inf. Syst. Q.*, 2001, 401–426.
34. W. H. Delone and E. R. McLean, The Delone and McLean model of information systems success: A Ten-year update, *J. Manag. Inf. Syst.*, 2003, **19**(4), 9–30.
35. See Chapter 13: M. Joselevich, P. D. Moro and M. A. Martínez, in *Digital Learning and Teaching in Chemistry*, ed. Y. J. Dori, C. Ngai and G. Szteinberg, Royal Society of Chemistry, United Kingdom, 2023.
36. K. Shraim and Z. N. Khlaif, An e-learning approach to secondary education in Palestine: opportunities and challenges, *Info. Tech. Dev.*, 2010, **16**(3), 159–173.
37. W. Hadad, *ICT-in-education Toolkit Reference Handbook*, 2007, InfoDev. From http://www.infodev.org/en/Publication.301.html, May, 2017.

CHAPTER 6

A Framework for Learning to Teach Chemistry on a Digital Platform: The Case of Chemical Equilibrium

E. MAVHUNGA*[a], M. ROLLNICK[a] AND D. VAN DER MERWE[a]

[a]Science Education, Wits School of Education, University of the Witwatersrand, Johannesburg, South Africa
*E-mail: Elizabeth.Mavhunga@wits.ac.za

6.1 Introduction

Humanity has become a digital society with major implications for all economic sectors including chemistry education. This move has created a need to seek for greater understanding informed by refined discipline-based digital teacher-knowledge theoretical frameworks. This is particularly urgent in the preparation of prospective chemistry teachers, where the traditional classroom pedagogies are increasingly becoming insufficient to meet the demands for online learning and teaching. As schools and higher education institutions try to mitigate various disruptions to the academic agenda by turning to remote teaching, more and more students find themselves physically separated from each other and their teachers. Even more concerning is the observed consequence of inequality in developing countries in accessing basic education. This experience is compounded by an increase in teacher anxiety as they navigate unchartered waters in trying to transform traditional

Advances in Chemistry Education Series No. 11
Digital Learning and Teaching in Chemistry
Edited by Yehudit Judy Dori, Courtney Ngai and Gabriela Szteinberg

Published by the Royal Society of Chemistry, www.rsc.org

classroom lessons into digital content. This chapter aims at contributing towards easing this physical–virtual tension by introducing a refined theoretical framework that links teacher knowledge for teaching science to the consideration for the isolated learner who learns virtually.[1] To illustrate the application of the newly conceptualized theoretical framework, we use an example of designing and creating a teaching video on chemical equilibrium. The conceptualization of the refined theoretical framework is informed by the understanding that the kind of teacher knowledge required to teach science content effectively over online platforms is complex. It is more than the mere integration of technology into a science lesson rather, requires deep knowledge transactions across multiple theoretical domains. We envisage that a refined theoretical framework would advance science teacher preparation by guiding the creation and delivery of pedagogically sound digital science lessons. It will serve as seed for growing our understanding of discipline-based digital teacher knowledge as technology developments progress. It will also serve as an analytical tool for framing ongoing research on the nature of teacher knowledge for creating sound digital science content for teaching through digital platforms. We begin with an account of our view of the centrality of Pedagogical Content Knowledge (PCK) as a discipline-based theoretical framework for teacher preparation.

6.2 The Preparation of Science Teachers through Pedagogical Content Knowledge

In science education, teacher preparation is widely regarded as been successfully conducted through the implementation of PCK. PCK is a framework based on the centrality of teachers' understanding and the transformation of the content knowledge of science as an academic discipline into versions that are comprehensible by learners. Ever since Shulman[2] introduced PCK, many science education researchers have empirically explored PCK's core element that enables such a powerful transformation. PCK is commonly understood to enable pedagogical transformation of content knowledge from a complex interaction among its components. At an operational level, PCK has been found more useful when considered at a topic level within a discipline. This is largely because of its topic specific nature, and the idiosyncratic way in which teachers apply it across topics. Mavhunga and Rollnick[3] have in previous studies developed a conceptual framework that operationalized the understanding of the topic specific nature of PCK, by enlisting the knowledge components that are content specific from which the transformation of content knowledge emerges, see Figure 6.1.

Having defined PCK at a topic level distinctively as Topic Specific Pedagogical Content Knowledge (TSPCK), has enabled the use of the construct to structure teacher education interventions, and to frame research on teacher knowledge for teaching science in various core topics of Chemistry.[4,5] We acknowledge that the use of the TSPCK construct has significantly

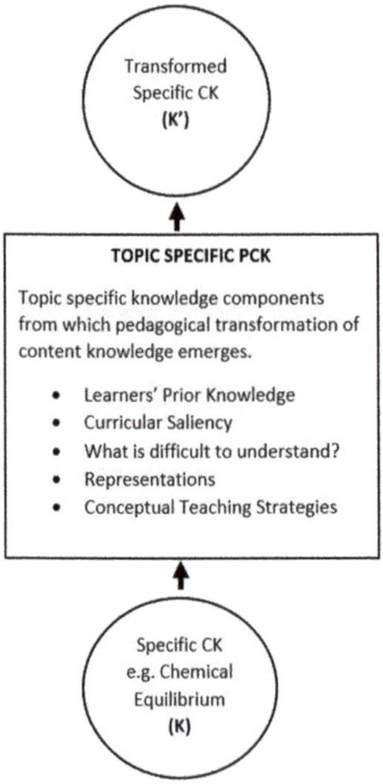

Learners' Prior Knowledge	Refers to the common misconceptions and alternative conceptions students may have. This also includes knowing about the correct knowledge they possess.
Curricular Saliency	Refers to identifying most important meaning to be established in a topic, as well as their sequencing for teaching.
What is difficult understand?	Refers to being able to identify which concepts are difficult to understand and be able to find different possible interventions to teaching them.
Representations	Refers to the understanding of the use of different representations of teaching a topic, including illustrations, analogies, simulations and models.
Conceptual Teaching Strategy	Refers to effective instructional strategies for specific concepts, misconceptions, areas that are difficult to understand. It is often a cumulative knowledge component embracing the use of the other knowledge components mentioned above.

Figure 6.1 Content specific knowledge components of TSPCK.

contributed to the improvement of the science pedagogy. However, as society moves into digitization, little is known about retaining and exporting the transformative benefit afforded by TSPCK when teaching is conducted through online teaching platforms. The purpose of this chapter is therefore to construct a refined theoretical framework for describing teacher knowledge needed for planning and teaching of digital pedagogically transformed science content through online education platforms. We have named the refined construct as digital-TSPCK (d-TSPCK), thereby distinguishing it from the traditional TSPCK theoretical construct.

6.3 Digital-TSPCK

The construct d-TSPCK is defined as the teacher's professional knowledge for digitally transforming content of science topics and teaching it efficiently through digital learning platforms for learner understanding. As with all other versions of PCK, the primary benefit to be derived from d-TSPCK is the pedagogical transformation of content knowledge for learner understanding. However, for d-TSPCK the process of transforming content knowledge of a topic is to happen digitally, resulting in the transformed content knowledge being in a digital form packaged in layouts such as a teaching video, simulation, animation, *etc.* As with traditional TSPCK, the process of pedagogical transformation of content knowledge continues to emerge from the consideration of the interactions among the knowledge components of the traditional TSPCK construct. However, in the case of d-TSPCK, these knowledge component interactions have been technologically enabled. By being technologically enabled, we mean the implementation of the TSPCK components, and their interactions have been made feasible through some form of teacher digital competence.

Consider a possible example of the interactions of the traditional TSPCK knowledge components in a lesson segment where a teacher is discussing shapes of molecules using the VSEPR model. It is important to note that simple recording of this lesson delivered in a typical traditional lesson using a video camera, would not constitute d-TSPCK even if the desirable traditional TSPCK component interactions are present. It will simply reflect traditional TSPCK captured using a digital tool, a video camera. The presence of d-TSPCK is demonstrated by the ability of the teacher to have some digital competence in creating purposefully designed digital teaching content reflecting the desirable transformative interactions among the traditional TSPCK knowledge components. A possible example could be a segment in a purposefully designed teaching video, where the teacher has made a strategic decision to use the electron molecular geometrical structural representations, instead of the version that shows no electrons to represent molecules with the same number of covalent bonds but different shapes *e.g.*, NH_3 *versus* BF_3. The purpose here would be to illustrate the effect of the electron pair repulsion. Hidden in this choice is the teachers' understanding of the common learner's misconceptions that all molecules with same number of bonds have the same molecular shape, and the difficulty in understanding a 3D conception from a 2D representation.

The teaching video segment begins with an image showing a static electron geometry structure of BF_3. The teacher then uses a digital highlighting function to illuminate the number and nature of the bonding electron pairs. The next shot adds a static electron geometry representation of the NH_3 molecule side by side with the BF_3. The teacher here also uses a digital highlighting function to point to the bonding electron pairs, drawing the learners' attention to the same number of covalent bonds as in the BF_3 molecule. The teacher then digitally highlights the location of a lone electron pair. The digitally highlighted parts are simultaneously presented with verbal explanations and labels while they are being described. These dynamic highlights are synchronized in space and time to achieve the intended emphasis on the presence of a lone pair thereby setting-up the scene for a discussion on its differentiating effect on the shape in comparison to BF_3. The emphasis on what is important and what is peripheral in this example is a crucial element of curricular saliency interacting with the consideration for common learner misconceptions as well as the well-known difficulty in perception of a 3D conception from a 2D representation. The point illustrated here is that the teacher's consideration for the interactions among the components of traditional TSPCK have been technologically enabled resulting in the digitally transformed content knowledge, balanced appropriately with considerations of the cognitive load for learning through multimedia.

It is important to note that the reference to technology when PCK is considered, particularly with reference to the process of transformation of content knowledge is not new. Looking back at the original descriptions of PCK, Shulman[2] in his conception of PCK considered the use of appropriate technologies when teachers are thinking about the representations of the concepts being taught, as included in PCK. Here we see Shulman[2] placing the emphasis on the knowledge component for which the technology is used (representation), rather than the technology itself. In a similar vein, looking at the original descriptions of the subsequent Technology, Pedagogy, and Content Knowledge (TPACK) construct, Cox and Graham[6] acknowledge this understanding in their reflection and elaboration on the TPACK model, and concede that "as the technologies used in those activities and representations become ubiquitous, TPACK transforms into PCK" (p. 64, 2009). These authors described their view on the meaning of the 'technology component' in the TPACK model to refer only to 'emerging technologies' (that is, new technologies as used in traditional education settings) rather than any common form of technology or technological device as commonly seen in many TPACK studies.[7]

For d-TSPCK, it is immaterial which type of technology (ubiquitous or emerging) is used if the Shulman's[2] principle of 'appropriateness' and the transformation of content knowledge are observed. Furthermore, the use of technology is not only in enhancing the component of representations, but as seen in the example above, technology is required to bring to view the entire pedagogical transformation process of content knowledge into the digital mode. This includes also revealing the wisdom behind teachers' choices of discipline-specific representations used; their mode whether static or dynamic, the synchronization of events, the quantity of information shared at a time, and the appropriateness of the sequence of the discussions. Thus, the critical role of technology in the case of d-TSPCK is to translate and bring to view the complex transactions of component interactions of traditional TSPCK in the emerging new digital content. This function is distinct from enhancing one or two of the individual components of traditional TSPCK, and even from the integration of technology into a traditional classroom setting. Teachers who can affect these interactions in a digital setting show the ability to import their traditional TSPCK with consciousness of learner cognitive load demands when learning through multimedia. So, put simply, traditional-TSPCK is an important pre-requisite for acquiring d-TSPCK, however not enough by itself to enable the building of proficient digital lessons. This observation alludes to a new blend of additional knowledge domains and competencies making up d-TSPCK rather than traditional TSPCK. For the construct of d-TSPCK, in addition to the traditional TSPCK, there is a priority to acquire specific teacher digital competences, and a learning theory describing the needs of a learner who is to receive the digital lesson separated from the traditional classroom. This observation is reflective of the new complexity in describing teacher professional knowledge for digitalization. In order to deal with the relationship between these different pieces of knowledge and competencies, and to harness the richness of the emerging new knowledge, a theory of knowledge becomes necessary.[8]

6.3.1 Positioning d-TSPCK Within a Theory of Knowledge

The digitalization of education has brought back the debate between the rationalist and practice perception of knowledge. Generally, this theory *versus* skills debate represents two historically clashing schools of thought about whether skills and competencies (knowing how) should be considered as similar knowledge to propositional knowledge (knowing that), and which form of knowledge should precede the other.[9] A 'third way' is to consider the idea of 'powerful knowledge'. Powerful knowledge is a construct that considers knowledge as a resource for the cultivation of human powers enabling critical thinking, exploration beyond context, in contrast to knowledge as an end to itself.[8] It is a view that acknowledges the relational character of knowledge emerging from the contextual grounding of knowledge in societal settings enabling learning by acquaintance (knowing what). Seeing knowledge as powerful knowledge, transforms the binary description of 'knowing that' and 'knowing how' into powerful 'knowings and knowns'[10] thus encompassing the context and associations inclusively.[9] In this view, the knowns are connected to human activity as knowings. Tacit forms of knowledge such as teachers' TSPCK, including the context in which the known is functional in some format (planning or teaching) are acknowledged in this view.[9] A theoretical framework based on the above premise enables a focus on capacity building and how people become cultivated through acquisition of knowledge in given contexts. Referring to knowledge as 'knowings' and 'knowns' thus appeals to the objective of this chapter as we explore possible ways of framing the long-established teachers' knowledge for teaching science in traditional settings and the new digital competencies required for adapting for teachability on digital platforms, seamlessly. In the next section, we look at the knowings and knowns for constructing d-TSPCK as a guiding framework.

6.4 Digital Teaching Competence

In science education the use of technology in classrooms has traditionally been largely integrated into lessons for purposes of generically enhancing learning. The use of technology in such settings has been guided largely by the theoretical framework of Technological Pedagogical and Content Knowledge (TPACK)[11] with a focus on introducing technology into traditional lesson settings. TPACK has been presented as attempting to identify the nature of knowledge required by teachers for technology integration in their teaching.

However, in practice, the operationalization of the TPACK construct has been criticized for the predominant research tendency to focus on the technological element being introduced and less on revealing the complex transactions between technology than with the content being taught, *e.g.*, Santos and Castro (2021),[12] and Cetin-Dindar and colleagues (2018).[13]

Apart from TPACK, other frameworks exist that attempt to define digital teaching competencies within disciplines in teacher education. These include the: critical digital literacy,[14] substitution, augmentation, modification, redefinition (SAMR)[15] and the UNESCO ICT competencies framework for teachers which has been gaining acceptance due to its specific focus on teacher development.

The UNESCO ICT competencies framework for teachers is of interest to the current chapter for its progressive nature in emphasizing usage of technology that goes beyond technology as an end by itself. The framework rather promotes the use of technology in engaging with the disciplinary content towards cultivating students to be collaborative, problem-solving, and creative participants in a digital society. Furthermore, it evokes teacher rationale behind technological choices and applications in line with the value of PCK of aiming for 'appropriateness' in technology usage.[2] The framework is organized into three distinct productivity levels that connect teachers' spheres of work to an increasing scope of situated engagement of technology with the disciplinary knowledge, as shown in Table 6.1. The first level is *Technology Literacy*, and it refers to enabling teachers to use ICT to ensure efficient teaching and learning. The second level is *Deepening Knowledge*, referring to the use of ICT to ensure learners acquire in-depth knowledge of their school subjects and apply it to complex, real-world problems. The third is *Knowledge Creation*, enabling both teachers and learners to create new digital content within their disciplines using technology.

The first column of the table highlights the teachers' selective spheres of work that are more applicable to the d-TSPCK construct. The second column refers to UNESCO's typical education spheres of work where a teacher may find their feet in using technology. The last three columns reflect the differentiated levels of sophistication in the engagement with technology. The UNESCO teachers' spheres of work are continually being refined but all the newer versions have retained the same description of the sophistication levels of usage of technology, which fit the purpose of this study, particularly the capacity to deepen knowledge and create digital content. The fit is seen in the descriptors for each level of sophistication that emphasizes the extent of smooth engagement of ICT use and content knowledge, making the act of 'knowing' and that of 'what is known' seamless. While the UNESCO framework does not specify the technology to be used (ubiquitous or emerging), it promotes appropriate technology use as directed by the differentiated levels of sophistication. We assert that the envisaged growth in teacher knowledge associated with d-TSPCK, will demand teachers to be competent across all three levels of sophistication as they design and create digital content for teaching proficiently through online platforms.

Table 6.1 An expanded illustration of The UNESCO ICT framework for teachers.[16][a]

		Three sophistication levels of ICT use		
		Technology Literacy (Basic)	**Knowledge Deepening** (Developing)	**Knowledge Creation** (Exemplary)
	Teachers' spheres of work			
	Understanding ICT in education	Policy awareness	Policy understanding	Policy innovation
	Curriculum & assessment	Basic knowledge	Knowledge application	Knowledge society skills
This is of interest for d-TSPCK.	Pedagogy	Access and integrate technology, *based on pedagogical rationale.*	Complex problem solving, *with enabled digital content modifications.*	Self-design, *delivery of teaching* and management.
	ICT	Basic Tools	Complex tools	Pervasive tools
	Organization and administration	Standard Classroom	Collaborative groups	Learning organizations
This is of interest for d-TSPCK.	Teacher professional learning	Digital literacy *for appropriateness.*	Manage and guide knowledge deepening *with enabled competencies to modify digital content.*	Teacher as a model learner, *demonstrating new learning (for self and the learner) acquired in each cycle of knowledge creation.*

[a]The italicized phrases are inserts making the links to d-TSPCK more visible.

6.5 Cognitive Theory of Multimedia Learning

In addition to using technology to create digital disciplinary content that is pedagogically transformed, science teachers must learn how to present information in a digital product such as a teaching video taking into account both learner and multimedia-platform contextual considerations. Mayer[17] asserted that the use of multimedia in ways that are in line with the working of the human brain are more likely to lead to meaningful learning. He explained a principle known as the "multimedia principle" that is based on three understandings with respect to learning with multimedia: Firstly, humans use two separate channels (auditory and visual) for processing information. Secondly, each channel has a finite capacity similar to the notion of cognitive load.[18] Lastly, learning is an active process of filtering, selecting, organizing, and integrating information based on prior knowledge. Humans can only process a finite amount of information in a channel

Table 6.2 Pedagogical principles of cognitive theory for multimedia learning.

Signaling	Signaling is a digital technique that can be used to place emphasis and reduce extraneous material such as redundant text, images, diagrams, *etc.*
Segmenting	Segmentation focuses on dividing instructional multimedia into smaller concepts of the topic that are split up across a lesson.
Weeding	Weeding refers to the awareness that the addition of redundant information in a piece of multimedia should be avoided.
Matching modality	This is the ability to make choices and apply the best technology for a specific purpose.
Content control	This principle advocates for learners to have control over the pace of the multimedia learning to become more meaningful and effective.

at a time and make sense of incoming information by actively creating mental representations. These three assumptions culminate into a set of twelve multimedia pedagogical principles for use in creating digital content that are sentient to the Cognitive Theory for Multimedia Learning. In this study, they are expressed interactively through a set of five principles shown in Table 6.2.

At the heart of the above principles, there emerges a nuanced guideline on how the use of ICT could facilitate and promote learning when multimedia is used. We also encourage the use of this guideline to include an open discussion on the ethics of observing copyrights of digital content drawn from the internet.

6.6 Pulling It Together: A Model Describing d-TSPCK

Drawing from the above-mentioned pieces of theoretical frameworks, d-TSPCK is illustrated as emerging from the interactions of the different theoretical constructs as shown pictorially in Figure 6.2.

The construct of d-TSPCK is conceptualized as an interactive blend of teacher 'knowings and knowns' for teaching transformed digital science content knowledge on digital platforms. The blend is drawn from the theoretical frameworks describing: (i) the digital teaching competences ranging from digital literacy to digital content creation, (ii) considerations of the pedagogical multimedia principles for learning through digital platforms, all grounded (iii) on traditional TSPCK as the teachers' province for transforming content knowledge into comprehensible versions for the learner. From this blend, a refined teacher knowledge construct that elucidates the deep complex translational transactions across topic specific content pedagogy, multimedia learner cognitive load considerations and technology emerges, as d-TSPCK. The overall nature of the interactions among these frameworks, like in the case of PCK and TSPCK, is postulated

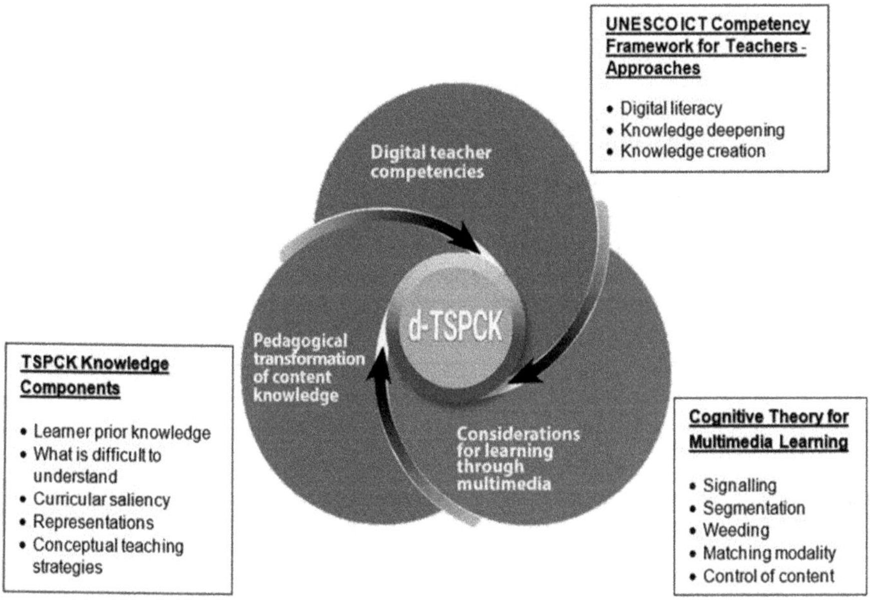

Figure 6.2 An illustrative model of d-TSPCK.

to be interactive and dynamic. The primary value to be derived from d-TSPCK, is the enabled digitalization of teacher's knowledge for teaching while retaining the centrality of the discipline's transformative pedagogical rigour and the associated topic specificity nature in the digitalization of PCK.

6.7 Demonstrating d-TSPCK Within Chemical Equilibrium

To illustrate the conception of d-TSPCK in operation, we draw on an example of teaching chemical equilibrium digitally. This example was extracted from an intervention for introducing d-TSPCK using chemical equilibrium in a methodology class of 4th-year pre-service chemistry teachers. The example illustrated here is from the teacher educator's first video segment introducing the topic by describing the pre-requisite conditions for the state of chemical equilibrium.

Example: *A representative extract.*

As noted before in the case of traditional TSPCK, not all activities in a teaching video are moments of d-TSPCK. However, there are specific moments when the interactions of traditional TSPCK are evidently digitally enabled. This is to happen with considerations for a balance in the learner's cognitive load when creating the digital content. The specific moments in this specific teaching video were sequenced shots delivering the message about the pre-requisite conditions for chemical equilibrium.

The opening visual shot had a title introducing the topic of chemical equilibrium as well as four hyperlinked statements that were about the Big Ideas of chemical equilibrium[19] *presented in chronological order as shown in* Figure 6.3.

The hyperlinks led to teaching videos that were a composite part of the topic. The educator clarified the segmented structure of the digital lessons as hyperlinked short videos to allow learner flexibility, differentiated learner prior knowledge and affording choice and control by the learner. This opening shot moved to the next uncluttered visual showing a symbolic representation of a chemically balanced equation for the forward reaction of sulfur dioxide (SO_2 (g)) reacting with oxygen (O_2 (g)) to form sulfur trioxide (SO_3 (g)), using a single forward arrow. The educator explained what this chemical equation was representing, using her voice and displaying a passport size self-portrait in the bottom corner of the screen. As she mentioned the single forward arrow and its meaning it was simultaneously highlighted zooming out in size and changing in colour, see Figure 6.4. *She then directed students to pay attention to the physical states of the reactants and products which dynamically zoomed out as she referred to each one. However,*

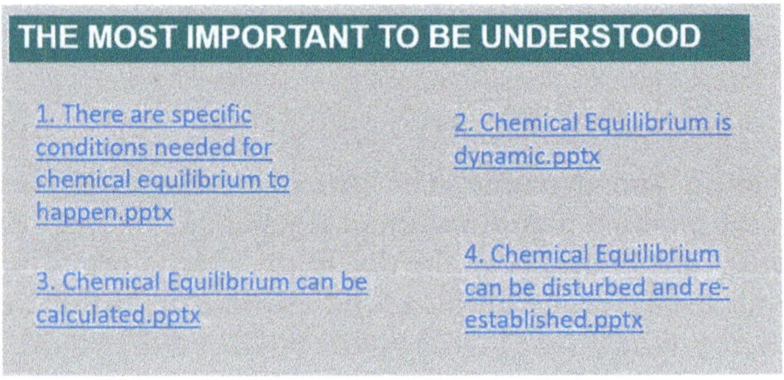

Figure 6.3 Illustrative interaction between segmenting and the Big Ideas of the topic.

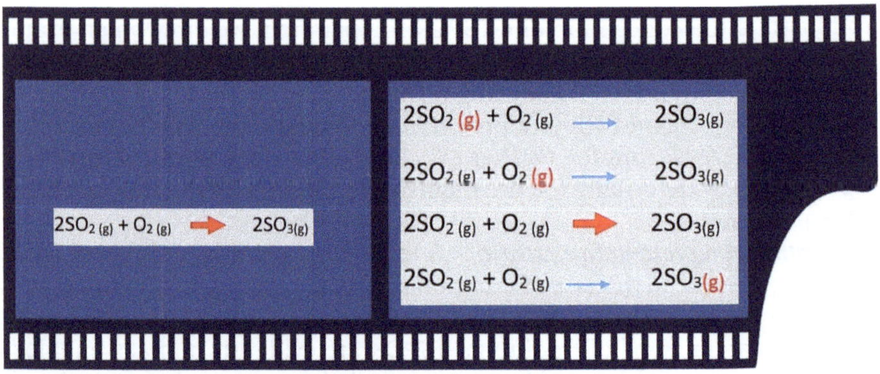

Figure 6.4 Illustrative dynamic highlighting.

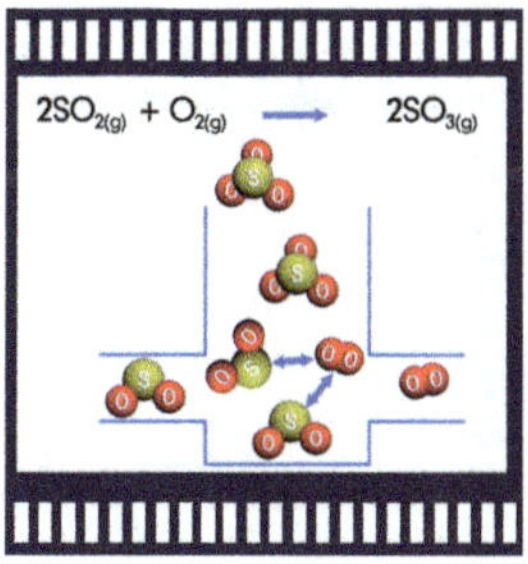

Figure 6.5 Illustrative use of multiple representations including sub-microscopic representations.

no further explanation was provided at this point. She then introduced a static sub-microscopic representation of the reactants and products as solid spheres in distinguishable colours on the same plane, see Figure 6.5. *These spheres entered a reaction vessel from each side, colliding to form products that were moving out of the container. She provided a description of this reaction pointing to the loss of products to the outside environment.*

In the next shot she asked students to consider the same reaction happening in the reverse order, where the SO_3 (g) was considered a reactant and the SO_2 (g) and O_2 (g) the resulting products. For this, she used a symbolic chemical equation also with a single forward arrow. She pointed out that if this reaction was happening in the same open vessel, it, too would go to completion and the products would be lost to the environment. The final scene was a visual animation of the same reaction using pre-counted sub-microscopic solid spheres to represent the reactants (SO_2 and O_2) in a closed container. She drew the attention of the students to the starting quantities of each reactant and asked them to pay attention to the reaction as it progressed. She used visuals of molecules of SO_2 (g) and O_2 (g) colliding and forming SO_3, and the product breaking back into reactants. After some time has elapsed (a few seconds), she posed a question in form of a poll (using Edpuzzle), asking students whether they think any of the reactions and their conditions if left undisturbed, would come to an end, and the reason for their answers. The Edpuzzle function enables students to access the correct explanation and proceed with the video on completion of their initial attempt. The correct answer was that neither reaction would come to an end, as the reactions are reversible in nature and occurring in a closed system.

Beyond the poll, the educator provided a summary of the most important content to be learned in this video segment. The visual shot following the Edpuzzle task was a narrated representation of the three types of thermodynamic systems with arrows distinguishing each by the allowed exchange of either matter or/and energy or none. The representation of the closed system was linked to the earlier example of the reaction in a closed container discussed above. The same representation was brought back and placed side by side with that of the closed system, see Figure 6.6. *The educator highlighted the conditions of a closed system as enabling the reverse reaction to happen in the same container. She added that a non-*

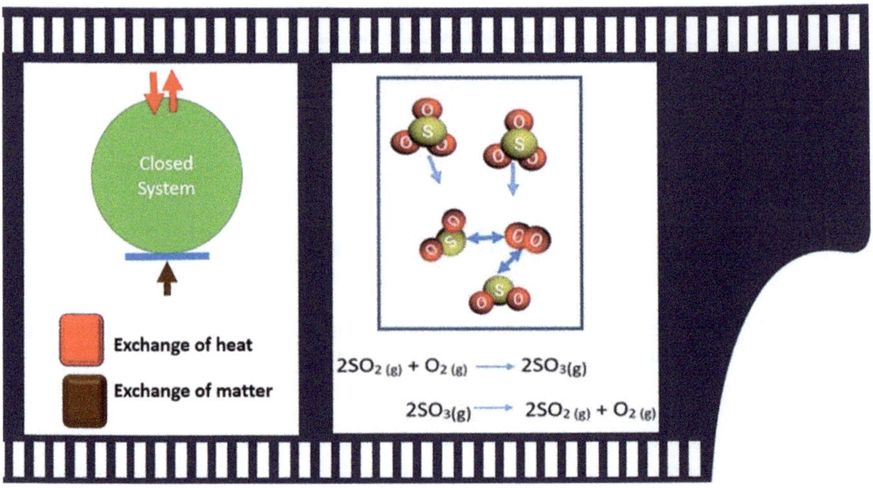

Figure 6.6 An illustration of calculated omission for the use of the phrase 'chemical equilibrium'.

reversible reaction in nature will remain as such even in a closed system. Her verbal statement was synchronized with a view of a symbolic representation of the forward chemical reaction of SO_2 (g) and O_2 (g) and the reverse reaction of (SO_3 (g)) placed on consecutive lines, each with single arrows still (without combining them to use the chemical equilibrium double arrow yet). The final shots were visual examples of closed systems in the aqueous media.

In the above example, there are several transactions of multiple component interactions across traditional TSPCK and the pedagogical principles of multimedia digitally enabled. The design of the teaching reflects segmentation into a set of short, interlinked teaching videos. The interlinked segments are purposely sequenced and given titles that reflect the core understanding to be mastered as the topic's Big Ideas. This reflects a transaction between the principle of multimedia called segmenting and the TSPCK's curricular saliency providing an informed rationale in knowing what is core in a topic and considered necessary for learners to develop a comprehensive understanding.[19] There is evidence of appropriate choice and scaffolding of meaning through the use of multi-level representations including symbolic and sub-microscopic levels deemed essential for providing powerful in-depth explanations in chemistry.[20] Overlaying the use of multiple-level representations is a repeated sense of curricular saliency emphasizing the most important features in a representation. This is seen in the signposting for later discussion, the importance of noting the physical states of the reagents. Such emphasis is enabled digitally through the teacher's competence in applying dynamic signaling. Dynamic signaling is a powerful pedagogical multimedia principle for selectively placing appropriate cognitive load on important features for the learners' attention. The sophistication of the transactions across the different constructs, ushers-in the emergence of the understanding of the pre-requisite conditions for chemical equilibrium. The teacher further blended to the observed transactions calculated omissions such as avoiding mentioning specific concepts such as 'complete' and 'incomplete reactions', preferring to use the term 'coming into an end'. This is seen in the writing out of the forward and the reverse reactions separately beneath each other while avoiding combining the reactions yet. These omissions are informed by the recognition of the difficulty in understanding the notion of incomplete reactions before going into the next Big Idea of dynamic equilibrium. It is also in calculating the timing for dealing with common learner misconceptions associated with thinking that reversible reactions go to completion at chemical equilibrium. Both these additional considerations are drawing from the components of the traditional TSPCK construct deepening the extent of the observed inter-construct transactions further. This example exhibits quite sophisticated interactive transactions across the three different theoretical constructs presented in Figure 6.2, with each construct bringing-in a distinct characteristic effect. It illustrated the possible meaning of experiencing seamless transactions across teacher's knowings and knows alive.

6.8 Conclusion

In this chapter we have addressed the theme of the book *Best Practices of Teaching and Learning Digitally* in two ways. First, we provided the conceptualization of a possible theoretical framework for digitizing teacher knowledge for teaching science, called digital-TSPCK. Second, we further illustrated the application of the construct in creating pedagogically transformed digital content in the topic of chemical equilibrium. Our effort is in line with Deng's[21] argument for an understanding of the rationale behind a school curriculum that is not devoid of the societal and community goals that embodies it. These include the development of teacher capabilities responsive to the society's current thinking, and the associated progressive technological developments. The newly conceptualized framework describing digital-TSPCK is born out of the realization of the growing lack of theoretical constructs sufficiently addressing teacher knowledge for teaching in the digital realm. The conceptualized construct of d-TSPCK draws on additional competences and knowledge to be mastered, that refines the traditional TSPCK framework into a more up-to-date version that is more responsive to teaching science in a digitized society. No less important than the digital topic specific teaching practices, is the specificity of the feedback given digitally during and after the topic specific learning process, as pointed out by Lawrie.[22] This framework is considered progressive as it builds on powerful pedagogical gains from our traditional understanding such as the centrality of pedagogical transformation of content knowledge, while allowing us to transit into the digital realm of teaching and learning. The human capacity building potential to be realized from this new framework suggests a new and refined discipline-based teacher disposition. A disposition that demands more than simply using technology to enhance learning, but to create and bring to light smooth transitions across multiple levels of digital teacher competencies and discipline-specific pedagogy with balanced learner cognitive load considerations. As stated previously,[6] the technology in use becomes translucent allowing the prominence of pedagogical transformation of content for learner understanding. We believe the refined digital-TSPCK construct will provide a possible vehicle for chemistry teacher educators and researchers for framing further research bringing deeper understanding for the digitization of teacher knowledge.

References

1. See Chapter 7: J. Huwer, C. Barth, A. Siol and I. Eilks, in *Digital Learning and Teaching in Chemistry*, ed. Y. J. Dori, C. Ngai and G. Szteinberg, Royal Society of Chemistry, United Kingdom, 2023.
2. L. S. Shulman, *Educ. Res.*, 1986, **15**, 4–14.
3. E. Mavhunga and M. Rollnick, *Afr. J. Res. Math. Sci. Technol. Educ.*, 2013, **17**, 113–125.
4. O. S. Akinyemi and E. Mavhunga, *Chem. Educ. Res. Pract.*, 2021, **22**, 282–302.
5. L. V. A. Hale, J. C. Lutter and G. V. Shultz, *Chem. Educ. Res. Pract.*, 2016, **17**, 700–710.
6. S. Cox and C. R. Graham, *TechTrends*, 2009, **53**, 60–69.
7. G. W. R. Fernandes, A. M. Rodrigues and C. A. Ferreira, *Res. Sci. Educ.*, 2020, **50**, 673–708.
8. Z. Deng, *Cambridge J. Educ.*, 2018, **48**, 335–351.
9. I. Carlgren, *J. Curric. Stud.*, 2020, **52**, 323–336.
10. J. Dewey and A. F. Bentley, *J. Philos.*, 1945, **42**, 225–247.
11. M. J. Koehler and P. Mishra, *Contemp. Issues Technol. Teach. Educ.*, 2009, **9**, 60–70.
12. J. M. Santos and R. D. R. Castro, *Social Sciences & Humanities Open*, 2021, **3**, 100110.
13. A. Cetin-Dindar, Y. Boz, D. Yildiran Sonmez and N. Demirci Celep, *Chem. Educ. Res. Pract.*, 2018, **19**, 167–183.
14. J. Hinrichsen and A. Coombs, *Res. Learn. Technol.*, 2014, **21**, 1–16.
15. R. R. Puentedura, Transformation, technology, and education [Blog post], 2006, retrieved from http://hippasus.com/resources/tte/.
16. UNESCO, *UNESCO ICT Competency Framework for Teachers*, UNESCO, 2011.
17. R. E. Mayer, in *The Cambridge Handbook of Multimedia Learning*, ed. R. E. Mayer, Cambridge University Press, New York 2005, vol. 1, pp. 1–12.
18. J. Sweller, P. Chandler, P. Tierney and M. Cooper, *J. Exp. Psychol.*, 1990, **119**, 176–192.
19. J. J. Loughran, A. Berry and P. Mulhall, *Understanding and Developing Science Teachers Pedagogical Content Knowledge*, Sense Publishers, Rotterdam, 2006.
20. A. H. Johnstone, *J. Chem. Educ.*, 1993, **70**, 701–704.
21. Z. Deng, *J. Curric. Stud.*, 2007, **39**, 503–535.
22. See Chapter 27: G. Lawrie, in *Digital Learning and Teaching in Chemistry*, ed. Y. J. Dori, C. Ngai and G. Szteinberg, Royal Society of Chemistry, United Kingdom, 2023.

CHAPTER 7

Learning With Digital Media About the Chemistry Behind the Recycling of Digital Hardware

JOHANNES HUWER[a], CATHERINE BARTH[a], ANTJE SIOL[b] AND INGO EILKS*[b]

[a]University of Konstanz, Universitätsstr. 10, 78464 Konstanz, Germany; [b]University of Bremen, Leobener Str. NW2, 28359 Bremen, Germany
*E-mail: ingo.eilks@uni-bremen.de

7.1 Introduction

Green and sustainable chemistry is a contemporary issue that is highly discussed, both in chemistry and in chemistry education.[1] Different concepts of sustainable chemistry were implemented and were discussed in terms of their meaning to chemistry education.[2] In general, the political visions outlined in the Agendas 21 and 2030, such as the Sustainable Development Goals (SDGs)[3] or the planetary boundaries concept by Steffen *et al.*,[4] are perhaps the best known. As interpreted in the Wedding Cake Model by Rockström and Sukhdev, these concepts often emphasize the priority of ecological sustainability over economic or social development.[5] Without wishing to challenge the importance of this priority, the following question must be asked: what about other sustainability considerations that focus more on economic and social developments, *e.g.*, the European Union (EU) critical raw materials concept[6] or related considerations in other countries?[7,8]

Advances in Chemistry Education Series No. 11
Digital Learning and Teaching in Chemistry
Edited by Yehudit Judy Dori, Courtney Ngai and Gabriela Szteinberg

Published by the Royal Society of Chemistry, www.rsc.org

Critical raw materials are those with a high economic importance for a certain region and a simultaneously high supply risk.[6,7] The best-known example is certainly the criticality of light and heavy rare earth elements which are in terms of supply risk at the top of the European Union critical raw material list published every three years.[6] Although some of these rare earth elements are quantitatively not that rare on earth, they can only be found in few economically accessible sources—primarily in China, in a region called Inner Mongolia. Another well-known example are the platinum group metals, which are indeed very rare with platinum occupying only about 0.0005 ppm of the Earth's upper crust.[9]

A source for some of these critical raw materials that has not yet been adequately developed would be the consistent recycling of electronic waste—especially in times of increasing digitization in economy, mobility, or education.[10] It is, however, that the chemical processes behind the recycling of electronic waste is neither a typical issue in traditional academic and school chemistry curricula, nor in the chemistry education literature. The project Rare Earths & Co is addressing this issue and focuses on learning with and about digital media.[11] The project is using a digital learning environment with augmented and virtual reality applications in combination with experiments for school classes and non-formal student laboratories.

The theoretical framework behind the project is the growing demand to integrate green and sustainable chemistry into chemistry curricula at all levels.[1] It is suggested from the literature to add content from green and sustainable chemistry to the curriculum and to start sustainability education in chemistry from a socio-scientific issues-based approach[2] with the aim to raise the relevance of chemistry education and students' perception thereof.[12] By this means, the objective of the project is to open the topic of recycling electronic waste in a new way and to focus on new learning experiences from disadvantaged to gifted students who study chemistry. The project aims to contribute to the body of knowledge by bringing a new topic into chemistry education, namely the recycling of electronic waste and evaluating the learning environments enriched by augmented and virtual reality. The question behind the project is: how can the chemistry behind the recycling of electronic waste, a problem intensified among others by the digitalization of education, can be included in chemistry teaching?

7.2 Electronic Waste, Sustainability and Chemistry

Sustainability has become a regulatory idea for international politics, set out in the Agendas 21 and 2030, among others.[3] It has thus become a guiding idea of national and international political processes. This is currently reflected in particular in the 17 Sustainable Development Goals (SDGs) issued by the United Nations in 2015, which the international community has committed to implement by 2030.[3] One of the goals (SDG 4) calls for quality education for everyone with a special focus on imparting skills that are necessary for sustainable development. By 2030, target 4.7 in particular asks the worldwide community to "ensure [that] all learners acquire knowledge and skills needed to promote sustainable development, including among others through education for sustainable development and sustainable lifestyles, human rights, gender equality, promotion of a culture of peace and non-violence, global citizenship, and appreciation of cultural diversity and of culture's contribution to sustainable development."

7.2.1 Quality Education and Digitization

Quality education includes creating the best possible learning opportunities for learners of all ages. One of these opportunities is to find the most promising ways that lie in the use of digital media in teaching-learning-arrangements. This led to a nationwide program in Germany called *Digitalpakt Schule.*[13] The program was launched in 2019 by the national Ministry of Education and Research and has led to massive investments in digital infrastructure in schools. 5.5 billion Euros were provided to update the digital infrastructure in schools. At the end of this process, among other investments, schools will have quality digital access and all teachers as well as learners will be equipped with up-to-date digital devices, primarily tablet computers.

Availability of digital devices for all teachers and learners will also impact the teaching of chemistry. Digital devices will provide easier access to worldwide information resources, being it information as such or to animations, visualizations, or virtual and augmented reality applications in class. These devices will be used in different forms to enrich chemistry learning. The different forms encompass new chances for practical work, *e.g.*, for data logging and easier analysis of data. They can help to better understand phenomena, *e.g.*, by slow-motion videos of chemical phenomena. Video tutorials, digital learning environments or digitally enriched textbooks will become available easier. Finally, the digital devices themselves and new information and communication forms can become an interesting topic for learning chemistry[14] both for students and teachers,[15] *e.g.*, about the chemistry behind the making and function of digital devices and what do to do with them physically after they are no longer usable.

7.2.2 A Different View on Digitization

Digitization in schools, however, can also be viewed from a different angle in terms of sustainability. The SDGs also focus on sustainability in other areas, more specifically issues that relate to the preservation of our natural environment. For example, the increasing air traffic is repeatedly criticized in the international discussion as one of the main causes of climate change. The approximately equally high energy consumption by digital technologies and the constantly growing energy demand for increasing data traffic are not being questioned to the same extent. For sustainability education in chemistry classes, chemistry education must focus on all chemistry related effects of digitization in terms of systems thinking.[16] Chemistry education needs to reflect the issue of digitization in terms of its impact on energy and resource consumption as well as the growing amount of electronic waste when the digital devices come to the end of their operation time. Digitization offers new learning opportunities, but it also requires reflection on resources consumption and potential environmental pollution. Thus, digitization is a multi-faceted sustainability issue, and raising the quality of education is just one of them.

Digitization requires investments in infrastructure from the digital device in hand to data lines, storage units, and data centers. All of this has an impact on the environment, from the extraction of raw materials, their processing and use, to disposal or recycling. Digitization has social and political effects as well, for example when it comes to the exploitation of raw materials and the often politically influenced trade with them.[3] Chemistry plays a central role in all of these steps.[17] Not only the environmental impacts on life, on land or under water (SDGs 14 and 15) or on climate (SDG 13) are to be considered. Global trade with minerals in different parts of the world applies to issues such as global justice (SDG 10), sustainable and clean energy use (SDG 7), or sustainable consumption (SDG 12). Chemistry class is linked in one way or another to all of the SDGs as discussed in the Global Chemical Outlook II issued by the United Nations Environment Program.[15] Competent and sustainability-conscious handling of digital media and the necessary hardware consumption is an emerging question, not only from an educational perspective. Geopolitical and economic interests also play an important role in the implementation of digitization. Risks for hardware production are increasingly predictable. Politically, there is discussion of how supply bottlenecks and dependencies on raw materials supply from different countries can be limited or avoided, for example from politically unstable countries and regions or those in which ores are extracted and processed under ecologically questionable and inhumane conditions. One of the often-discussed examples is Coltan mining for the extraction of tantalum in the Congo.[18]

When it comes to many raw materials that are important for digitization, Europe is almost completely dependent on ore imports as primary source.[6] For many raw materials there are no economically viable sources to be exploited in Western Europe, as it is for many other countries in the world. Such raw materials are considered critical for economic and societal development in Europe. Since 2011, the European Commission has regularly published assessments of non-energetic and non-agricultural resources under the concept of critical raw materials. In 2020, 30 currently critical raw materials were identified to have high economic importance and, at the same time, to have a high supply risk.[6] China is often named here as the most influential country for the global supply for many critical raw materials, *e.g.*, rare earth elements, magnesium, tungsten, antimony, gallium and germanium. Other countries dominate the supply in other areas, such as Brazil (niobium) or the USA (beryllium and helium). Platinum group metals are concentrated in Russia (palladium) and South Africa (iridium, platinum, rhodium and ruthenium) and Chile is named as the main exporter of lithium. Chile currently accounts for 44% of the global lithium production; the EU covers 78% of its lithium needs with imports from Chile. The list of critical raw materials includes heavy rare earth elements, light rare earth elements, and metals of the platinum group. This means there are not just 30, but actually well over 40 raw materials with a critical supply status. In addition to the risks associated with the provision, in many cases critical raw materials can, in fact, not be substituted at all or only with difficulty in their areas of application and the recovery rates of their recycling are inadequate. Similar assessments exist in other countries.[7,8] The EU furthermore started stronger regulations concerning the trading of minerals, and a regulation was issued on the handling of minerals from conflict areas in May 2017, which came into force at the beginning of 2021. Its aim is to create a system for compliance with the duty of care in the supply chain. The possibilities for armed groups and security forces to trade tin, tantalum, tungsten, gold, or other corresponding ores are to be restricted.[19]

7.2.3 Digitization and the Need for Investment in Better Recycling

Those reasons create a need for action for more sustainable electronic waste recycling. For example, about 20 kg of electronic waste are generated each year on average per person in the case of Germany. Although in Germany old devices can be returned to communal collection points and dealers free of charge and the manufacturers take over the disposal, only 45% of the devices get back into the cycle.[20] In response to the disposal problem, the European Commission issued guidelines in 2002/2003 with the aim of reducing the use of hazardous substances in production (including particularly toxic heavy metals, plasticizers, and halogenated flame retardants) and to reduce the proportion of electronic devices in household waste. The controlled disposal is intended to combat the illegal export of electronic scrap and to recycle rare and valuable raw materials. The guidelines have been implemented in Germany since 2015 through the Recycling Management Act and the Electrical and Electronic Equipment Act.

The content of precious metals per personal computer (PC) or mobile phone is often only a few milligrams. But, given the total volume of old devices, there is an economically relevant potential. For example: To extract 1 g of gold in African mines, about 2 tons of rock have to be extracted from great depths and processed. Recycling just five PCs can yield the same amount. Old electronic devices, so-called "end-of-life" material, are usually dismantled manually at first and the recyclable fractions are sorted according to materials and ingredients. The collected components are then mechanically shredded, magnetically separated and sent to the respective refurbishment processes. Gold, copper and platinum, among others, are recovered from so-called shredder fine fractions by electrolytic processing.

In the public discussion in Europe and beyond, when recycling electronic waste is concerned, special attention is paid to the rare earth elements known as technology metals.[21] None of these metals occur naturally in pure form, but always as a mixture of different compounds. Many of the 16 stable rare earth elements are part of electronic components, magnets, batteries, energy-saving lamps, or LEDs. Their primary extraction harbors major environmental risks. The remaining residues mostly contain toxic waste (heavy metals, acids, fluorides) and are stored in artificial ponds, which cannot always be considered safe due to the often-inadequate environmental regulations. In addition to the danger to soil and groundwater, there is also a permanent risk of radioactive exposure since many of the rare earth element ores are accompanied by radioactive substances (thorium, uranium).

Once installed, rare earth elements are difficult to recover from electronic waste. The amounts are small, the processes complicated and, so far, hardly affordable on an industrial level. According to the current state, less than one percent of the processed rare earth elements are recycled in the case of Germany.[22] Due to the steadily growing demand, however, the production of some rare earth elements has risen sharply in recent years, especially when focusing on neodymium, dysprosium and praseodymium. A good example for this is the strongest permanent magnets currently available, made from a neodymium-iron-boron alloy with the composition $Nd_2Fe_{14}B$. Such magnets are found in wind energy and electrical systems, but also in data storage media, headphones, or vibrating elements in cell phones. The economic prognosis for recycling also looks favorable as the prices of rare earth metals and many other raw materials are constantly rising.

7.3 Electronic Waste Recycling as an Issue for Chemistry Education

7.3.1 The Project Rare Earths & Co

Digital electronic devices are ubiquitous, but knowledge of their structure and ingredients is often only available to experts. In the education sector, it is mainly tablets that have spread almost exponentially in recent years. The reasons are their chance of mobile and simple use as well as the broad spectrum of potential applications. Another factor for the development was certainly the COVID-19 pandemic and the associated need to quickly make devices (and software) available for distance learning, leading to an even faster growth in the purchase of digital equipment with its associated potential effects—a challenge foreseen, however, even before the pandemic.[23]

The tablet itself not only offers the opportunity to learn through it, but it can also be considered a relevant topic for chemistry education. Learning about the use as well as the consumption of digital devices has individual, societal and vocational relevance both in the students' present and future, as outlined in the dimensions of relevant science education by Stuckey and colleagues.[12] Tablets contain almost 40 different metals and their compounds, various plastics, dyes, adhesives, and glass. The project Rare Earths & Co develops educational offers on the recovery of important raw materials from old tablets as well as a focus on their use and substitution potential,[11] which was recently also suggested for mobile phones.[24] The project pursues two complementary development goals: first, the development of learning environments for secondary chemistry students that can be carried out in non-formal educational laboratory environments and, second, the development of digital learning content that can be used independently of a well-equipped laboratory (*e.g.*, in distance learning or in less equipped schools).

The focus on non-formal learning and digital environments as an inclusive approach to education was chosen for two reasons. The first reason is the greater freedom for curriculum development outside the formal school sector along with the chance of immediate implementation. The other reason is to allow students different learning experiences outside the regular classroom practice. Non-formal learning environments offer chances for more in depth learning experiences in terms of inquiry-based learning and provide room for better supporting both gifted students and those with differing interests or educational disadvantages.[25] The combination with an ICT learning support also allows learning for students who do not have access to a corresponding non-formal laboratory, be it for physical, practical, or geographical reasons.

7.3.2 The Digital Learning Environment

The digital learning environment combines presentation of information on the practical level with augmented reality (AR; Figure 7.1) and virtual reality (VR; Figure 7.2) applications. With regard to distance learning, both lines of

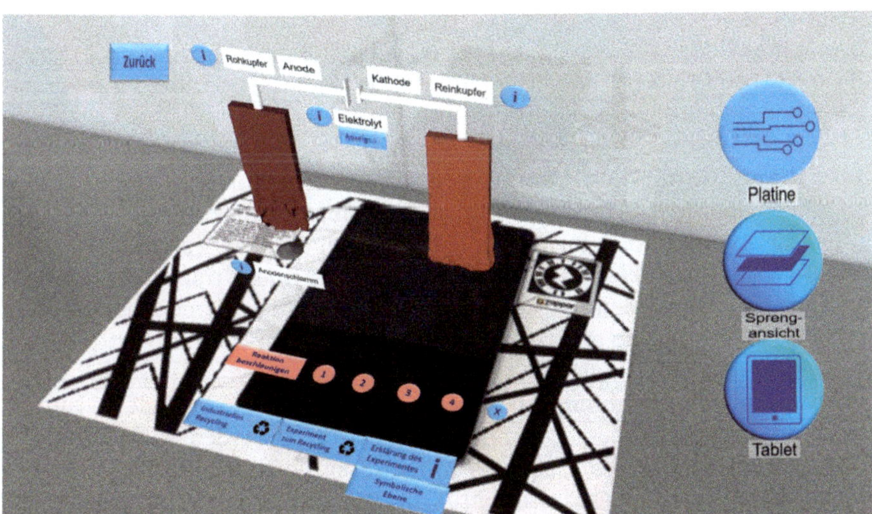

Figure 7.1 Learning by augmented reality about electrolysis for copper recycling. The real-world tablet is augmented by an animation about the electrolytic process of refining copper.

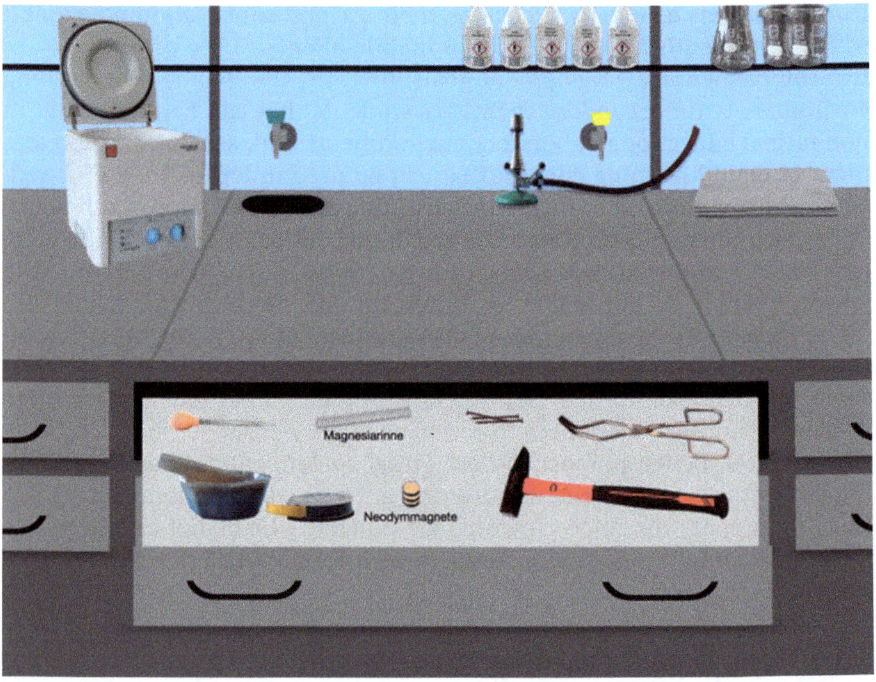

Figure 7.2 Virtual laboratory on the recycling of neodymium. A virtual space is provided where students can virtually conduct the experiment for recycling of neodymium from waste magnets.

development offer the chance that at least virtual practical and experimental experience is offered which might otherwise not be possible for practical issues or reasons of cost and risk.

The access to the chemistry content starts with the dismantling of an iPad in the digital learning environment based on AR (Figure 7.3). AR was chosen since it allows to enrich natural reality (objects, places, *etc.*) with digital (static or dynamic) information. In our case, a proper (physical) iPad is overlaid with a digital learning environment using image tracking. To design the learning environment, all components of an iPad were recreated to scale with a 3D CAD program and then added using ZapWorks Studio (https://zap.works/studio/).

The learning environment has a modular structure so that it can be expanded and updated. The components can be provided with functions and other learning content, such as animations. The hardware requirement for the digital learning environment is an iPad and a second mobile device (student's own mobile phones or a second tablet). To start the learning environment, the iPad is placed on a trigger image and scanned with the second device using the free ZappAR app (https://www.zappar.com/) so that the interactive AR learning environment is activated.

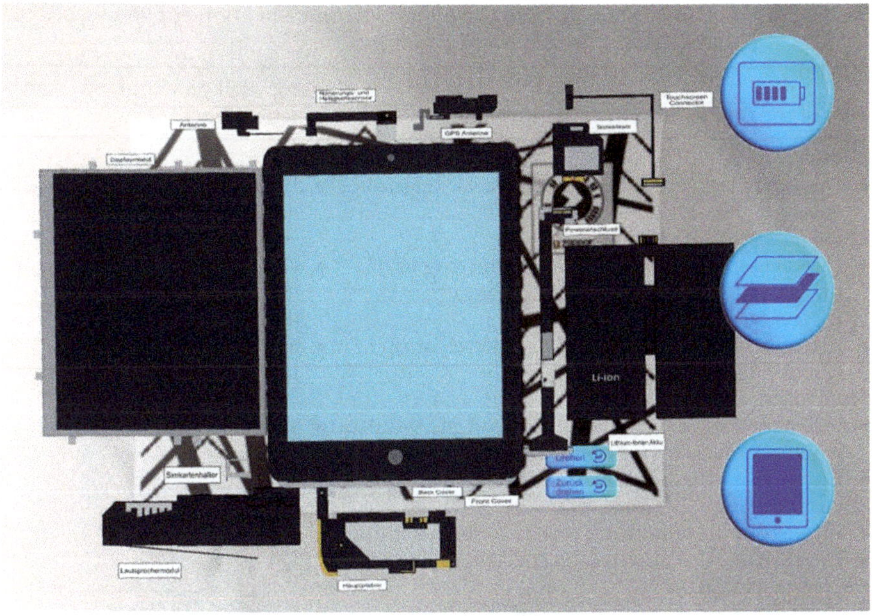

Figure 7.3 Disassembling an iPad in an augmented reality environment.

From this point, the different features of the learning environment can be worked on in a self-regulated way. The different features can also lead to several further activities in the classroom or in the non-formal laboratory. Table 7.1 provides an overview on the foci of the digital learning environment and associated potential practical work, *e.g.*, concerning recycling of copper, the function of OLEDs, or the specific type of glass used for the tablet screen.

In using augmented and virtual reality applications the student can enrich the physical experience of touching the iPad (or doing observations along experiments in class) by finding further pictures or animations. The move within the virtual and augmented reality is scaffolded by explanations, both for the macro- and submicroscopic levels. With respect to the students and learning goals, students can be allowed to move freely in the digital environment or can be guided by questions and tasks. Thus, there can be learning in a combination, *e.g.*, on copper recycling, by finding out where and for what purpose copper is used in the iPad, doing the copper electrolysis in class, and finding an animation of copper electrolysis on the particle level with corresponding explanations in the augmented reality application. In doing so, the project aims both to provide a basis for the learning of what underlies these chemical concepts, *e.g.*, copper recycling, but also to raise awareness among students for the importance of responsible consumption of electronic products and the need for proper waste collection and recycling. In the end there is hope that students also start sustainable action by releasing their electronic waste to the corresponding recycling collection points.

Table 7.1 Exemplary topics from the digital learning environment and associated potential for practical work.

Foci of the digital learning environment	Associated potential for practical work
Complete iPad	
• Digitally disassembling an iPad *via* AR (Figure 7.3).	• Physically disassembling an old iPad.
Li-ion battery	
• Interactive 3D animation of the charging and discharging processes of a Li-ion battery at the molecular level.	• Charging and discharging a dual graphite Li-ion battery pack.
• Interactive 3D animation of the intercalation of electrons into the graphite grid.	• Model experiments on Li-ion batteries.
Mainboard (gold recovery)	
• Video on the dissolution of gold and the Wohlwill electrolysis on laboratory scale.	• Wohlwill electrolysis.
• Interactive 3D animations on the Wohlwill electrolysis.	
• Interactive 3D animation on the redox reaction.	
• Interactive 3D animation on the Miller process.	
Cables (copper recovery)	
• Video on the dissolution of copper and the experiment of copper refinement by electrolysis on laboratory scale.	• Copper refinement by electrolysis.
• Interactive 3D animation on copper electrolysis.	
• Interactive 3D animation of the redox reaction.	
Magnets (raw earth metals recovery)	
• Interactive virtual laboratory on the recycling of neodymium from magnets.	• Recycling of neodymium from magnets.

7.4 Conclusion and Outlook

Due to the COVID-19 pandemic, selected elements of the learning environment have so far only been tested in distance teaching of smaller groups and in chemistry teacher training seminars. The feedback was generally very positive. The AR learning environments were perceived by the students and student teachers as a learning opportunity with high potential for self-regulated and autonomous learning.[11] In addition, we interviewed four chemistry teachers from different schools about the learning environment. All the four teachers stated that the topic of electronic waste recycling is considered "very important" (T1–4) for chemistry lessons, since "the students are themselves consumers of digital devices" (T1) and "valuable raw materials are built into them [the device]" (T2). Most of the teachers indicated that the learning offer should be used both in school lessons and as an offer in non-formal student laboratory learning (in German: *Schülerlabor*). Special opportunities were seen in the support of "making invisible processes visible" (T3). Likewise, it was emphasized that "students can be cognitively supported in their understanding with these media by being able to look into the particle level" (T3). Furthermore, it was stated that students could "learn independently and self-directed" (T4). The general evaluation of the learning environment was positive in all feedbacks (*e.g.*, "motivating for the students" (T4)). The teachers, however, also wished for more components to be explored and an even more detailed insight for upper secondary level chemistry students.

The virtual laboratory was considered as a good opportunity to understand experiments at least virtually when having no access to a real laboratory. The case presented here so far shows an example on how the important and relevant topic of electronic waste recycling can be integrated into chemistry teaching. There are initial indications that the teaching and learning activities may contribute to students' awareness and engagement for sustainability. It presents a case how to use innovative digital tools—augmented and virtual reality applications—to enrich chemistry learning and demonstrates that the suggested learning materials and environment may be applied both in distance and face-to-face teaching and learning.

After full implementation of both, the digital learning environment and the non-formal laboratory, both parts of the project will be subject to more thorough evaluation. The process of evaluation will focus on the most promising scenarios to include the topic and the teaching activities into chemistry education, both for formal education in schools and non-formal education.

Acknowledgements

We would like to thank the Deutsche Bundesstiftung Umwelt for the extensive funding of the project (DBU, no 34467).

References

1. V. G. Zuin, I. Eilks, M. Elschami and K. Kümmerer, *Green Chem.*, 2021, **23**, 1594.
2. M. Burmeister, F. Rauch and I. Eilks, *Chem. Educ. Res. Pract.*, 2012, **13**, 59.
3. UN 2015, www.un.org/ga/search/view_doc.asp?symbol=A/RES/70/1&Lang=E, accessed March 30, 2021.
4. W. Steffen, *et al.*, *Science*, 2015, **347**(6223), 736.
5. J. Rockström and P. Sukhdev, 2016, www.stockholmresilience.org/research/research-news/2016-06-14-how-food-connects-all-the-sdgs.html, accessed March 30, 2021.
6. EC (European Commission), https://eur-lex.europa.eu/legal-content/EN/TXT/?uri=CELEX:52020DC0474, accessed March 30, 2021.
7. S. M. Hayes, E. A. McCullough and E. A. Resour, *Policy*, 2018, **59**, 192.
8. K. J. Schulz, J. H. DeYoung Jr, R. R. Seal II and D. C. Bradley, *Critical Mineral Resources of the United States – and Environmental Geology and Prospects for Future Supply*, US Geological Survey, Reston, 2017.
9. M. L. Zientek, P. J. Loferski, H. L. Parks, R. F. Schulte and R. R. Seal II, in *Critical Mineral Resources of the United States – Economic and Environmental Geology and Prospects for Future Supply*, ed. K. J. Schulz, J. H. DeYoung Jr, R. R. Seal II and D. C. Bradley, US Geological Survey, Reston, 2017, ch. N1.

10. Z. Sun, H. Cao, Y. Xiao, J. Sietsma, W. Jin, H. Agerhuis and Y. Yang, *ACS Sustainable Chem. Eng.*, 2017, **5**, 21.
11. J. Huwer, C. Bath, A. Siol and I. Eilks, *CHEMKON*, 2021, **28**, 235–240, (in German).
12. M. Stuckey, A. Hofstein, R. Mamlok-Naaman and I. Eilks, *Stud. Sci. Educ.*, 2013, **49**, 1.
13. BMBF (Federal Ministry for Education and Research), 2019, www.digitalpaktschule.de/, accessed March 30, 2021. (in German).
14. See Chapter 8: N. Belova, A. Heckenthaler, J. M. Tietjen and C. Zowada, in *Digital Learning and Teaching in Chemistry*, ed. Y. J. Dori, C. Ngai and G. Szteinberg, Royal Society of Chemistry, United Kingdom, 2023.
15. See Chapter 15: M. Tal, D. Zreke, M. Hugerat and A. Hofstein, in *Digital Learning and Teaching in Chemistry*, ed. Y. J. Dori, C. Ngai and G. Szteinberg, Royal Society of Chemistry, United Kingdom, 2023.
16. P. G. Mahaffy, A. Krief, H. Hopf, G. Mehta and S. A. Matlin, *Chem. Int.*, 2018, **40**(3), 34.
17. UNEP (United Nations Environment Program), 2019, www.unep.org/resources/report/global-chemicals-outlook-ii-legacies-innovative-solutions, accessed March 30, 2021.
18. S. Behrendt *et al.*, *Rare Metals Measures and Concepts for the Solution of the Problem of Conflict-aggravating Raw Material Extraction – the Example of Coltan*, UBA, Berlin, 2007.
19. EC (European Commission), 2017, https://eur-lex.europa.eu/legal-content/EN/TXT/?uri=CELEX:32017R0821, accessed March 30, 2021.
20. UBA (German Environment Agency), 2020, www.umweltbundesamt.de/daten/ressourcen-abfall/verwertung-entsorgung-ausgewaehlter-abfallarten/elektro-elektronikaltgeraete#sammlung-und-verwertung-von-elektro-und-elektronikaltgeraten-drei-kennzahlen-zahlen, accessed March 30, 2021. (in German).
21. T. G. Goonan, *Rare Earth Elements – End Use and Recyclability: U.S. Geological Survey Scientific Investigations Report 2011-5094*, US Geological Survey, Reston, 2011.
22. B. Bensmann *et al.*, 2020, www.vdi.de/ueber-uns/presse/publikationen/details/vdi-statusreport-strategische-nutzung-von-rohstoffen-in-deutschland, accessed March 30, 2021. (in German).
23. B. Hileman, *Chem. Eng. News*, 2002, **80**(26), 15.
24. M. Barceló-Oliver, *et al.*, *J. Chem. Educ.*, 2020, **98**, 454.
25. F. Affeldt, S. Tolppanen, M. Aksela and I. Eilks, *Chem. Educ. Res. Pract.*, 2017, **18**, 13.

CHAPTER 8

Chemistry-based Information in Social Media in Light of Scientific Media Literacy—Teachers' Views and Classroom Implementation in Secondary Education

N. BELOVA*[a], A. HECKENTHALER[b], J. M. TIETJEN[c] AND C. ZOWADA[a]

[a]University of Bremen, Leobener Str. NW2, 28359 Bremen, Germany; [b]Cato Bontjes van Beek-Gymnasium, Bergstr. 26, 28832 Achim, Germany; [c]Wilhelm-Kaisen-Oberschule, Valckenburghstr. 1-3, 28201 Bremen, Germany
*E-mail: n.belova@uni-bremen.de

8.1 Introduction

In our everyday lives, we are surrounded by a dizzying number of media messages. The media landscape is evolving and changing at an incredibly fast pace. In the last decade especially, social media platforms have gained massive popularity. Recent data show that the number of social media users worldwide has increased from roughly a billion in 2010 to 3.6 billion in 2020.[1] Currently, media aspects are addressed in various school subjects—primarily

Advances in Chemistry Education Series No. 11
Digital Learning and Teaching in Chemistry
Edited by Yehudit Judy Dori, Courtney Ngai and Gabriela Szteinberg

Published by the Royal Society of Chemistry, www.rsc.org

the humanities and social sciences—while in the science classroom the application is still quite spotty and mostly limited to traditional media types like newspapers.[2] Different definitions of the notion of media literacy can be traced in the literature. Two main dimensions of media literacy were outlined in the media and information literacy curriculum for teachers published by UNESCO: "reading" and "writing" media.[3] On the one hand, students need to understand the different forms of communication used in the media, while on the other, they must be able to create their own media. According to Scheibe and Rogow,[4] the common definition of media literacy is described as individuals having capabilities which enable them to navigate their way through a media-dominated environment. They explicitly name four dimensions that define the concept of media literacy: accessing, analyzing, evaluating and creating media.

Many topics discussed in social media have scientific underpinnings, yet only in very rare cases are they discussed by science experts. Information can be spread by anyone, and it is often impossible or at least challenging to distinguish its credibility or factual correctness (for an overview of criteria for the analysis of YouTube videos of chemical topics see Chapter 13 by Joselevich *et al.*).[21] Understanding the use (and abuse) of scientific information is an important competency for critical citizens, however, learning through and about the media has not yet played a significant role in science education. Different types of social media, such as social networks (*e.g.*, Instagram or Facebook) or blogs are a popular source of information for young people. They not only fulfill communication purposes, but also involve the creation and distribution of content. The contents may contain chemical-related information which in many cases may not be based on scientifically proven sources. Generally, misinformation influences others in their opinion, which in turn can lead to a change in consumer behavior and even influence the views on the nature of science.[19] In general, there have been recent suggestions for a more thorough implementation of social media platforms such as Facebook[5] or Instagram[6] in the lower secondary science classroom, but there is still a lack of research and development of teaching materials especially on learning about social media. This chapter presents the results of an exploratory interview study on the use of social media in the science classroom, which revealed a rather skeptical assessment of the teachers interviewed. Based on these results, we have created a teaching module that can serve as an example for the use of social media and is presented in the following sections.

8.2 Scientific Media Literacy for Everyone?

Merging scientific literacy and media literacy in school science courses should form part of citizenship education in the modern age that can be named "scientific media literacy"[7] or "scientific media education" (SME).[8] Scientific media literacy is a concept that encompasses many different goals, skills and possible classroom activities connected to the learning of science and forms a justification for more intense media education in the science classroom. Reid and Norris recently outlined certain research areas for the next decade in scientific media education, emphasizing that skills related to SME are equally important for all citizens in modern societies.[8]

A general theoretical justification for a more thorough societal focus in science education can be derived from the central European tradition of *Bildung* or *Allgemeinbildung.*[9] *Allgemeinbildung* (which can be literally translated as "general knowledge") inseparably bounds education to a democratic understanding of society. It defines all objectives of education under consideration of a societal perspective, for education in general but also for all school educational domains in particular—among them science education. Within *Allgemeinbildung*, one part of the word, *Allgemein* (which can be translated as 'general') has two dimensions. The first dimension means to achieve *Bildung* as a specific set of skills for all citizens in a society independent from their ethnic, cultural, or socio-economic backgrounds. The second dimension aims at *Bildung* in all human capacities under inclusion especially of those skills necessary for coping with life in the society the students live and operate.[10] The part of the word *Bildung* is a more complex construct. In the 1950s and 1960s the German scholar Wolfgang Klafki tried to define and conceptualize a modern understanding of *Bildung* in the means of *Allgemeinbildung.* He suggested understanding *Allgemeinbildung* as the preparation of the young generation to live a self-determined life in a democratic society.[10] Klafki described the overarching goal for any education that the young generation becomes able to recognize and follow their own interests in society and to behave within society as responsible citizens. The theoretical framework of *Allgemeinbildung* gives a strong justification for an inclusion of societal questions and issues in contemporary science education. Skills developed along these lines will also be important for societal debates and decisions concerning the development of students' future participation as responsible citizens. In this sense the theory of *Allgemeinbildung* supports the point of view of a Scientific Literacy for All instead of science learning only for selecting and preparing future scientists and engineers.[11] Referring back to the media, Reid and Norris argue that even adults are not able to understand media messages that include scientific information,[8] which shows the need to equip citizens the related competency for scientific media literacy early in school, and the related research-based practices are required. However, the use of media in science classroom and the related students' learning outcomes are still limited in research. Thus, from our point of view, a critical reflection of media messages should be part of education available for everyone, regardless of their socio-economic or migratory background, in all school subjects. Developing capabilities to cope with information related to science in the media is indispensable if young people wish to become scientifically and media literate citizens.

8.3 Reflecting Upon Media Messages—Not for Everyone? Teachers' Views

To promote such competencies in school, teachers need to address them. To answer the question of the extent to which science teachers recognize the usefulness of using social media and show the willingness to use them not only as a communication platform in their lessons, an exploratory interview study comprised of semi-structured interviews 40 and 60 minutes each was carried out, with eight teachers from the federal state of Bremen, Germany. These teachers all work at secondary comprehensive schools, 'Oberschule', with their very heterogeneous classes encompassing a variety of nationalities, backgrounds and levels of performance. Four male and four female chemistry teachers with different levels of professional experience were interviewed. The teachers ranged in age from 28 to 61, with varying work experience—four teachers with under 10 years of professional experience and four with over 10 years (ranging from 2.5–35 years). The data were transcribed and evaluated with qualitative content analysis according to Kuckartz.[12] Table 8.1 shows the core categories and the respective number of codes (number of times the respective category could be assigned to a unit in the interview transcripts).

The teachers presented a rather diverse picture of media literacy and described it as something multidimensional and as a combination of learning with and about media:

> Media literacy ... on the one hand, it is the use of different media, for example the mobile phone or the iPad. [...] But also the critical reflection of platforms like Facebook, WhatsApp or Instagram [...], that one can be influenced and manipulated there and that you should be aware that people without a professional background can spread information there. [I.8]

The teachers mostly recognized the relevance of media literacy (a total of 113 codes) due to the omnipresence of the media, but highlighted problematic aspects (106 codes) more frequently compared to opportunities and strengths (64 codes) when using social media in school. A frequently mentioned problematic aspect was the uncertainty of passing on information in social networks. Often a general potential of these media types was emphasized, but not related to one's own teaching. Especially, the three teachers who were over 40 years old showed a generally rather weak self-concept with regard to social media in the science classroom, for which the lack of personal use was mentioned as the main reason: *"You would have to be theoretically familiar with these platforms to be able to apply them. If you don't use them yourself, I would say it's difficult to work with them. [...] Because if you don't know how these networks work, you can't teach students any critical reflection"* [I.5]. Four of the eight teachers pointed out that at some point one will probably no longer be able to shut oneself off against the influence of social media but criticized the lack of one's own authenticity. Quite often we could encounter a fairly negative attitude towards social media, which can be linked to a lack of own experiences: *"[...] colleagues [...] use social media [...] for their*

Table 8.1 Core categories and selected sub-categories from the evaluation of the interviews.

Main category - Selected sub-categories	Description	Number of codes
Media literacy	Definitions of media literacy, aspects related to its potential, role of media literacy in school, competencies related to media literacy	163
- Own definition		59
- Competencies		32
- Potential		10
Media usage	Usage of different types of media (esp. social media) both in and privately	162
- No social media usage		50
- Negative perception towards social media usage in school		24
- Positive perception towards social media usage in school		11
Self-concept of using modern media	Assessment of own perceived competency in the use of modern media	20
Strengths/opportunities of using social media in the science classroom	Aspects that were connected to a positive impact of using social media in the science classroom	64
Problematic aspects of using social media in the science classroom	Aspects that were connected to a negative impact of using social media in the science classroom	106
- Accessibility		23
- Uncertainty		9
- Students' competencies		14
Relevance	Relevance of fostering media literacy *via* social media in the school context, *e.g.*, for participation in society	113
- Relevance for the students: participation in society (literate citizens)		41
- Relevance for the teachers: teachers as "agents of change"		8
Skills related to social media	Skills which students and teachers must either possess or can acquire wile dealing with social media	
- Of the students		14
- Of the teachers		17
School factors influencing the use of social media	Factors influencing the use (or lack of use) of social media in the science classroom, *e.g.*, lack of teaching material	61

assignments and thereby try to be closer to the students. I don't know if that's so good. I haven't had any experience with it and of course I won't, because I don't know enough about these media" [I.6].

Surprisingly, three teachers saw some kind of "cognitive barrier" which prevents their students (who often stem from socially difficult backgrounds) from questioning scientific content in social networks:

> It's probably stupid to say, but in the end, there is an intellectual limit for the students, maybe because they belong to a particular social class. They deal with all kinds of information so differently in their peer groups and

their families and they only spread bold statements [...] and they won't get out of there. Trying to discuss it, that's like fighting against windmills." [I.2]

Interestingly, according to half of the teachers, critical reflection of media messages is declining in the schools, although the opposite should be the case nowadays: *"I believe that reflecting media is no longer an issue in schools today, that's why they [the students] find such tasks particularly difficult and do not even know what is required from them."* [I.6]. At various points in the interviews, the teachers said that they require the students to undertake this critical reflection of social media, without, however, making themselves responsible for guiding this process.

Overall, two types of teachers could be identified based on the data, which we call "the open type" and "the skeptical type". The first type includes two teachers and the latter type includes the remaining six. "The open type" includes teachers aged 28 and 32 years who use social media on a daily basis and recognize that "social media literacy" should also be promoted in science classes: *"So if I [...] can bring them [the students] to the point where they have the knowledge to critically reflect and evaluate everything they see on social media, I would say I have accomplished a huge educational goal"* [I.7]. "The skeptical type", on the other hand, does not use social media and does not recognize the potential of such networks for teaching and learning: *"Not much can be learned when using social media"* [I.5]—this answer referred both to content knowledge as well as competencies related to media literacy. This typification corresponds to the classification found by Belova and Eilks in the context of implementing advertising into the science classroom.[2]

Overall, the attitudes of the surveyed teachers towards social media in chemistry classes can be described as careful caution or even open rejection, although the teachers recognize the multidimensionality of media literacy as an educational goal of all school subjects. The study findings may suggest that the use of social media is a generational issue. However, the acceptance of the necessity of a critical media literacy for the students does not seem to be compatible with this rejection from part of the teachers. Apparently, some teachers feel that there is a cognitive barrier for some students in using and reflecting social media. This is very much at odds with our view of science media education for all as described above. We believe that with suitably prepared teaching material on relevant media-related topics, it is possible to reach all learners. To illustrate this with an example, we present below a teaching module on parabens in social media for upper secondary school students (grades 8–12).

8.4 Curriculum Development: Parabens in Cosmetics

The teaching module "Parabens in Social Media" was developed based on the results of the interview study described above, where teachers highlighted the lack of teaching examples. Thus, the module is meant to function as a kind of "bast practice example", as the trial showed a very time-efficient use and positive feedback from the students. The central learning objective of the module was to show the students, using the example of the substance class of parabens, that the presentation of scientific facts in social media does not always have to go hand in hand with scientific consensus, and thus to enable them to assess such facts. A subordinate goal was, among other things, learning about the substance class of parabens as well as the functioning of preservatives in general. In the following, some technical aspects around parabens are presented and then the module and its evaluation are outlined.

8.4.1 General Information on Parabens

Parabens are a collective name for the ester of the *para*-hydroxybenzoic acid. For almost a hundred years they have been used in pharmaceuticals, tobacco products, cleaning agents, cosmetics as well as food, and their preservative effect prevents these products from germs, mold, fungi and their spores. Parabens are chemically stable and act over a wide pH range, which is an advantage over other preservatives. Despite their effectiveness, parabens have a rather bad image which is especially prominent in social media. They are said to promote breast cancer, trigger allergies and have high hormone activity.[13] But, contrary to the opinion that is common in the social networks about parabens, parabens are reliable and harmless preservatives. Interestingly, despite the lack of studies proving that (especially ethyl and methyl) parabens are not safe to use, their bad image is actively promoted *via* social media. A natural preservative which is sometimes traded as an alternative to parabens and also finds its place in the teaching unit is grapefruit seed extract (GSE). Grapefruit seed extract is obtained through the processing of waste products from the production of grapefruit juice.[14] As described above, it is advertized that the extract is also highly effective compared to parabens. However, there are no studies that confirm this effectiveness.[15] For people with an allergy to citrus fruits, the extract causes rashes or reactions in thc oral cavity. Taking in a high concentration can cause a headache.[11] The developed material deals with the contradiction between the presentations of parabens in social media and the actual state of research as well as with the comparison of parabens and grapefruit seed extract from a sustainability perspective.

8.4.2 Structure of the Lessons

Table 8.2 shows an overview of the module's structure.

As mentioned in Table 8.2, the evaluation is based on a spider web diagram. This was developed by Zowada, Zuin, Belova, and Eilks[16] and is shown in Figure 8.1 in a form already completed for parabens and GSE. The criteria were developed in part using the twelve principles of green chemistry[17] and are intended to provide learners with conversation starters about the differences and similarities of various substances in the context of a

Table 8.2 Structure of the module "Parabens in social media".

Phase	Content
Introduction, raising awareness for the discussion on parabens in social media	A collage of several excerpts on the topic of parabens to illustrate the topicality and controversy is shown to the students First impressions and questions are collected
Social media research	The students conduct a search on parabens in social media and write down their first impressions (mostly negative)
Initial positioning on an opinion scale	Based on the search, the students mark their position on parabens on an opinion scale (between "my opinion on parabens is negative" and "my opinion on parabens is positive")
Background information on parabens and GSE (science-based)	The students read texts with information on parabens/GSE
Evaluation/comparison of parabens and GSE	The students compare parabens and GSE using a spider web diagram[16] and discuss discrepancies (see Figure 8.1)
Summary *via* an "Instagram post"	The students summarize their insights in the form of an "Instagram post" (see Figure 8.2)
Final positioning on the opinion scale	The "Instagram posts" are attached to the opinion scale from the beginning of the lessons and the potential shift is discussed (see Figure 8.3)
Optional content (after the comparison of parabens and GSE)	*An experiment on the production of a cream and another worksheet on the chemistry of parabens*

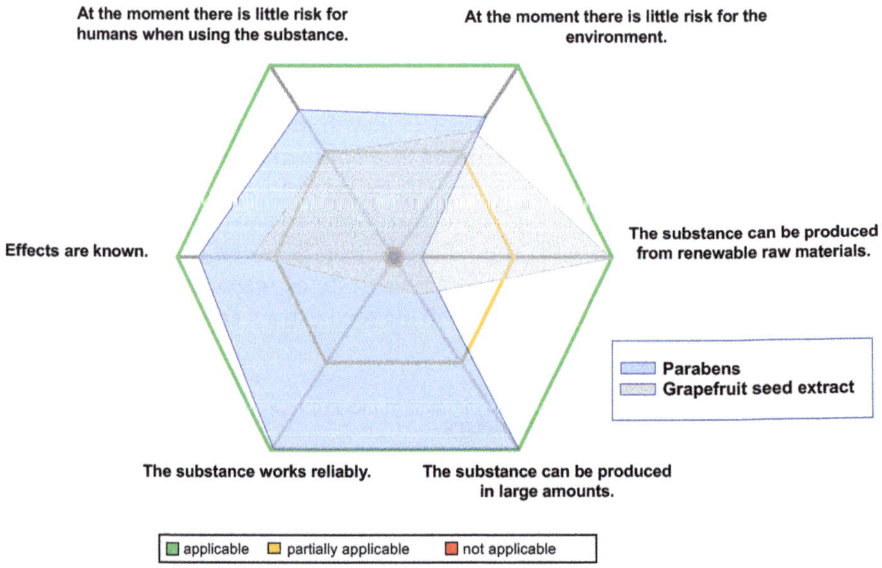

Figure 8.1 Simple spider chart for evaluating the use of chemicals filled out for parabens and grapefruit seed extract.

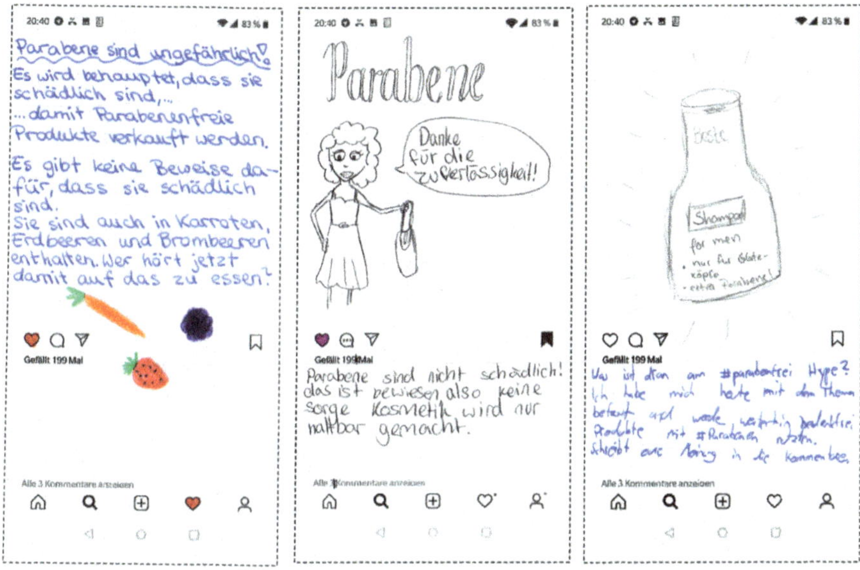

Figure 8.2 "Instagram Posts" created by the students. Translations (from left to right): (1) Parabens are harmless. They are said to be harmful in order to sell paraben-free products. There is no evidence that they are harmful. They are also found in carrots, strawberries and blackberries. Who stops eating now? (2) Parabens – thanks for the reliability! Parabens are not harmful. That's proven, so don't worry, cosmetics are only made to last. (3) What's up with the paraben-free hype? I have learnt about the topic today and I will continue to use products with parabens. Write your opinion in the comments.

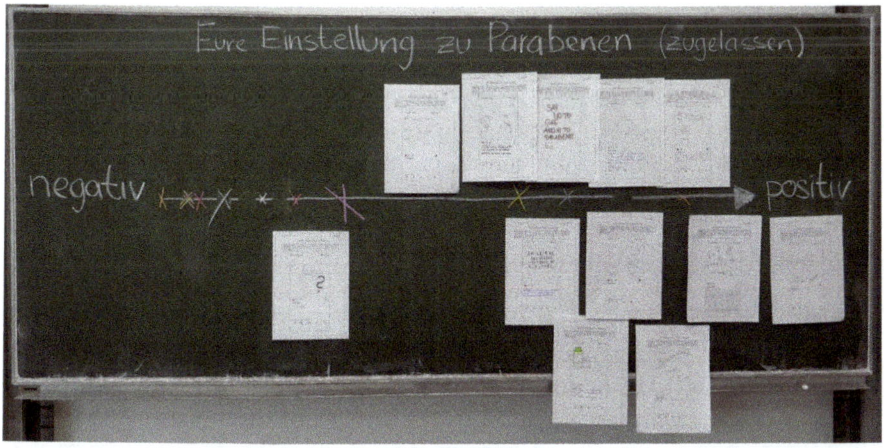

Figure 8.3 Final opinion scale with the attitudes towards parabens at the beginning (crosses) and at the end of the module ("Instagram posts").

sustainability assessment based on a three-point Likert scale (agree–partially agree–disagree). In addition to criteria aligned with green chemistry principles, the focus is on economic issues, such as functionality or the potential for mass production. With the help of the material, the students evaluate the level of each criterion and mark their evaluation on the chart. The six marks are then connected, and the resulting area is hatched. The completed diagrams can then be compared, and it becomes clear that parabens are to be evaluated more positively than the initial picture suggests. This view is reinforced by the more negative evaluation of grapefruit seed extract.

8.4.3 Research Design and Reception of the Module

The module was developed by means of action research[20] and tested with 168 students (f = 91, m = 63, other or no answer = 14) of an eighth-grade class of a grammar school in Bremen (without the additional content), whereby the module could be implemented in two teaching hours (90 min). It was evaluated using a questionnaire containing three open-ended questions (aspects the students liked/disliked as well as their opinion on whether the EU should increase the use of parabens, and 17 Likert-scaled items). The evaluation of the results of the questions with an open response format was carried out according to Kuckartz[12] in the form of a qualitative content analysis and the Likert-scaled items were evaluated using descriptive statistics. After the first research, impressions were formulated about the presentation of parabens in social media, which (as anticipated) were negative. The comparative evaluation of GSE and parabens with the aid of the spider web diagram provided a good basis for discussion due to the visualization.

8.4.4 Findings

The active engagement of the students was evident in the "Instagram posts" that were created, examples of which can be seen in Figure 8.2. These were made with great zeal and offered room for creativity. Familiarity with the design may have played a crucial role. Often, additional small changes were made to the material, such as painting the heart of the Instagram post red or increased the number of likes in it. In addition to the good reception of the material, it became apparent during the processing that the students were able to take up and present the controversy presented to them in their own way. Thus, references were made to the supposed alternative GSE, the intentions of the cosmetics industry with the advertising of paraben-free products were addressed, or the advantages of parabens, such as their mode of action, were presented as an argument. Figure 8.3 also shows the change of opinion in the course of the teaching unit. The "Instagram posts" are placed here on the opinion beam (see Figure 8.3) and suggest an overall more positive assessment of parabens, which also gave rise to a discussion.

The Likert-scaled items led to the following results: while the majority of the students (62%) rated the topic of parabens as interesting, they emphasized the general use of social media even more positively (87% agreed with the statement that social media made the lessons more interesting). Building on this, there is a desire for social media to play a greater role within the school in the future (85%). In addition, feedback on the structure and design of the lesson was also positive. The design of the lesson received an approval rating of 94%, which is one of the highest values in the survey. The meaningfulness of using the sustainability network (88%) and the fun of working with it (85%) ties in with the positive feedback regarding the material. Tables 8.3 and 8.4 summarize the results of the open-ended questions. Table 8.4 clearly shows that much more positive than negative aspects were mentioned by the students.

Some students expressed very differentiated positions on the use of parabens, as the following example shows:

> *According to current research results, I would find [the use of parabens] okay and it would not bother me any further. However, if studies that are adequately conducted would harden the suspicion of the British study [linking paraben use to cancer], I would look into the issue more and then make my personal decision.* [S.30.4]

Overall, it can be deduced from the results of the questionnaire that the teaching unit on parabens in social media, including the material developed, was received positively and that almost all students profited from learning with and about the media implemented. It is interesting to see that after the lessons the students gained a more balanced and reflexive view on chemistry in general and parabens in particular. It can even be claimed that the lessons helped to convey a different idea of the role of chemistry in the public discourse and increased the perception of relevance for many students: *"Finally something that enriched chemistry classes. [...] Knowledge of ingredients is important. And that social media often exaggerate things and manipulate us"* [S.83.3]. In this context, another striking feature is the answers to the questions related

Table 8.3 Results of the first open-ended question.

Open-ended question	Answer	Percentage (if applicable)	Anchor example
Comment on the statement *"The cosmetics industry plans to use parabens in more products in the future"*	Yes	53%	*"I would "support" it because parabens are not toxic in the amounts we use. After all, the toxic ones are banned."*
	No	24%	*"I still think the risks are too high, so no."*
	Neutral	23%	*"I don't really care about it because I don't use cosmetics anyway."*

Table 8.4 Results of the third and the second open-ended questions.

Core category	Sub-categories	Number of codes
Positive aspects		
Lesson design		69
	Student-centeredness	18
	Variation of methods	12
	Room for creativity	10
Material		57
	Spider-web diagram described positively	16
	Design of worksheets described positively	17
Methods		54
	Internet research described positively	24
	Group work described positively	11
Social media use		42
Topic		31
	Interesting	16
	Relevant	10
High learning outcomes		13
Critical reflection		11
Negative aspects		
Material		32
	Too long texts	9
	Too much writing	9
	Too many worksheets	6
Lesson design		15
	Pacing (too fast)	9
Methods		9
	More group work desired	4
Non-interesting topic		9
Content too difficult		9

to the influence of social media. While almost all respondents assume that social media influences many people, they exclude themselves from this potential influence. This phenomenon is called the "third-person effect". It describes that the assessment of one's own influence through mass media is far below the assessment for other people.[18]

8.5 Conclusions and Implications

From the theoretical point of view, this chapter indicates the importance and potential of teaching science through social media while developing the competency of critical thinking about information posted on social media in order to raise a generation of smart, critical consumers while also contributing to empowering the relevance of science teaching and learning. From a practical perspective, this chapter demonstrates the application of the learning module on the topic of parabens in cosmetics, and students' views on this module. We suggest that teaching through and learning about social media in the science classroom of the future can be a potentially fruitful issue in science education, in parallel with a change in teachers' attitudes towards such issues. This includes bettering the contribution of science to the development of scientific media literacy in the next generation of learners as a prerequisite for their societal part as critical citizens, while at the same time contributing to the increase of the perception of relevance of science learning. This goes in line with the tradition of *Bildung*—the preparation of the next generations to live an independent lifestyle and to contribute to a democratic society. Due to the crucial role science in general and chemistry in particular plays in our everyday lives, we must keep up the debate about the role of a societal perspective for a contemporary and relevant science education.[11] However, application in the classroom is still neglected all over the world.[11] In the sense of *Bildung*, a re-orientation of the objectives of science education is necessary both in its content and pedagogy. The teaching sequence presented in this chapter shows in an exemplary way for how chemistry education can contribute to a more critical view on the way chemical content and chemistry as a domain are depicted in social media. The students realize that the image of certain chemistry-based products or compounds presented in social networks is not in line with the actual scientific background. Such insights then in turn lead to a more critical reflection on messages in media *per se*, and ultimately to a more critical consumer and empowerment to participate in societal debates about chemistry—based on *Allgemeinbildung*.

It is now crucial to develop further teaching materials which can help teachers to move beyond the news media and consider using social media more frequently in the science classroom. Available teaching strategies, pedagogies and examples are quite limited so far. In terms of educational policy, learning about mass media in general and social media in particular need to be more systematically incorporated into modern science education. This is a clear task for science curriculum development and accompanying research on its effects.

It is also of great importance to start from the students' side asking which types of social media they find relevant and which factors contribute to their decisions. Educators should keep an eye on the social media habits of their students. Effectively, it is not that important through which platform scientific information is spread as long as we have clear criteria to evaluate it—for instance, using evaluation tools such as the spider chart diagram presented in this chapter.

With regards to the implementation of such measures into classroom practices, it seems advisable to include both learning with and about social media in science education. This includes integrating learning about corresponding pedagogies and teaching examples into university programs for pre-service science teacher education and into in-service, continuous professional development programs. Teachers still seem to be reluctant to incorporate social media in science lessons, which can be counteracted preventively by such activities. The idea that reflection skills should be made available to all students (in the sense of scientific media literacy for all) should find greater acceptance among teachers. The aim here is not to completely turn teachers' attitudes around in a short period of time, but to show, by linking science and media education in an exemplary way, that this can be done in an activating way, without much effort and in a time-efficient way. Thus, the skepticism that became apparent in the interview study can at least be gradually reduced. There is need for further development of suitable pedagogies and illustrative examples as well as in-service and pre-service teacher training. It also needs evidence-based curriculum development that leads to credible teaching scenarios that can serve as exemplary patterns and useful illustrations.

Speaking about the concrete implementation of the teaching module described in this chapter, especially the introduction of the spider chart diagram, received positive feedback. The completion of the diagram and the subsequent discussion fostered communication and assessment skills of the students. All other aspects of the lessons were also assessed positively. All students were able to complete the worksheets without major problems, the tasks or texts were not too difficult. For higher grades, a discussion at the meta-level (about the general role of scientific information in the media) may also be possible, as it took place with some students after the lessons.

In terms of empirical research, there is a need for more evidence, both on the prerequisites for as well as on the long term effects of including social media in science classes. This includes factors such as considering learners' knowledge gains, developing higher-order cognitive and communication skills, meta-cognition, and attitudes towards science.

Acknowledgements

We gratefully acknowledge the valuable guidance of Dr Rea Lavi from Massachusetts Institute of Technology, MA, USA, throughout the writing process.

References

1. Number of social network users worldwide from 2017 to 2025, https://www.statista.com/statistics/278414/number-of-worldwide-social-network-users/, accessed April 2021.
2. N. Belova and I. Eilks, *Int. J. Sci. Math. Educ.*, 2016, **14**, 1233.
3. Media and information literacy. Curriculum for teachers, https://unesdoc.unesco.org/ark:/48223/pf0000192971, accessed December 2020.
4. C. Scheibe and F. Rogow, *The Teacher's Guide to Media Literacy*, Corwin, Thousand Oaks, 1st edn, 2012.
5. R. Blonder and S. Rap, *Educ. Inf. Technol.*, 2017, **22**(2), 697.
6. F. Affeldt and I. Eilks, *Sch. Sci. Rev.*, 2018, **99**(369), 90.
7. S. N. Chang Rundgren and C.-J. Rundgren, in *Science Education Research and Education for Sustainable Development*, ed. I. Eilks, S. Markic and B.

Ralle, Shaker, Aachen, 1st edn, 2014, SSI pedagogic discourse: embracing scientific media literacy and ESD to face the multimedia world, pp. 157–168.
8. G. Reid and S. P. Norris, *Cult. Stud. Sci. Educ.*, 2016, **11**(1), 147.
9. J. Sjöström, N. Frerichs, V. G. Zuin and I. Eilks, *Stud. Sci. Educ.*, 2017, **53**, 165.
10. W. Klafki, in *Teaching as a Reflective Practice: The German Didaktik Tradition*, ed. I. Westbury, S. Hopmann and K. Riquarts, Lawrence Erlbaum, Mahwah, 1st edn, 2000, Teaching as a reflective practice: the German Didaktik tradition, pp. 85–108.
11. A. Hofstein, I. Eilks and R. Bybee, *Int. J. Sci. Math. Educ.*, 2011, **9**(6), 1459.
12. U. Kuckartz, *Qualitative Inhaltsanalyse. Methoden, Praxis, Computerunterstützung*, Beltz, Weinheim, 4th edn, 2018.
13. D. Błędzka, J. Gromadzińska and W. Wąsowicz, *Environ. Int.*, 2014, **67**, 27.
14. Grapefruit und Arzneimittel, https://www.pharmazeutische-zeitung.de/inhalt-31-2005/pharm1-31-2005, accessed March 2020.
15. M. Eryılmaz, D. Cicek Polat, K. Akalın and M. Coskun, *Hacettepe Univ. J. Fac. Pharm.*, 2018, **38**, 1.
16. C. Zowada, N. Belova, V. G. Zuin and I. Eilks, *Chem. Educ. Res. Pract.*, 2020, **21**, 141.
17. P. T. Anastas and C. J. Warner, *Green Chemistry: Theory and Practice*. Oxford University Press, Oxford, 1st edn, 1998.
18. I. Huck and H. B. Brosius, *Publizistik*, 2007, **52**(3), 355.
19. K. Dhingra, *J. Res. Sci. Teach.*, 2003, **40**(2), 234–256.
20. I. Eilks and B. Ralle, in *Research in Chemical Education - what Does This Mean?* ed. B. Ralle and I. Eilks, Shaker, Aachen, 1st edn, 2002, Participatory action research in chemical education, pp. 87–98.
21. See Chapter 13: M. Joselevich, P. D. Moro and M. Agustina-Martínez, in *Digital Learning and Teaching in Chemistry*, ed. Y. J. Dori, C. Ngai and G. Szteinberg, Royal Society of Chemistry, United Kingdom, 2023.

CHAPTER 9

Digital Learning Platforms: Digital Platforms for Increasing Inclusion in Chemistry Education

R. BLONDER(0000-0003-4796-4678)*

Department of Science Teaching Weizmann Institute of Science, Rehovot, Israel
*E-mail: ron.blonder@weizmann.ac.il

9.1 Digital Platforms for Increasing Inclusion in Chemistry Education

The transition to online teaching, due to the COVID-19 pandemic, has become a reality for all education systems worldwide. It has raised the need to better understand the web-based online platforms and the pedagogy of using them. Puentedura[1] suggested the SMAR model, which includes four levels of selecting, using, and evaluating technology in educational settings. The four levels that create the SMAR acronym are substitution, augmentation, modification, and redefinition. At the substitution level, teachers transfer their chemistry teaching pedagogy, which they used before the pandemic, to the digital space. For example, teachers who previously taught chemistry by lecturing now prepare a PowerPoint presentation and do it *via* Zoom (or other platforms for online meetings). Another example of transferring the

Advances in Chemistry Education Series No. 11
Digital Learning and Teaching in Chemistry
Edited by Yehudit Judy Dori, Courtney Ngai and Gabriela Szteinberg

Published by the Royal Society of Chemistry, www.rsc.org

"old" pedagogy to digital space is group work pedagogy. In the classroom, teachers used to split the class into small groups and actively work in collaboration on a given assignment. However, this pedagogy was transferred to digital space by using the "breakout rooms" option in Zoom. To keep track of students' learning the same way they used to before the pandemic, teachers and schools have applied learning management systems (LMS) such as Moodle, which can address this need.

However, the redefinition level of using technology in teaching involves a transformative approach.[2] In this approach, teachers use technology for teaching chemistry in a way that transforms their pedagogies. For example, LMS systems provide real-time data about students' learning. Using this information, the teacher can track students' mistakes and misconceptions during the lesson and can address them in real time.[3] Learning analytics can provide the teacher with suggestions on how to customize learning paths for different students by analyzing the data of many students who are not in the teacher's own class. This feature can be a "game changer" when an inclusive chemistry learning environment is designed. Another example is the use of animations and other technological resources that can be used to visualize macroscopic, microscopic, and symbolic levels of understanding.[4–6] When all students use their computer, the learning process and the inquiry of the connections between the different levels of understanding shift from being teacher centred to student centred.[7] The use of animations and a variety of online resources for chemistry teaching can address the needs of diverse learners and can lead to an inclusive approach to chemistry teaching.[8] To progress up the SMAR levels, teachers should develop their professional knowledge[9] and integrate technological knowledge into their pedagogical content knowledge, namely, develop technological pedagogical content knowledge (TPACK).[10] However, the web-based online platforms can either support or hinder teachers to reach a certain level on the SMAR ladder.

Online learning environments that are commonly used (*e.g.*, Moodle, Blackboard, Schoology, and Canvas) and even social network sites (*e.g.*, Facebook, YouTube) can be used at the four previously mentioned levels. However, to design an online learning environment that supports the higher levels of using technology in chemistry teaching, one that can be modified and redefined by the designer of the environment, using a transformative approach, the designer should develop and incorporate an understanding of technology with pedagogical content knowledge. Several environments are presented in this section; they present how, with a careful design, the environment can lead to a different type of chemistry learning and teaching based on online technology. Interestingly, the environments provide platforms for including a variety of learners when teaching chemistry.

In Chapter 10, Usher and Barak focus on the novel abilities of online environments to connect diverse learners in a MOOC course about nanotechnology.[11] They explore how the level of a group's diversity, which they defined as "the distribution of differences among members of a group [that] includes bio-demographic variables, such as gender, mother tongue, and age, or task-related variables such as academic discipline and academic level" is correlated with aspects of innovative thinking that they were able to measure. Whereas in a regular academic course the diversity of the participants is relatively low, the MOOC environment provides a unique platform that connects learners from different disciplines, countries, academic levels, and languages. Usher and Barak found that some of the diversity characterizations are positively related to the innovative thinking developed by the learners. However, a higher diversity of mother tongue was negatively associated with innovation of the group project. In terms of equity, the last result should be considered in order to support the inclusion of learners from different societies in group work involving diverse learners.

In Chapter 11, Marchak, Kesner, and Frailich describe the online environment "General Chemistry and Industrial Chemistry for the Service of Mankind" throughout its 20 years of development and adaptation.[12] This chapter demonstrates how the environment supported the needs of formal chemistry teaching twenty years ago by using animations about chemical bonding. However, this teacher-focused approach was gradually replaced by a student-centred pedagogy when the environment was accompanied by a non-formal competition of chemistry projects. The environment organized resources from a different nature, for example, repositories of chemical terms and concepts, as well as video tutorials of soft skills (how to communicate scientific research). The environment was planned to be used during the years of the competition. However, during the COVID-19 pandemic when the competition was canceled, students and teachers kept using the materials for self-learning of chemical concepts. These results constitute evidence of the need for open access environments to support inclusion of diverse learners in the process of students' self-regulated learning of chemistry.[13]

Chapter 12, "The Next Level in Inclusive Chemistry Education: A Model Approach Using a Multitouch Learning Book",[14] describes a model for a three-level design approach of an adaptive environment for chemistry learning. Kranz and Tiemann suggest that the MiC (model for inclusive chemistry teaching) supports inclusive chemistry teaching. In this study, they critically evaluate traditional teaching of chemistry concepts and present how they were able to reorient teaching in a unit about "Fire and Flame" in order to make the digital learning environments effective for inclusive chemistry teaching. They found that students having different characteristics and achievement levels differentially used various features of the e-Book. Nevertheless, all kinds of students used the interactive components of the e-Book and completed the final course requirements. Namely, the MiC system provided a suitable framework for developing an inclusive online environment for chemistry learning. This result emphasizes the possible advantage of technology that can transform teaching and learning towards achieving a high level of personalization.

In Chapter 13, "Can YouTubers provide powerful tools for addressing heterogeneity in the classroom? An analysis of videos about the Periodic

Table using the TPACK framework",[15] Joselevich, Moro, and Martínez take an innovative approach to improving chemistry teaching. They methodically examined YouTube videos about the Periodic Table that were created by YouTubers with a high number of followers in order to distill principles that can be transferred to school chemistry teaching and that can improve students' interest. Using the TPACK theoretical framework,[10] they evaluated the pedagogical aspects of the videos as well as the technologies and contents. They revealed that none of videos helped to overcome the epistemological obstacles and common misconceptions in learning the Periodic Table and did not help identify predominant traditional teaching models. However, they found a wide range of communication styles that include formal, informal, colloquial, traditional, and comical or humoristic language as well as the use of youth slang, songs, and body language. They suggest that this approach can be adopted by chemistry teachers in order to better address the variety of students' preferences while they learn chemistry.

In Chapter 14, "A formalized conceptual model-based approach for fostering and assessing students' systems thinking in undergraduate chemistry education",[16] Lavi presents a unique tool that can support system thinking in chemistry learning. In the last years, system thinking has received a lot of attention by the community of chemistry educators, and it has also influenced teacher professional development.[17] Lavi presents the use of a modelling tool used by scientists and engineers called Object-Process Methodology (OPM). The OPM tool supports pre-service teachers and helps them to better understand the connections between the components of a complex system, the way the system is influenced by them as well as reveals emerging phenomena in a real-world biochemical phenomenon. It is interesting to see that in this study a technological tool used by professionals, which can be learned in an academic MOOC, can be applied to foster system thinking in a course for prospective teachers. This tool enables students to overcome a variety of barriers in learning the complex skill of system thinking in chemistry.

In Chapter 15, "Chemistry teachers' awareness of sustainability through social media: cultural differences",[18] Tal *et al.* recognize that social media is prevalent across the world, and may be an effective tool for education. They identify sustainability, particularly through chemistry, as a relevant topic worldwide, and investigate pre- and in-service teachers' awareness and understanding of sustainability. Similar to Blonder and Rap (2017),[19] they investigated participants' perspectives on using social media for environmental and chemistry communication and education. The researchers situated their findings within cultural differences noted by the Arab and Jewish participants as well as with respect to whether the participants were pre- or in-service teachers. Cultural differences included differences in prioritization of the sustainability issues, potentially due to differences in lifestyle and living conditions, as well as differences in social media usage. They found that pre-service teachers reported a lower awareness of sustainability issues, and for those who were aware of them, a majority learned about them through social media when compared to other sources. Facebook, Instagram, and YouTube were the favoured social media platforms for the participants, suggesting that these may be leveraged for educational purposes.

In recent years, adaptive learning environments have been developed and implemented (*e.g.*, Apoki, 2021,[20] Peng *et al.*, 2019[21]). These environments use artificial intelligence tools to adapt the learning route of the learners to their performance in the system, without interference from the teacher. The chapters in this section highlight the high potential of digital technology in supporting a more inclusive approach in teaching chemistry. However, these examples also emphasize the important role of the teacher in personalizing the learning process toward more inclusive chemistry teaching. Integrating these environments leads to a change in the role of the teacher, which should include the digital literacy of data-based informed teaching. Teachers that can interpret data gathered by the LMS about the learners to improve their decision-making process[22] can harness the advantages of digital environments for more personalized inclusive teaching.[3] More research about the use of digital environments by chemistry teachers from the teachers' perspective is needed. A better understanding of the teachers' stance and needs will improve implementation of digital environments for personalization, and it will greatly improve chemistry teaching and learning.

References

1. R. Puentedura, SAMR, http://hippasus.com/resources/tte/.
2. R. Puentedura, SAMR, http://www.hippasus.com/rrpweblog/archives/2014/08/22/BuildingTransformation_AnIntroductionToSAMR.pdf.
3. E. Aviran, E. Easa, S. Livne and R. Blonder, in *Early Warning Systems and Targeted Interventions for Student Success in Online Courses*, IGI Global, 2020, pp. 90–111.
4. A. H. Johnstone, *J. Comput. Assisted Learn.*, 1991, **7**, 75–83.
5. Y. J. Dori and M. Hameiri, *J. Res. Sci. Teach.*, 2003, **40**, 278–302.
6. B. S. Dorfman, B. Terrill, K. Patterson, A. Yarden and R. Blonder, *Chem. Educ. Res. Pract.*, 2019, **20**, 772–786.
7. S. Rap, Y. Feldman-Maggor, E. Aviran, I. Shvarts-Serebro, E. Easa, E. Yonai, R. Waldman and R. Blonder, *J. Chem. Educ.*, 2020, **97**, 3278–3284.
8. E. Easa and R. Blonder, *Chem. Teach. Int.*, 2022, **4**, 71–95.
9. E. R. Hamilton, J. M. Rosenberg and M. Akcaoglu, *TechTrends*, 2016, **60**, 433–441.
10. P. Mishra and M. J. Koehler, *Teach. Coll. Rec.*, 2006, **108**, 1017–1054.
11. See Chapter 10: M. Usher and M. Barak, in *Digital Learning and Teaching in Chemistry*, ed. Y. J. Dori, C. Ngai and G. Szteinberg, Royal Society of Chemistry, United Kingdom, 2023.
12. See Chapter 11: D. Marchak, M. Kesner and M. Frailich, in *Digital Learning and Teaching in Chemistry*, ed. Y. J. Dori, C. Ngai and G. Szteinberg, Royal Society of Chemistry, United Kingdom, 2023.

13. Y. Feldman-Maggor, R. Blonder and I. Tuvi-Arad, *Internet Higher Educ.*, 2022, 100867.
14. See Chapter 12: K. Kranz and R. Tiemann, in *Digital Learning and Teaching in Chemistry*, ed. Y. J. Dori, C. Ngai and G. Szteinberg, Royal Society of Chemistry, United Kingdom, 2023.
15. See Chapter 13: M. Joselevich, D. P. Moro and M. A. Martínez, in *Digital Learning and Teaching in Chemistry*, ed. Y. J. Dori, C. Ngai and G. Szteinberg, Royal Society of Chemistry, United Kingdom, 2023.
16. See Chapter 14: R. Lavi, in *Digital Learning and Teaching in Chemistry*, ed. Y. J. Dori, C. Ngai and G. Szteinberg, Royal Society of Chemistry, United Kingdom, 2023.
17. R. Blonder and S. Rosenfeld, *J. Chem. Educ.*, 2019, **96**, 2700–2703.
18. See Chapter 15: M. Tal, D. Zreke, M. Hugerat and A. Hofstein, in *Digital Learning and Teaching in Chemistry*, ed. Y. J. Dori, C. Ngai and G. Szteinberg, Royal Society of Chemistry, United Kingdom, 2023.
19. R. Blonder and S. Rap, *Educ. Inf. Technol.*, 2017, **22**, 697–724.
20. U. C. Apoki, *Computers*, 2021, **10**(5), 59.
21. H. Peng, S. Ma and J. M. Spector, *Lect. Notes Educ. Technol.*, 2019, 171–176.
22. T. Nazaretsky, M. Ariely, M. Cukurova and G. Alexandron, *Br. J. Educ. Technol.*, 2022, **53**, 914–931.

CHAPTER 10

Group Diversity and Innovative Thinking: Lessons Learned From a MOOC on Nanotechnology

MAYA USHER*[a] AND MIRI BARAK[a]

[a]Technion—Israel Institute of Technology, Israel
*E-mail: oshermaya@gmail.com

10.1 Literature Review

10.1.1 Nanotechnology Courses and Innovation

Within chemistry education, nanotechnology is an interdisciplinary subject that integrates chemistry, engineering, biology, and medicine. To understand the importance and powerful use of nanotechnology, and nanomaterials in particular, learners must understand the surface chemistry of the particles. Chemistry plays a significant role in nanotechnology as it involves the makings of new forms of matter.[1] Academic courses on nanotechnology present fundamental principles of nanomaterials and their vital role in the design of novel sensing properties and applications.[2,3] Courses on nanotechnology discuss the chemical and physical mechanisms of molecules ranging from 1–100 nm. They present students with theoretical and practical understanding of the principles behind the creation and/or manipulation of nanomaterials, either by scaling up from single groups of atoms or by reducing bulk

Advances in Chemistry Education Series No. 11
Digital Learning and Teaching in Chemistry
Edited by Yehudit Judy Dori, Courtney Ngai and Gabriela Szteinberg

Published by the Royal Society of Chemistry, www.rsc.org

materials into nanoparticles. In higher education, nanotechnology courses are usually delivered to advanced students with a strong science and engineering background.[2]

The field of nanotechnology requires innovation as it is continuously evolving in response to human needs. Innovation is conceptualized as the cognitive and social processes that lead to the application of new or significantly improved ideas. It is perceived as the outcome of the cognitive process of innovative thinking.[4,5] Innovation and innovative thinking are essential competencies in the academia and the workplace of the 21st century.[6–8] Training engineers to be innovative is a nationwide priority as it advances society and enables successful competition in today's global and competitive economy.[9,10] The importance of innovation in the engineering industry led to the examination of this competency, and its promotion, in the context of science and engineering higher education.[2,9]

The literature on innovation in science and engineering education can be divided into two main research strands. First, studies that examine innovation through the analysis of students' learning outcomes (*e.g.*, Barak & Usher, 2020;[11] Henderson & Clark, 1990).[12] Second, studies that examine innovative thinking through self-report questionnaires (*e.g.*, Barak *et al.*, 2020;[4] Fila *et al.*, 2015).[13] With regards to measuring the innovation level of students' learning outcomes, an accepted method is the distinction between four types of innovation: incremental, modular, architectural, and radical.[12] The distinction between the four types of innovation is made based on whether a moderate or a drastic change was made in the core components of the product and in the links between the components. Incremental innovation refers to the improvement of existing products while using existing knowledge; modular innovation refers to making drastic changes in the components of the existing technology, while leaving the links between the components unchanged; architectural innovation refers to making drastic changes in both the core components of the technology and the links between them; and radical innovation refers to the introduction of novel products, previously unknown technologies, and/or revolutionary business models. In a recent study, the four types of innovation outcomes were used to examine the level of innovation in team projects of engineering students who studied in a hybrid course *vs.* a MOOC. The findings indicated that the hybrid students produced team projects with a higher level of innovation, as they created projects with a higher potential for radical type of innovation.[14]

With regards to measuring innovative thinking through self-report questionnaires, Dyer and colleagues[5] identified four consistent behavioural tendencies as the catalysts to the generation of innovative ideas: (a) questioning—asking questions that challenge the status quo; (b) observing—intensely examining the world around while seeking new ideas; (c) experimenting—engaging in active investigations to explore options and generate novel designs; and (d) idea networking—creating networks of individuals with diverse backgrounds and perspectives that could lead to new ideas and insights. The four constructs were used as a framework for the generation of the 'innovative thinking questionnaire',[11] which was applied in a recent study on international engineering students.[7] The international students who studied with peers from their home country, reported higher gains in obtaining innovative ideas through observing, questioning, exploring, and idea networking, compared with their counterparts. Other studies that examined the distribution of the four constructs among engineering students found that students tend to score high on the component of experimenting and significantly low on idea networking.[9,13] The studies described above were conducted in a classroom setting. However, as the number of university courses offered online is increasing fleetly,[15] behaviors that were identified in the literature as promoting students' innovation might be expressed differently in online learning environments.

10.1.2 Massive Open Online Courses

Over the last two decades, online education has seen the emergence and adoption of massive open online courses (MOOCs). MOOCs are web-based learning environments in which participants from all over the world, regardless of religion, race or gender, can acquire knowledge for free without any commitment or prior requirements.[11,16,17] While sharing many characteristics with online courses, MOOCs are somewhat different as they facilitate massive participation, attracting thousands of learners, from a wide range of countries, cultures, and backgrounds.[3,18,19] Due to the free access they provide for a global audience of learners, MOOCs were documented as a means for fostering diversity, equity, and inclusion.[20] The assumption is that if the online courses are free and available online, barriers to access could be overcome. Yet, the promise of free education to the masses has not always been fully realized since MOOCs do not adequately address to the learning traditions, contexts, and needs of learners from developing countries.[21] According to recent studies, one way in which MOOC designers could address this issue is by encouraging collaborations among MOOC learners, preferably in small groups. This could create learning communities that interact beyond social, cultural, and political boundaries.[11,22]

Indeed, given the greater diversity in student population, MOOCs provide the opportunity to collaborate with learners from different backgrounds and to work on learning products in diverse groups.[3,23] Such learning was documented as promoting knowledge construction, motivation, and creativity of MOOC learners.[3,16,24] Moreover, being exposed to individuals from diverse backgrounds has the potential to advance the cognitive processes that generate innovative ideas.[7,11]

Despite its possible advantages, group work in MOOCs is not without challenges. First, MOOCs often end with dysfunctional groups due to the high drop-out rates or insufficient matching between group members.[3,24] Second, due to the geographical distance, interpersonal interactions among students, and among them and the teaching staff, are more limited, reducing the ability to receive real-time support.[16,24,25] Third, online learners are less exposed to peers' non-verbal behaviour, which might result in misunderstanding or misinterpreting peers' ideas, leading to frustration and disruption of performance.[23] The unique characteristics of MOOCs, as expressed above, may influence the nature of interactions among learners, and consequently, their learning outcomes, in general and innovation capabilities, in specific. Investigating the characteristics of these emerging learning environments is now more relevant than ever since many science and engineering courses have shifted from frontal teaching to online learning due to the COVID-19 outbreak.

10.1.3 Group Diversity in Science and Engineering Education

Working in groups on authentic industrial problems is a well-known practice in science and engineering education.[26,27] Group work provides students with a realistic and complex industry-like experience that they are likely to encounter during their professional careers.[26,27] Group work was documented as having positive effects on students' academic performance and motivation (Esparragoza and colleagues, 2015.[28] See also Chapter 29 in this book by Fung and Kim).[29] Moreover, it was reported as stimulating new ideas and perspectives among science and engineering students.[10,11,20] However, the application of effective group work has many challenges, one of which is working with members from diverse backgrounds that have yielded varying educational outcomes.[10,30]

Group diversity refers to the distribution of differences among members of a group.[31,32] This includes bio-demographic variables, such as gender, mother tongue, and age, or task-related variables such as academic discipline and academic level.[8,32,33] Group diversity is central to science and engineering education, since the ability to work effectively as an engaged member of a small group is among the top most important behavioural attributes for university graduates.[34] In this context, group diversity has yielded inconsistent results with regards to students' learning outcomes. For example, group diversity in terms of gender and culture was reported as fostering students' academic performance.[28,35] Conversely, other studies have reported gender and age diversity as factors that inhibit the promotion of novel and creative ideas.[30,31] More recently, diversity researchers have begun to pay attention to the aspect of innovation—one of the prominent skills required in the 21st century workplace, and hence, a required skill for science and engineering students.[2,33]

10.2 Research Goal and Questions

The goal of the current study was to identify the diversity level of groups formed by MOOC learners and to examine its association with learner's perceived innovative thinking and the innovation level of their group project. This goal raised the following research questions:

1. What is the diversity level of groups formed by learners enrolled in a MOOC on nanotechnology?

2. What is the perceived innovative thinking and the innovation level of group projects of science and engineering MOOC learners?
3. What are the correlations between group diversity, perceived innovative thinking, and group project innovation?

10.3 Methods

10.3.1 Research Setting and Participants

The study was undertaken in the context of a MOOC on nanotechnology, which integrates broad, interdisciplinary domains within science, technology, engineering, and mathematics (STEM) education. The MOOC teaching staff included one instructor, a worldwide expert in nanotechnology, and two teaching assistants who were Ph.D. students at the time.

The MOOC included a series of short, pre-recorded video lectures, an e-book, and discussion forums. It also included three types of graded assignments: ten weekly quizzes (30% of the course grade), individual assignments (20% of the course grade), and a final project (50% of the course grade). The final project engaged learners in designing an innovative nanosensor that might help humans in their daily lives and contribute to the good of human society. The project was conducted in small groups, and learners were encouraged to form groups of three or four members.

The study included 109 science and engineering undergraduate and graduate learners who enrolled in the MOOC on the Coursera platform. All participants were asked to sign an informed consent form. They were informed that participation is voluntary, and they were given the choice to withdraw at any time. The study received the approval of the institutional review board (IRB). The participants represent a diverse group of learners, from more than 20 countries, with various cultural, academic, and professional backgrounds. In terms of gender, this group constituted of 65% males and 35% females, ranging in age from 18 to 45 years (mean = 29.5, SD = 8.2). As written above, the participants were required to work on the final project in small teams of three or four members. They were not assigned into teams by the teaching staff, but rather they were encouraged to find suitable partners and form their own teams by using a designated online forum, where they introduced themselves and their academic background. Since the team members came from different geographical locations, they worked on their projects remotely *via* online systems such as Google Docs and Dropbox.

10.3.2 Research Methods and Tools

The study applied a one-group quasi-experimental design,[36] with one independent variable (*i.e.*, the level of group diversity) and two dependent variables (*i.e.*, perceived innovative thinking and group projects' innovation level).

The level of group diversity refers to the heterogeneity among the members of the small project groups in terms of bio-demographic and task-related differences. It was calculated for each small group of three or four MOOC learners who worked together on the final project. This was done using Blau's index of heterogeneity, which ranges between 0 for a perfectly homogeneous population, to 1 for a perfectly heterogeneous population.[37] The index was separately calculated for each of the following four variables: gender, mother tongue, academic discipline, and academic level. Learners' background information was collected *via* a pre-questionnaire, which was distributed to all participants before the beginning of the course.

The participants' perceived innovative thinking was evaluated using online pre- and post-questionnaires.[4] The questionnaire includes 13 items, formatted on a Likert-type scale, ranking from 1 (strongly disagree) to 5 (strongly agree). Based on the work of Dyer and colleagues,[5] the items were grouped into four constructs: observing (three items), questioning (three items), idea networking (three items), and experimenting (four items). See Appendix A for the modified innovative thinking questionnaire.

The innovation level of group projects was evaluated according to the innovation taxonomy, which is used to identify and measure the innovation profile of nanotechnology group projects.[11] It includes four categories: (1) innovation type—whether the newly designed nanosensor presents incremental, modular, architectural, or radical type of innovation; (2) product necessity—the extent to which the nanosensor addresses a problem that requires immediate action or attention; (3) STEM interdisciplinarity—the extent to which the nanosensor is presented from a wide range of science, technology, engineering, and mathematics (STEM) perspectives; and (4) market readiness—the extent to which the nanosensor has the potential to produce a tangible product with commercial potential. Each of the four categories is assigned with a numerical score on a scale of 1 to 4, with the highest possible score for project innovation being 16.

Data were analyzed using IBM's Statistical Program for the Social Sciences (SPSS). We applied a paired-samples *t*-test to compare the mean scores on the pre- and post-questionnaires, and Pearson correlation tests to examine potential associations between the three dependent variables.

10.4 Findings

10.4.1 The Diversity Level of Groups Formed in a Nanotechnology MOOC

To answer the first research question and gain a deeper insight into the diversity level of groups formed by MOOC learners, we examined four variables: gender, mother tongue, academic discipline, and academic level. Gender was distributed according to male, female, or other. Mother tongue was distributed according to English, Russian, Hebrew, Spanish, Mandarin, and other. Academic discipline was distributed according to engineering faculties (chemical engineering, mechanical engineering, *etc.*) or science faculties (physics, chemistry, *etc.*), medicine, mathematics, and other. Academic level was distributed according to undergraduate students, M.Sc. students, and Ph.D. students. The overall group diversity level was calculated for each project group of three to four learners. The level of group diversity was calculated for each of the four variables separately, using Blau's index, which ranges between 0 to 1.[38] Table 10.1 details the frequency distribution for each of the four variables of group diversity, each column amounts to 100%.

The average Blau's index for the project teams formed by the MOOC learners was 0.45 (SD = 0.25). Results show that the project groups were the most diverse in terms of mother tongue (mean = 0.51, SD = 0.25) and academic discipline (mean = 0.59, SD = 0.25). Apparently, learners did not hesitate to carry out projects with people from different countries and from various fields of science and engineering. Results further show that the project groups were the least diverse in terms of gender (mean = 0.33, SD = 0.20) and academic level (mean = 0.41, SD = 0.31). Apparently, the MOOC learners were looking to carry out group projects with people from the same level of education and the same gender, whether it was done intentionally or not.

10.4.2 Learners' Perceived Innovative Thinking and the Innovation Level of Nanotechnology Group Projects

To answer the second research question, we analyzed MOOC learners' responses to the pre- and post-questionnaires. A paired-samples *t*-test indicated a significant increase in participants' perceived innovative thinking over the time of the course for each of the four subscales: questioning, observing, experimenting, and idea networking (Table 10.2).

Table 10.2 shows that at the end of the course, the participants expressed higher ratings on the behaviours of observing (mean = 3.94, SD = 0.50) and experimenting (mean = 3.92, SD = 0.51), and lower for idea networking (mean = 3.83, SD = 0.53) and questioning (mean = 3.82, SD = 0.45). A series of paired samples *t*-tests indicated statistically significant differences between the four subscales' scores. For example, observing was found as significantly higher than questioning ($t(108) = 2.52$, $p = 0.01$, $d = 0.25$) and networking ($t(108) =$

Table 10.1 Group diversity according to the four variables.

	% Of group projects ($n = 30$)			
Blau's index	Gender	Mother tongue	Academic discipline	Academic level
<0.29	25	11	6	29
0.30–0.39	33	15	6	10
0.40–0.49	25	22	11	11
0.50–0.59	11	7	26	15
0.60–0.69	6	35	31	26
0.70<	0	10	19	9

Table 10.2 MOOC learners' perceived innovative thinking, pre- *vs.* post-course results ($n = 109$).

	Pre		Post			
Category	Mean	SD	Mean	SD	*t*	Sig
Questioning	3.63	0.46	3.82	0.45	7.46	0.00
Observing	3.77	0.51	3.94	0.50	4.73	0.00
Experimenting	3.73	0.51	3.92	0.51	5.86	0.00
Idea networking	3.69	0.58	3.83	0.53	9.80	0.00
Total mean	3.70	0.38	3.88	0.37	8.76	0.00

1.82, $p = 0.05$, $d = 0.03$). Questioning was found as significantly lower than experimenting ($t(108) = 2.01$, $p = 0.04$, $d = 0.21$) and observing ($t(108) = 2.52$, $p = 0.01$, $d = 0.25$).

To analyze and grade the innovation level of group projects, we used the Project Innovation Taxonomy[2] that includes four categories: innovation type, product necessity, STEM interdisciplinarity, and market readiness, on a 1 to 4 scale. An example can be seen in a project entitled 'Polygraph Based on Nanotechnology'. The rationale of the project assumes that the operation of the traditional polygraph requires heavy and complex equipment that may disturb both the users and the subjects. Hence, the group members developed a miniaturization of the traditional polygraph device in the form of a tiny, flexible electronic human skin-like sensor that can be connected to the user's body for the maximum convenience of the user. The project addressed a problem that has the potential to impact a limited range of consumers, hence it received 2 out of the 4 possible points for *product necessity*. The suggested sensor was presented in the project mainly from the technological aspect, and not enough information was given regarding the fabrication process that will allow this nano-based sensor to be released to the market. For these reasons, the project received 2 points for *STEM interdisciplinarity* and *market readiness*. Lastly, the fundamental change proposed in this project is in the core components of the product, while no change will be done in the links between them. Hence, the project was classified as presenting *modular innovation* and received 2 points for *Innovation type*. Overall, this project received a score of 8 points out of the maximum of 16 points.

Results indicated that the MOOC learners received the highest scores for product necessity and STEM interdisciplinarity (mean = 3.19, SD = 0.85; mean = 3.04, SD = 0.10, respectively), and lowest scores for innovation type and market readiness (mean = 2.07, SD = 1.10; mean = 2.67, SD = 0.60, respectively). A series of paired samples *t*-tests indicated statistically significant differences between the four categories. Product necessity was found as significantly higher than innovation type ($t(108) = 12.18$, $p = 0.00$, $d = 1.15$), STEM interdisciplinarity ($t(108) = 2.25$, $p = 0.03$, $d = 1.25$), and market readiness ($t(108) = 7.14$, $p = 0.00$, $d = 0.71$).

10.4.3 Associations Between Group Diversity, Innovative Thinking, and Project Innovation

To answer the third research question, we examined the correlations between group diversity, perceived innovative thinking, and group project innovation. Out of the four observed variables of group diversity (gender, mother tongue, academic discipline and academic level), only one was significantly correlated with participants' perceived innovative thinking. Findings indicated a statistically significant correlation between group diversity in terms of *academic discipline* and perceived innovative thinking ($r(109) = 0.22$, $p = 0.02$). This finding raises the possibility that the more diverse the project group was in terms of academic discipline, the greater they perceived themselves to be innovative thinkers at the end of the course.

Out of the four observed variables of group diversity, two variables were significantly positively correlated with group project innovation: academic discipline ($r(109) = 0.58$, $p = 0.00$) and academic level ($r(109) = 0.50$, $p = 0.00$); while one variable was negatively correlated: mother tongue ($r(109) = -0.43$, $p = 0.00$). Group diversity in terms of "gender" did not correlate significantly with project innovation. Meaning that the more diverse the project group was in terms of academic discipline and levels, the higher the score was on innovation. Conversely, the more diverse the project group was in terms of members' mother tongues, the lower the score was on project innovation.

10.5 Discussion

The literature shows mixed results regarding the influence of group diversity on the innovation capabilities of science and engineering students (*e.g.*, Yueh *et al.*, 2015[27]). We approached this gap by examining group diversity in a more granular and precise manner, by analyzing different variables of group diversity (bio-demographic and task-related). Overall, three key findings result from this study.

The first finding refers to the diversity level of MOOC learners who worked on a project in a small group. The project groups received an overall average Blau's index of 0.4 out of 1, which indicates relatively low heterogeneity. Even though MOOC learners had the opportunity to form project groups with participants from diverse countries and backgrounds,[3,19] it seems they were looking to carry out projects with people who are similar to them, whether it was done intentionally or not. This behavior is referred to in the literature as homophily—the tendency to associate with others who share similar traits, such as gender, ethnicity, and level of education, which has been reported to be even more salient in MOOCs.[38]

Findings further indicated that out of the four variables of diversity that were observed, group diversity in terms of *academic discipline* received the highest scores, while group diversity in terms of *gender* received the lowest scores. It seems that the MOOC learners did not hesitate to carry out nanotechnology projects with people from various fields of science and engineering, but still preferred to collaborate with people from the same gender. This result coincides with the work of Kellogg and colleagues,[38] who reported that if students share the same gender, it increases the likelihood of a tie between them. Moreover, this result could be attributed to the interdisciplinary nature of the nanotechnology field, which encompasses multiple disciplines that draw knowledge from diverse sources.[39] According to Eto,[40] the multidisciplinarity in nanotechnology focuses mainly on chemistry, extends to physics and material sciences and, to a lesser extent, biology and instrument technology. Hence, it is possible that the reason for the high group diversity in terms of the academic discipline is related to the research population which was very heterogeneous in terms of fields of study.

The second finding refers to MOOC learners' significant increase in their perceived innovative thinking. This coincides with studies that reported innovative thinking as a skill that can be taught and promoted, and hence should be integrated into the science and engineering curriculum.[7,33] Out of the four innovative behavioural constructs, learners scored significantly higher for observing and experimenting, and lower for idea networking and questioning. These results are in line with previous studies, which reported that engineering students scored highest in the innovative behaviour of experimenting, and significantly lower on idea networking.[9,13] With regards to group projects' innovation, results indicated that learners received the highest scores on the categories of product necessity and STEM interdisciplinarity, and lowest scores on innovation type and market readiness. This corresponds to a study by Thayer and colleagues,[6] which addressed the need to place more attention on idea implementation to develop quick-to-market products.

The third finding refers to the correlations between the distinct characteristics of group diversity, learners' perceived innovative thinking, and group project innovation. Findings revealed that out of the four observed variables of group diversity, only *academic discipline* was significantly correlated with both perceived innovative thinking and group projects' innovation. This empirical evidence complements previous research that reported group diversity in terms of academic discipline as contributing to the development of creative and innovative projects.[8,41,42] A contradictory finding emerged from the examination of *mother tongue* diversity, which was found to be significantly negatively correlated with group project innovation. This suggests that heterogeneity in group members' mother tongues may serve an obstacle to MOOC learners' efforts to develop innovative projects.

A possible explanation may be related to the way the online learners communicated with each other while working on the final projects. They used mostly asynchronous technology as an intermediary tool that provides learners with indirect interactions and limited support, which may result in misunderstanding or misinterpreting of peers' ideas.[14,25,43] This may have impeded their ability to communicate directly and explain their ideas when collaborating with their group members.[23,30] Hence, we believe that providing online learners with more options for direct and timely support from their instructors and peers may increase their ability to communicate efficiently and produce innovative learning outcomes. The use of synchronous technological tools, such as Zoom, Google Meet or Microsoft Teams, will enable them to interact directly with their instructors and peers, to ask questions, and receive support in real-time.

To sum up, it seems that diversity in online learners' mother tongues made it difficult for them to communicate openly and express their ideas clearly, which resulted in lower scores on innovation in their group projects. Thus, better mechanisms should be developed to help science and engineering online learners to overcome language barriers and be able to effectively communicate novel ideas with their peers. To promote innovation among learners, we recommend developers and instructors of online science and engineering courses, and MOOCs in particular, to encourage collaborative learning in small teams that are heterogeneous in terms of academic discipline and homogenous in terms of mother tongue, and to communicate by using both asynchronous and synchronous technological tools. In the context of MOOCs, this may increase equity and inclusion since learners who are not fluent in English would still be able to take part in the English MOOC and have the opportunity to benefit from the knowledge of peers who have the same native language but come from different disciplines. In such manner, MOOCs would meet their goal of providing accessible high-quality education to the masses, including non-native English-speaking learners.[2,18]

Appendix A

Table 10.3 The modified innovative thinking questionnaire.

Category	Item #	Statement
Questioning	1	I regularly ask questions that challenge other learners' assumptions
	2	While learning, I often ask questions that challenge the status quo
	3	Other learners are challenged by my questions
Observing	4	While learning, I think of new ideas that come through observing the world
	5	New ideas often come to me when observing how other learners interact and do things
	6	I have a history of taking things apart in order to reexamine them
Experimenting	7	I actively search for new ideas through experimenting
	8	When learning, I actively search for new ideas through experimenting
	9	I experiment to understand how things work and create new ways of doing things
Idea networking	10	I have a network of individuals with which I interact to get new ideas
	11	I have a network of learners whom I trust to bring new perspectives and refine new ideas
	12	I initiate meetings with other learners to spark new ideas
	13	I participate in diverse professional and/or academic forums

References

1. G. M. Whitesides, *Small*, 2005, **1**, 172–179.
2. M. Barak and M. Usher, *Comput. Educ.*, 2019, **137**, 1–11.
3. M. Barak, A. Watted and H. Haick, *Comput. Educ.*, 2016, **94**, 49–60.
4. M. Barak, A. Watted and H. Haick, *Assess. Eval. Higher Educ.*, 2020, **45**, 212–223.
5. J. H. Dyer, H. B. Gregersen and C. Christensen, *Strategic Entrepreneurship J.*, 2008, **2**, 317–338.
6. A. L. Thayer, A. Petruzzelli and C. E. McClurg, *Am. Psychol.*, 2018, **73**, 363.
7. M. Barak and S. Yuan, *Thinking Skills Creat.*, 2021, **39**, 100766.
8. M. Usher and M. Barak, *Comput. Educ.*, 2020, **144**, 103702.
9. J. L. Hess, N. D. Fila and S. Purzer, *Int. J. Eng. Educ.*, 2016, **32**, 1236–1249.
10. N. D. Fila and S. Purzer, *Int. J. Eng. Educ.*, 2014, **30**, 1405–1418.
11. M. Barak and M. Usher, in *Active Learning in College Science*, Springer, 2020, pp. 639–653.
12. R. M. Henderson and K. B. Clark, *Adm. Sci. Q.*, 1990, 9–30.
13. N. D. Fila, J. L. Hess, P. D. Mathis and S. Purzer, in *2015 ASEE Annual Conference & Exposition*, 2015, pp. 26–338.
14. M. Barak and M. Usher, *Eur. J. Eng. Educ.*, 2022, **47**, 299–313.
15. See Chapter 5: R. Eidelman and Y. Shwartz, in *Digital Learning and Teaching in Chemistry*, ed. Y. J. Dori, C. Ngai and G. Szteinberg, Royal Society of Chemistry, United Kingdom, 2023.
16. C. Kulkarni, J. Cambre, Y. Kotturi, M. S. Bernstein and S. Klemmer, in *Proceedings of the 18th ACM Conference on Computer Supported Cooperative Work & Social Computing*, Vancouver, BC, Canada, 2015, pp. 1116–1128.
17. J. Zhu, M. Mo and C. Mok, *Educ. Psychol.*, 2018, **38**, 1106–1128.
18. T. Dillahunt, Z. Wang and S. D. Teasley, *Int. Rev. Res. Open Distance Learn.*, 2014, **15**, 177–196.
19. M. Zhu, A. R. Sari and M. M. Lee, *Educ. Technol. Res. Dev.*, 2020, **68**, 1685–1710.
20. L. Carver and L. M. Harrison, *Lib. Educ.*, 2013, **99**, n4.
21. R. F. Kizilcec, A. J. Saltarelli, J. Reich and G. L. Cohen, *Science*, 2017, **355**, 251–252.
22. C. B. Chandler, R. M. Quintana, Y. Tan and J. M. Aguinaga, *J. Appl. Instr. Des.*, 2021, **10**, 1–14.
23. M. Usher, M. Barak and H. Haick, *Thinking Skills Creat.*, 2021, **42**, 100965.
24. V. Kovanović, S. Joksimović, O. Poquet, T. Hennis, I. Čukić, P. de Vries, M. Hatala, S. Dawson, G. Siemens and D. Gašević, *Comput. Educ.*, 2018, **119**, 44–58.
25. M. Liu, W. Zou, Y. Shi, Z. Pan and C. Li, *J. Comput. High. Educ.*, 2020, **32**, 307–329.
26. N. Wengrowicz, Y. J. Dori and D. Dori, in *Cognition, Metacognition, and Culture in STEM Education*, Springer, Cham, 2018, pp. 191–216.
27. H.-P. Yueh, Y.-L. Liu and W. Lin, *Int. J. Eng. Educ.*, 2015, **31**, 220–228.
28. I. E. Esparragoza, S. L. Farak, J. R. Ocampo, J. N. Segovia, R. Viganò, J. Duque-Rivera and C. A. Rodriguez, *Int. J. Eng. Educ.*, 2015, **31**, 1255–1269.

29. See Chapter 29: F. M. Fung and Y. Kim, in *Digital Learning and Teaching in Chemistry*, ed. Y. J. Dori, C. Ngai and G. Szteinberg, Royal Society of Chemistry, United Kingdom, 2023.
30. H. W. Tseng and H.-T. Yeh, *Comput. Educ.*, 2013, **63**, 1–9.
31. L. E. Coursey, B. C. Williams, J. B. Kenworthy, P. B. Paulus and S. Doboli, *J. Creat. Behav.*, 2020, **54**, 253–266.
32. L. H. Pelled, K. M. Eisenhardt and K. R. Xin, *Adm. Sci. Q.*, 1999, **44**, 1–28.
33. J. Andrews and R. Clark, *Train. Eng. Innov.*, 2018, 177–197.
34. EBET, *Criteria for Accrediting Engineering Programs, 2016-17*, Baltimore, MD, 2016.
35. K. Lau, S. L. Beckman and A. M. Agogino, *Int. J. Eng. Educ.*, 2012, **28**, 293.
36. G. Campbell and N. Guibert, *Br. Food J.*, 2006, **108**, 233–242.
37. P. M. Blau, *Inequality and Heterogeneity: A Primitive Theory of Social Structure*, Free Press New York, 1977, vol. 7.
38. S. Kellogg, S. Booth and K. Oliver, *Int. Rev. Res. Open Distrib. Learn.*, 2014, **15**, 263–289.
39. A. L. Porter and J. Youtie, *J. Nanoparticle Res.*, 2009, **11**, 1023–1041.
40. H. Eto, *Scientometrics*, 2003, **58**, 5–33.
41. N. D. Fila, S. Purzer and P. D. Mathis, in *2014 ASEE Annual Conference & Exposition*, 2014, pp. 24–831.
42. S. T. Bell, A. J. Villado, M. A. Lukasik, L. Belau and A. L. Briggs, *J. Manage.*, 2011, **37**, 709–743.
43. R. Hussein-Farraj, M. Barak and Y. Judy Dori, *Interdiscip. J. e-Skills Lifelong Learn.*, 2012, **8**, 115–135.

CHAPTER 11

Integrating Web-based Learning to Make Industrial and Everyday Life Chemistry Accessible to High-school Chemistry Students

D. MARCHAK[a], M. KESNER*[a] AND M. FRAILICH[a]

[a]Department of Science Teaching, The Weizmann Institute of Science, Rehovot, Israel
*E-mail: mirikes2@gmail.com

11.1 Introduction

Since the nineties, the influence of Information and Communication Technologies (ICT) on the field of education has been growing fast. ICT offered new possibilities to enrich learning environments, content, and teaching strategies. Web-based learning environments gradually entered science education as the internet became available in schools and teachers gained the necessary techno-pedagogical knowledge. ICT enabled access to a wide variety of resources for enrichment and new tools for visualizing abstract phenomena, thus supporting the understanding of complicated scientific topics. It also promoted constructivist approaches to learning such as active learning, where the students are at the centre of the learning process,

Advances in Chemistry Education Series No. 11
Digital Learning and Teaching in Chemistry
Edited by Yehudit Judy Dori, Courtney Ngai and Gabriela Szteinberg

Published by the Royal Society of Chemistry, www.rsc.org

individually or in groups, as active builders of their own knowledge while the teacher functions more as a mentor or guide. Active learning can be understood as learning that strengthens student involvement in the learning process by asking students to engage in guided practice tasks. It was found to have a positive impact on student attitudes and achievements.[1] Meaningful learning, namely when content is retained and retrieved later in future learning scenarios to integrate new content,[2] is an expected outcome of active learning.

In this chapter we present the implementation of an ICT chemistry learning environment initiated by the Israeli Chemical Industry-Education Link Centre that operated at The Weizmann Institute of Science through the website "*General Chemistry and Industrial Chemistry at the Service of Mankind*". We also describe the evolution that this platform underwent through the years to adapt to the ever-changing demands that arose from curriculum modifications and teachers' and students' needs. We start by describing the development process that led to the original form of the website. We then present an example of how the website served as a free-for-all platform for supporting the teaching and learning of abstract chemical concepts in connection to real-world issues, and whose efficiency was validated by research. Lastly, we present how the website was reshaped into its current format and contents to support non-formal chemistry learning at the national scale. The COVID-19 pandemic, and related emerging needs regarding remote teaching and learning, strengthened the inclusive vision of our work that emphasized the importance of stimulating motivation by addressing students' individual interests through the connection of chemistry studies to the real world in a personalized way. We also discuss how issues of equity and inclusion in chemistry education were addressed and advanced through this ICT platform.

11.2 Developing the "*General Chemistry and Industrial Chemistry at the Service of Mankind*" Website

11.2.1 Website Rationale

Since the eighties there has been a growing understanding that chemistry should be presented in class not only as a scientific discipline, but also as a subject that influences the student's personal life and affects society. We found that integrating industrial and everyday life chemistry topics into the formal curriculum proved effective to increase the relevance and applicability of chemistry as learnt by high-school students in Israel.[3] In parallel, it was shown that when teachers used varied teaching methods, they were more successful in motivating their students, in creating activities that support the relevance and applicability of chemistry, in presenting occupational possibilities in chemistry-related professions, and in raising students' awareness regarding chemistry-related environmental issues. In addition, evidence accumulated demonstrating the educational power of ICT in helping to design varied contexts and methods for teaching and learning that encourage active and social learning.[4]

Notwithstanding this evidence, teaching resources in Hebrew that could allow web-based learning to be brought to class by chemistry teachers in Israel were very scarce, and in fact non-existent regarding the connection to chemical industry. For this reason, the "*General Chemistry and Industrial Chemistry at the Service of Mankind*" website was developed. The main goals for launching the website in 2002 were (1) to develop new teaching and learning materials suited to the novelty that ICT offered (visualization tools, communication facilities, access to varied reading materials, miscellaneous software like applets, simulations and games, *etc.*); (2) to introduce ICT to chemistry teachers and help them overcome any implementation reluctance by facilitating their introduction through open-access teachers' guides and courses; and (3) to create our own databases in Hebrew that combined chemistry-related, free resources about local industrial chemistry plants, their products and raw materials, and relevant technological, economical, and environmental issues. The latter goal was intended to advance equity and inclusion of Hebrew-speaking students as being part of a worldwide community of chemistry learners.

11.2.2 Website Development Process and Resulting Contents and Format

The website was developed as a curriculum-based interactive website for high school chemistry planned to highlight industrial and everyday life chemistry.[5] Its URL at present (2023) is: https://stwww1.weizmann.ac.il/learnchem. Both form and content of the earliest version of the website resulted from two years of development in collaboration with experienced chemistry teachers who participated in a series of development workshops. The website original components are presented in Table 11.1.

Regarding content, the web-based activities were available through the different website sections and were designed to attain one or several of the following aims: complementing the chemistry syllabus to emphasize the relevance that chemistry has in the industry and in daily life; enhancing students' investigative and thought processing skills through web-based active learning environments; and raising students' interest in chemistry and motivation to learn through enrichment activities and social learning. The wide range of learning environments and activities developed for the website was planned considering that most chemistry classes are heterogeneous and include diverse populations of students with varied backgrounds and abilities. From the start, the website offered different kinds of materials, such as dynamic learning units composed of several sub-units that spanned different aspects of specific chemistry topics and promoted equity among students by allowing them to make personalized decisions about their learning.

Table 11.1 Website components and sub-components.

Website components	Website sub-components
About the Chemical Industry-Education Link Centre	General information, publications, activities, courses, contact information
Link to the chemistry syllabus	Topic-relevant learning materials and activities
Glossary	General and industrial chemistry terms listing, including short explanations
Chemical industry in Israel	Databases on main industrial chemistry plants, including location maps
Chemical products in Israel	Databases on 100 chemical products, related to the syllabus, including visual models
Selected subjects in daily life chemistry	Medicine and health, foods and drinks, clothing, plastics, forensic and environmental chemistry
Students' corner	Games, home experiments, questions of interest, an open line to experts, "Why chemistry?" and "It is interesting to know that..." entries
Teachers' corner	Open line to experts, a collection of exam papers, guides for field trips to chemical industries, subjects for discussion and suggestions of discussion stages, interesting articles
Forums (students, teachers)	Questions and dilemmas discussed in class, support forums for in-service training courses, experimental groups of teachers
Guides	Student and teacher guides
From the news	"Hot" newspaper clippings
Page of the month	Selected up-to-date articles pertaining applications of chemistry, environmental and industrial chemistry
The people behind chemistry	Scientists, researchers, Nobel Prize winners, industry employees, prominent figures in the history of chemistry
Gallery	Syllabus and industrial chemistry related articles, pictures, presentations, videos, production processes flow charts
Related links	Organizations, museums, chemical industries abroad, governmental links, internet resources
Search	Internal search engine for the whole site and its databases

Eventually, more emphasis on laboratory work emerged from the chemistry inspectorate in the Ministry of Education, urging us to develop laboratory mini projects based on industrial and everyday life issues. Additional website sections were developed and incorporated to serve the needs of teachers who implemented the "Industrial Initiatives" approach:[6] a new teaching model for bringing industrial chemistry issues to class by varied learning activities assessed through diverse evaluation tools and a portfolio. All these approaches, and others, were gradually uploaded to the database of the website, free for use.

Through our collaborating teachers, we learned that presenting the website contents along with suitable teaching strategies and pedagogical guidance made the teachers more enthusiastic about using them with their students and helped them get more effective results. Thus, we promoted professional development programs that were directly connected to website implementation, from short workshops to long-term courses for teachers from all over the country, and we published a teachers' guide for using the website.

To know more about the educational effectiveness of the website's contents, we followed its implementation though several research projects using different research tools. The most important and large-scale research focused on learning the topic of chemical bonding through web-based activities. In the following section we briefly present some parts of a three-year study and discuss them in terms of the impact it had on the evolution of the website.

11.3 Assessing the Effectiveness of Web-based Learning

Chemical bonding is considered a key subject that forms the basis for understanding many phenomena in chemistry. It is a highly abstract concept that is generally difficult for students to visualize,[7] and it is usually taught with minor connections to applied chemistry. Thus, four web-based activities pertaining the subject of chemical bonding as taught in 10th grade, were developed within the framework of our website, and researched to assess their educational effectiveness. We also investigated students' perceptions of the classroom learning environment, their attitudes regarding the relevance that learning chemistry has in everyday life, and their interest in chemistry studies.[8]

The activities dealt with the following curricular sub-topics: (1) the atom—models of the atomic structure; (2) metals; (3) ionic substances; and (4) molecular substances. Each of these activities (except for models of atomic structure) consisted of four sections, each concerning a different aspect of content: (a) the microscopic structure of substances, (b) the physical properties of substances, (c) the relationship between the structure of the substance and its properties (macro–micro relationships), and (d) the relevance of the substance in the real world. The activities were based on tasks that the students had to perform in pairs (cooperative learning involving speech and negotiation), building on previous knowledge (constructivist approach involving associative connection and integration between new input and long-term memory of content), demanding the operation of several visualization tools and varied molecular models (active learning involving trial-and-error, learning-by-doing), accompanied with enrichment tasks connected to the uses and applications of the substances (promoting interest by realizing chemistry's significance in real-world scenarios).

This study was conducted during two academic years (2003–2004 and 2004–2005) and two groups participated in it: an experimental and a comparison group. The two groups had similar student–teacher profiles and followed the same syllabus. In both groups, the teachers first explained the topics in class using their own traditional tools such as static molecular models and pictures. The difference between the groups resided in the closing activity: while the comparison group students performed a traditional task, the experimental group students were exposed to the website and engaged in the above-mentioned web-based activities in pairs while the teacher assisted them. Overall, both groups dedicated the same amount of time to study the subject.

Data analysis in the study involved both quantitative and qualitative methodologies. The quantitative tools were administered to both groups regarding scores achieved in each curricular sub-topic. These tools included an achievement pre-test, to examine the students' previous chemistry knowledge, and an achievement post-test that assessed their knowledge and understanding of the concept of chemical bonding, together with the perceived relevance of chemistry. Qualitative data were gathered from interviews with students and teachers, and observations in the classrooms.[9]

Quantitative data analysis revealed no initial significant differences between the groups in achievement pre-test (Table 11.2). However, in the achievement post-test the experimental group outperformed the comparison group significantly (Table 11.3 and Table 11.4).

As shown in Table 11.4 the mean scores on the sub-topics examined in the post-test for the experimental group were significantly higher than those of the comparison group, including the perceived relevance of the subject. These results were supported by the insights from the qualitative analysis of the interviews with students and teachers, as well as students' feedback questionnaires and spontaneous comments collected from the experimental group during classroom observations. These gave us a deeper understanding of the advantages of this web-based learning approach.

Table 11.2 Achievement pre-test results.[a]

Experimental group $N = 161$ Mean (S.D.)	Comparison group $N = 93$ Mean (S.D.)	t value and significance level
73.30 (19.52)	72.15 (19.00)	0.45 N.S.

[a]N.S.: Non-significant.

Table 11.3 Achievement post-test results.

Experimental group $N = 145$ Mean (S.D.)	Comparison group $N = 88$ Mean (S.D.)	t value and significance level
76.11 (17.00)	63.41 (16.24)	5.7***

***$p \leq 0.001$.

Table 11.4 ANOVA results for student achievement post-test.

Sub-tests	Experimental group $N = 145$ Mean (S.D.)	Comparison group $N = 88$ Mean (S.D.)	F value and significance level
Atomic structure	78.68 (29.41)	60.54 (35.20)	17.81***
Metals	74.33 (21.28)	61.35 (20.70)	20.63***
Ionic compounds	76.70 (16.88)	64.45 (17.79)	27.53***
Molecular compounds	71.90 (19.51)	58.31 (19.04)	26.86***
Relevance (applications/ uses in everyday life)	74.93 (23.80)	57.28 (22.71)	30.59***

***$p \leq 0.001$.

Analysis of the interviews[9] conducted with teachers and students in the experimental group revealed the influence that integrating the web-based activities had on learning and understanding chemical bonding and its context. From these interviews four key factors appeared as having a positive contribution to the learning process: (1) identification of students' difficulties regarding *chemical bonding*, since while dealing with the activities cognitive conflicts came up that would otherwise remain un-noticed; (2) implementation of a constructivist, active learning approach that enhanced the learning experience; (3) visualization of abstract content by employing computer-based visual models and animations that lead to greater understanding; and (4) association between the chemistry concepts learned in class and their application in daily life and industrial scenarios, thus increasing students' awareness about the relevance of chemistry and its importance to the individual and the society. Table 11.5 includes some quotes from the interviews exemplifying these key factors.

Teachers in the comparison group reported in the interviews to have taught all the required content in a highly organized manner, using various static models like graphic depictions and ball-and-stick models, in alignment with the "traditional", teacher-centred format of instruction where the teacher is accountable for the instruction process.

One should not overlook that the development of the website and the web-based activities occurred when ICT was in its infancy in Israel. Its implementation in class required adapting to a new learning environment, to new modes of teaching and learning, and to the employment of digital skills not yet developed by teachers nor students. The importance of this study is reflected in the implications that can be drawn from its findings regarding the effectiveness of such a learning environment and its assimilation in chemistry classrooms. First, the use of web-based visualization tools, including dynamic models instead of static models, assists students' understanding of chemical bonding and helps them understand the transition between macroscopic and microscopic aspects of content. These tools can also help identify students' misconceptions. In addition, learning by the web-based

Table 11.5 Excerpts from the interviews of students and teachers in the experimental group.

Factor	Students' quotes	Teachers' quotes
Identification of students' difficulties	*The models and animations are a visual way of demonstrating the subject matter, it is better when you see it, sometimes it is difficult to understand it only from the teacher's explanation.*	*I observed that they did not understand well the contents that were taught in class.*
Implementation of a constructivist, active learning approach	*It's a matter of adding one thing to another, and in the end, everything adds up. It is better understood when you must formulate an answer rather than just listening in the classroom.*	*While engaging in the Web-based activity, they ask questions, and they need me... in the classroom I am the one who usually asks the questions; it is a major difference...*
Visualization of abstract content	*You really do not understand how the ions and electrons really move and what really happens... Then you see it in animation, so you understand it more.*	*When students see the micro and the structures in three dimensions and the electron cloud on the computer, they see that the models of molecular compounds have volume, that there is motion... and how one atom is arranged opposite another atom, and they see the connection between them. I think that this contributes greatly to learning concepts in chemical bonding.*
Association between classroom chemistry and real-world scenarios	*It's important to note everyday life in chemistry, because it's things you see in everyday life, and you do not know it's like that.*	*In the web-based activities, they were more exposed to the relevance of chemistry, in the book almost not.*

constructivist approach can lead to meaningful learning. Active, cooperative learning is required when students work in pairs to solve the tasks at their own pace, directing their questions to one another and to the teacher as they arise. In this aspect, the web-based activities significantly contribute to the teacher's explanations in class, providing scaffolding to support the learning process. Moreover, implementation of the web-based activities can cause teachers to reflect on their instruction strategies and on how to improve them. Teachers reported being satisfied with the activities since they allowed them to be more aware of students' difficulties, and that the web-based visual tools and interactive simulations helped them to demonstrate abstract phenomena to their students.[8] Lastly, the web-based learning environment can be suitable to connect chemistry content as taught in class with the world as perceived by students. Students reported[8] being satisfied with the web-based activities and the website, they enjoyed the process of learning chemistry and showed significantly higher awareness to the relevance of chemistry to daily life than the comparison group. These results are in line with other findings[9] that emphasize the contribution of technology-based chemistry representations that help to successfully promote visualization of content at the macroscopic, microscopic, and symbolic levels through constructivist knowledge integration.

11.4 Moving to a Web-based, Non-formal Learning Environment

Until 2002 industrial chemistry was a compulsory topic in the curriculum for students who major in chemistry. The decision made by the chemistry inspectorate to include this chapter within other selective chapters in the chemistry curriculum, led us to a gradual modification of the website content to adapt to new requirements. New web-based activities were developed for each chapter of the curriculum and adjusted to suit internal assessment requirements. On a larger scale, this decision led us to move on to a non-formal learning approach and initiate our next major initiative. In 2008 we launched a national projects competition by the full name of "*We Have Chemistry!—Chemistry, Industry, and the Environment in the eyes of the individual and society*". The project intended to bridge the gap between school chemistry and out-of-school learning by addressing students' individual interests and inviting them to make personal decisions about the chemistry-related content that they want to learn and the kind of project that they choose to prepare. Thus, the competition format was designed by combining three theoretical and practical frameworks: project-based learning (PBL), context-based learning (CBL) and science competitions. These were merged into a non-formal yet well-structured educational framework where students voluntarily and independently could decide to participate, in their free time, to work on any phenomena of their choice.

The main goal of the initiative was to have students experience independent and peer chemistry learning. The teacher functioned mainly as a mentor in cases where students wanted their help. In the beginning, only a small number of students joined the competition on their own and there were many teachers who avoided participating in the competition with their students because of being afraid of the additional time and effort the project might demand. Thus, following our objectives of equity and inclusion, we set to reach more chemistry students from more varied backgrounds and encourage them to partake in the competition regardless the availability of their teacher. To make this possible, the website was extended to include a competition section where students could connect directly to get support from the competition trained staff and access helpful information about the competition.

The original competition format comprised five project categories: "film", "poster", "newspaper article", "photography" and "laboratory inquiry", each named after the kind of learning item that students chose to submit. The main requirement regarding chemistry content was that the context of all projects should pertain to the chemical industry or daily life chemistry and emphasize explicitly the influence that chemistry has on the individual, the society and/or the environment. The first four categories were open to all chemistry students and meant to simulate the field of science communication. The inquiry-lab category, which was a part of the formal curriculum, was open only to students who majored in chemistry. Lately, the competition was expanded to include two additional categories, "podcast" and "comics", which have become popular in science communication.

Since the beginning of the competition initiative, we established an organizing committee that included industry people, inspectorate delegates and our staff composed of chemistry teachers and science educators. Our staff was specifically trained for the project and qualified to offer direct support for students and teachers throughout the different stages of their work (from registration to tips for improving the project before submitting the final version). Upon request, the staff helped students to find experts from the academy and industry with whom to work or interview, as well as professionals in the varied areas of communication media. The competition annex to the original website included sub-sections about each project category containing explanations about specific demands, assessment rubrics with the criteria by which the judges evaluate the project, and examples of projects that won first places of that category in previous years. Originally, we also organized students' regional conferences that would mark the beginning of the competition each year. These conferences included masterclasses with communication media experts and each student could choose to participate in one of them according to their chosen category. An evaluative study showed that participation in this competition increases the motivation to study chemistry compared with the preparation of similar projects within the framework of a school set-up,[10] thereby validating our decisions about the competition format.

Towards 2018 we moved a step further and decided to prepare short tutorial movies delivered by our collaborating communication media experts including explanations on how to prepare each specific kind of project. These tutorials were planned to substitute the conferences' masterclasses for there were always limiting factors regarding these conferences: (1) they took place on specific dates and hours so not all students could attend; (2) the masterclasses happened in parallel sessions so each student could participate only in one of these; and (3) teachers had to arrange for a bus and accompany the students after school hours so budget considerations were an issue for some populations. To overcome these limitations and encourage equity and inclusion, in 2018 we introduced the filmed tutorials through the competition website section, available online for all to access in their own time. Through the competition staff, students could connect with the professionals who gave the masterclasses. This approach proved particularly useful during the pandemic period, when the competition and its final conference unexpectedly had to switch to an online platform.

Following new accessibility regulations, the last upgraded version of the competition website was launched in September 2019. These regulations emphasized the inclusion of all students, embracing those with disabilities. The site was revised to be more user-friendly and easily manageable with bright colours, bigger and more clearly defined sections, easier registration, and shorter routes to content.

The number of students who enrolled in the competition grew gradually over the years from 300 in September 2008, to 1445 students from 54 schools in 45 different cities, from all around Israel, in September 2020. Between 2010 and 2018 there was a steady average of 570 participants per year. The steepest increase in participants was recorded between the years of 2018 and 2019 when their number rose from 540 to 1095, respectively. We believe that this increase in the number of participants can be partly attributed to the upgrading of the website and its accessibility to all since we did not change our advertising strategies.

The latest version of the website became possible through an advanced platform that also allowed the collection of data about its activity. Upon analysis of the number of website pages displayed as a function of time we could discern interesting patterns of website visitors' behavior. Figure 11.1 shows a plot of the data including marks of significant events during the time plotted.

Due to the COVID-19 pandemic and its restrictions, the last round of the competition took place during the 2019–2020 school year, during which the number of website pages displayed per day before summer recess averaged 123 and peaked around important dates matching the different competition

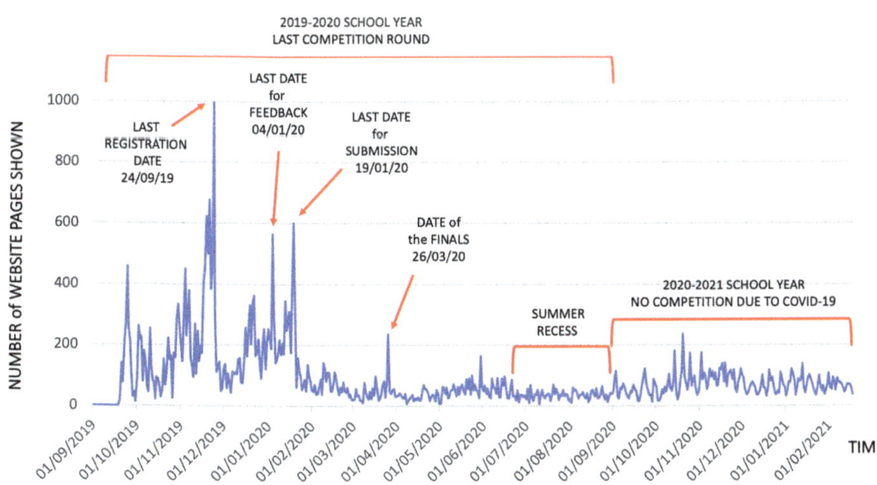

Figure 11.1 Plot of the number of website pages displayed over the period between 1 September 2019, and 18 February 2021, including significant dates.

stages (Figure 11.1). During the summer recess, as expected, this average decreased to 37. Our attention was caught by the fact that there were visitors entering the website's pages although the competition was not launched in September 2020, and that this number was higher than that seen during the summer recess. For example, the average number of pages shown per day between 1 September 2020, and 18 February 2021, was 73.

We investigated the website activity report to understand which pages visitors generally visited when accessing our website (Table 11.6). We found that during the competition year of 2019–2020 (12 months) the total number of pages of any kind that were shown was 35 374. The main activity on the website was connected to the competition, as expected. Meanwhile, only the glossary remained as non-competition-related content, and yet it proved to form a significant part of this activity, with the glossary portal being shown on demand 1116 times. This number does not include glossary pages that were shown when a visitor typed a specific term in the main page search field. Yet, it indicates the number of times visitors purposely reached for the glossary portal first. During the observed 5.5 months from the year of 2020–2021, the total number of page displays of any kind was 13 540. The main activity on the website was connected to the website glossary, with the glossary portal being shown on demand 1292 times, 176 times more than in the previous year. For instance, the term "titration" was specifically searched 253 times during the 2019–2020 year and 391 times during the 5.5 months of 2020–2021.

From comparing the website activity during a competition and a non-competition year, we can conclude that the website still attracts visitors and still offers a platform for independent learning. Of special use seems to be the database of chemical terms in Hebrew (the Glossary). Several teachers still contact the competition staff, to date, asking diverse questions about the availability of the competition section of the website. We know that in these cases teachers ask their students to engage in PBL tasks, such as the preparation of posters, and provide their students with detailed instructions sheets, tutorials, and assessment rubrics that appear on our website. Teachers then assess their students' work by these rubrics and the scores are reported as

Table 11.6 The five most-visited website pages with the number of times each page was displayed (frequency).

19 September 2019, to 31 August 2020		1 September 2020, to 18 February 2021	
Page name	Page frequency	Page name	Page frequency
Main portal	8213	Main portal	1625
Competitions portal	5020	Glossary portal	1292
Poster competition	3100	Glossary: Molar volume	585
Competition registration	2489	Competitions portal	504
Photography competition	1417	Glossary: Lewis structure	485

part of the alternative assessment required for certain topics of the chemistry curriculum (comprising 30% of the final grade in the discipline). This might explain the 271 entries to the "Poster Competition" page, or the "Poster Preparation" tutorial movie being specifically searched 122 times during the 5.5 months of the non-competition year.

11.5 Discussion

The original "*General Chemistry and Industrial Chemistry at the Service of Mankind*" website was a pioneering initiative to promote ICT in chemistry education in Israel at its early stages. Teachers could support, enrich, and vary their teaching practices through the website contents. Although initially teacher-oriented, it eventually became clear through its implementation and research that the website benefited students mainly by efficiently supporting understanding of abstract chemical content through simulation-aided visualization tools, offering a web-based, collaborative active learning experience, emphasizing the connection between chemistry and aspects of industry and everyday life, thus increasing their perceived relevance of and motivation to study chemistry.[8] Other research findings in following years also showed that when teachers used varied teaching methods, they were more successful in eliciting understanding and increasing students' motivation to learn.[11]

In a period of more than 20 years since the launching of the original website, and with the changes in chemistry curriculum, in pedagogical approaches and in ICT platforms, there was a constant need to renew the website and its components. Part of this evolution led to the "*We Have Chemistry!*" national projects competition, which supported non-formal learning, PBL and CBL. The resulting website format emerged as an inclusive platform not only for advancing the competition but also for encouraging independent learning among students from all social and demographic backgrounds in Israel. It was the outcome of a conceptual adaptation that the website underwent regarding its function and the role it played in chemistry education. It became an accessible, student-oriented platform for learning through enquiry, providing students access to the information needed to learn chemistry through working on their personalized projects.

The unusual situation that arose due to the COVID-19 pandemic gave us the opportunity to compare the website's activity between a competition and a non-competition year. Analyzing the data collected from the website's usage reports, we found that in addition to serving as a site of reference regarding the competition, the website can also serve as a free resource of varied chemistry contents for all.

One of the differences between privileged and underprivileged populations can be understood based on having access, or not, to better resources and know-how. We see the web-based platform that we presented in this chapter as a digital learning environment that proved to be appropriate for continuous renewal and adaptation to suit the needs of chemistry teachers and learners from varied backgrounds by offering rich contents, a wide database in the local language, and flexible techno-pedagogical tools while making industrial and applied chemistry widely available. Our next initiative is to reach all chemistry teachers in Israel to make them aware of the website open contents and how these can be used to advance meaningful, active learning through PBL tasks under the alternative assessment requirements. Around platforms like these[12] communities of educators can be established to take advantage of what ICT can offer in terms of allowing equal access for all to good quality education.

Acknowledgements

We would like to thank our partners in the Ministry of Education and in Industry, especially ICL—Israel Chemicals Ltd, for their continuous funding and support. We also wish to thank the chemistry teachers in Israel for their long-lasting trust and cooperation.

References

1. R. M. Felder and R. Brent, *ASQ Higher Educ. Brief*, 2009, **2**(4), 1.
2. R. E. Mayer, *Theory into Practice*, 2002, **41**(4), 226.
3. M. Kesner, A. Hofstein and R. Ben-Zvi, *Int. J. Sci. Educ.*, 1997, **19**(6), 725.
4. T. Myers, A. Blackman, T. Andersen, R. Hay and H. Gray, *J. Learn. Des.*, 2014, **7**, 38.
5. M. Kesner, M. Frailich and A. Hofstein, *Technology-Rich Learning Environments, A Future Perspective*, World Scientific, New Jersey, London, Singapore, Hong Kong, 2003.
6. A. Hofstein and M. Kesner, *Int. J. Sci. Educ.*, 2006, **28**(9), 1017.
7. See Chapter 18: D. F. Donnelly-Hermosillo, L. Gerard and M. Linn, in *Digital Learning and Teaching in Chemistry*, ed. Y. J. Dori, C. Ngai and G. Szteinberg, Royal Society of Chemistry, United Kingdom, 2023.
8. M. Frailich, M. Kesner and A. Hofstein, *Res. Sci. Technol. Educ.*, 2007, **25**, 179.
9. M. Frailich, M. Kesner and A. Hofstein, *J. Res. Sci. Teach.*, 2009, **46**, 289.
10. Y. Sharaabi-Naor, M. Kesner and Y. Shwartz, *SISYPHUS J. Educ.*, 2014, **2**(2), 100.
11. R. Blonder and S. Sakhnini, *Chem. Educ. Res. Pract.*, 2012, **13**(4), 500.
12. See Chapter 31: B. DeKorver and D. Herrington, in *Digital Learning and Teaching in Chemistry*, ed. Y. J. Dori, C. Ngai and G. Szteinberg, Royal Society of Chemistry, United Kingdom, 2023.

CHAPTER 12

The Next Level in Inclusive Chemistry Education: A Model Approach Using a Multi-touch Learning Book

J. KRANZ[a] AND R. TIEMANN*[a]

[a]Department of Chemistry, Humboldt-Universität zu Berlin, Unter den Linden 6, Berlin, 10099, Germany
*E-mail: ruediger.tiemann@hu-berlin.de

12.1 Introduction

Internationally, school education is undergoing rapid change because, as socially recognized new framework conditions, inclusion and digitalization require learning progressions from teachers in establishing an inclusive, digital world in school.[1] For chemistry education in Germany, the challenge is in delivering the Conference of Ministers of Culture's strategy on "Education in the Digital World", whose characteristic feature is the connection of inclusion—the participation of all in access to education—to digitalization.

Accordingly, digitalization should be embedded in an integrative way in inclusion, which is a recognized, overriding principle of society as a whole. The prerequisite for planning inclusive teaching with digital tools is therefore a model of inclusive teaching that takes aspects of digitization into account in a goal-oriented way (see also Chapter 10 by Usher and Barak).[2]

Advances in Chemistry Education Series No. 11
Digital Learning and Teaching in Chemistry
Edited by Yehudit Judy Dori, Courtney Ngai and Gabriela Szteinberg

Published by the Royal Society of Chemistry, www.rsc.org

For the international as well as for the German-speaking area, hardly any suitable models can be found as instruments for planning inclusive teaching with the integration of digital media. The paradigm shifts for teaching from exclusive to inclusive and analogue to digital are usually considered in isolation and rarely in their entirety. However, only in the synthetic consideration of these aspects can the inclusion process be optimized.[3] One task of research should therefore be to envisage the development and validation of a model for inclusive chemistry teaching.

We developed a "**m**odel for **i**nclusive **c**hemistry teaching"[4] (MiC) and, building on this, tested the transfer of the model to a practical, inclusive learning environment using an interactive learning book. The model was based on, among other aspects, research findings on inclusion, good teaching and teaching in heterogeneous groups.[5–7]

12.2 Model of Inclusive Chemistry Teaching

The MiC architecture is designed such that teachers can derive concrete planning-guiding assistance from it. MiC is characterized by three levels: the lowest level is determined by the state through the curricula and their subject-specific requirements; the second level looks at the individual, taking decisions after diagnoses, which, in addition to the educational reduction, concern the various approaches; and the third level concerns the decisions regarding the selection and how the problem is formulated. These levels can be followed step by step (see Figure 12.1).

12.2.1 Domain Level

The objectives, competences and standards as well as the subject areas and contents were formulated in the curricula. The teachers can select subject areas and contents and make further subject-related preliminary decisions. These introductory statements on level 1 also apply predominantly to traditional teaching. For inclusive teaching, two further premises are style-forming: joint work on the common subject and an open teaching approach (see Figure 12.2).

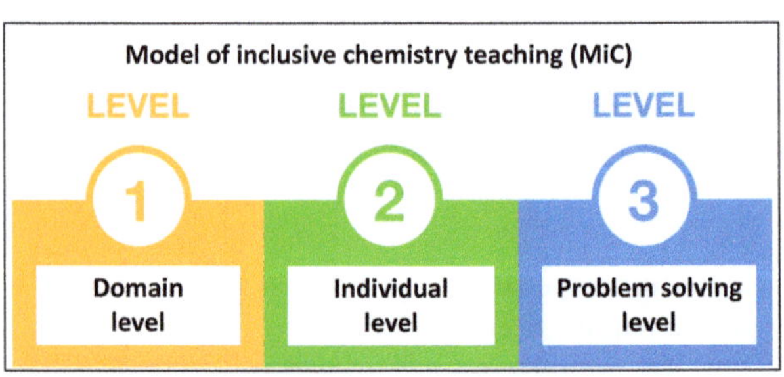

Figure 12.1 The three levels of MiC.

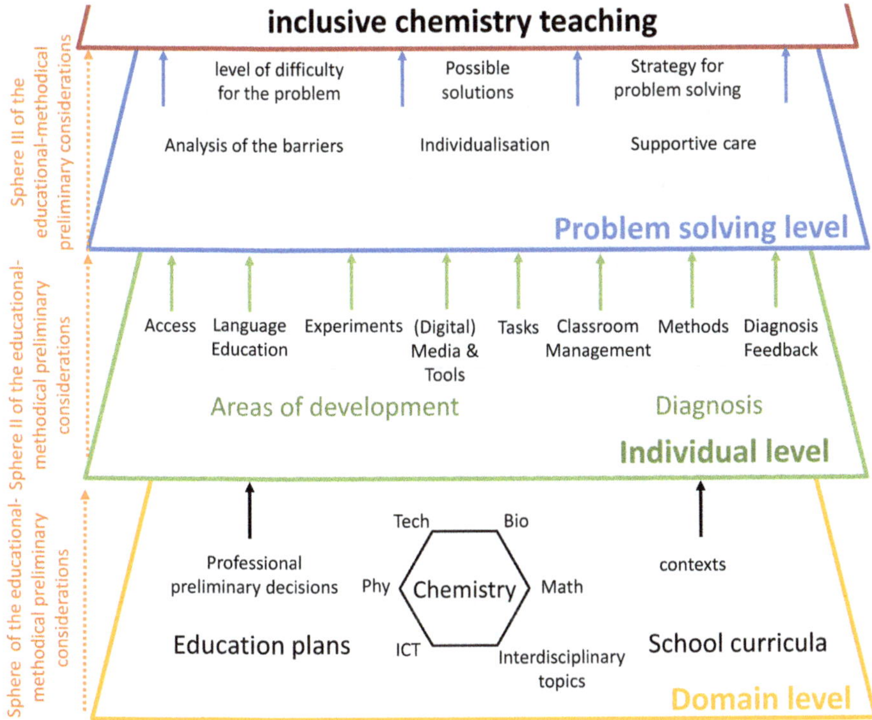

Figure 12.2 The MiC-levels in detail.

12.2.2 Individual Level

One of the key points of school inclusion and the MiC approach is a stronger focus on the individuality of learners in lesson planning, reflected in the second level. The diagnosis of learners in lesson planning includes the teacher's analysis of the individual learning status and progress, although it goes far beyond this. The other characteristics of the learners, *i.e.*, physical limitations, learning speed, haptic or intellectual learning types, must also be assessed.[8] Diagnosis and feedback also play a role in the implementation of lessons because it is the teacher's task to keep an eye on the progress of the current acquisition of competences.[9] The "encounter phase" of the lesson can lead to an individual motivation boost for all students through different approaches. Points of contact that involve all senses, including aesthetic or tension-inducing ones, are a characteristic of inclusive teaching.

Digital media can support the aforementioned aspects at the level of the individual, *e.g.*, by enabling self-evaluation through learning games and surveys, by deepening and clarifying models and model experiments (see Chapter 11 by Marchak *et al.*,[10] and Chapter 14 by Lavi[11]). The use of digital tools in an inclusive learning environment is determined by their added value for learning in heterogeneous learning groups. On the one hand, this can be determined *via* Puentedura's SAMR model[9] (substitution–augmentation–modification–redefinition, see Chapter 9 by Blonder[12]) with regard to the innovative potential of the application. On the other hand, it is necessary to consider interactive or adaptive elements. They allow individually clickable branches (clickables), which ensure media adaptation to the differently structured learning processes of the students. First assessments of interactivity have been formulated by Musch[13] and also by Niegemann,[14] who name as essential features the possibilities of intervention and control on the part of the learner who is able to actively intervene in the learning process and thus control the learning path and learning speed. Interactivity can thus be seen as an indication of the individualization of the learning process; at the same time, interactivity is a step towards the design of inclusive teaching (see Figure 12.2).

12.2.3 Level of Problem Solving

The high degree of support that is necessary during the problem-solving process in inclusive chemistry lessons requires intensive engagement with the formulation and the degree of difficulty of the problem.[15] Due to the traditional problem solving theory, abstract thinking ability is often a reason for either success or failure. In traditional teaching, the analysis of a problem's difficulty and the subsequent problem-solving process is rather the exception, whereas in inclusive teaching it is a fundamental condition for anticipating possible barriers and obstacles on the part of the students. Thus, it is a necessary prerequisite for the conception of learning aids and further support measures. Under the premise of open teaching, it is also important for the teacher to generate problems that allow multiple solutions and different problem-solving strategies (see Figure 12.2).

The spheres between the levels describe the spaces in which decisions are made regarding the design of the aforementioned aspects.

12.3 Transfer of MiC into an Inclusive Learning Environment

As part of the evaluation of a model for inclusive chemistry teaching (MiC), a learning environment on the topic of "fire and flame", a subject area of the Berlin-Brandenburg framework curriculum, was designed.[16] The orchestration of the learning environment with real experiments, an interactive Multitouch Learning Book[17]—a dynamic, web-based platform that allows the instructor to combine, *e.g.*, texts, videos or learning games—and a paper-based "researcher" booklet leads to a realistic design. It takes into account research-based design features for problem solving, motivational aspects, self-regulated learning with digital media, the sequence and type of experiments, and the use of prompts. For the transfer, the three levels of the model were gone through step by step, especially for the level of problem solving, the barriers of the problems were considered, and corresponding aids were designed.

Many experiments in chemistry lessons require a heat source and as such the gas burner is an indispensable laboratory and working tool that all students should be proficient in using. This topic requires a sophisticated, proactive classroom management in terms of accessibility to the subject rooms, equipment, chemicals and learning aids, as well as an adapted hazard potential.[16]

However, this subject requires an indispensable degree of instructional support, partly because of the potential danger but also to ensure that the competences required to work on more demanding problems can be acquired with the help of instructions. In the media-supported learning environment with the interactive textbook, attention is paid to a balance between instruction and construction. The four learning sequences each started with an instructional, verbal introduction. This is followed by a self-directed as well as cooperative learning phase with the digital textbook. This in turn is followed by a short phase of instruction immediately before the start of the experiment.[18] The experiments and phenomena are suitable for supporting both less gifted and highly gifted students in the sense of a broad understanding of inclusion.

12.3.1 Multi-touch Learning Books—A Forward-looking Digital Tool on the Way to Inclusive Teaching

Multi-touch learning books offer the opportunity for new learning formats. Thus, in addition to individualizing learning paths, linear and branched procedures can be designed. Furthermore, learning paths can be supported by assessment or diagnostic tasks, learning games and specific assistance. Multi-touch learning books are therefore a forward-looking digital building block on the way to inclusive teaching.[19]

Figure 12.3 shows the cover page, the icons structuring the learning book and the QR code with the link to the multi-touch learning book. The concrete design features of interactive learning environments are listed by Hesse and Mandl.[20] These include the following features: graphic navigation, for example *via* icons; buttons to click on (clickables) with the aim of active exploration of information; simulations and video tutorials to explain and deepen complex structures, processes and interrelationships; and feedback on the solution of tasks and exercises, including information on any errors that may have occurred and how to correct them.[20] These requirements are met in the interactive textbook on "fire and flame". Within the interactive learning book, after the introduction, the students go through the four learning sequences up to the transfer. Within the learning sequences LS1–LS4, the students can follow individual learning paths through the interactive branches (see Figure 12.4). The multi-touch learning book takes into account the three levels of the model *via* the topic, the development areas adapted to the grade level and learning group, and the difficulty levels of the problem with the assigned differentiating support.

Figure 12.3 A multi-touch learning book with a selection of the icons used, some of them interactive (information, tips, language aids and exercises), with QR code (link to the textbook).

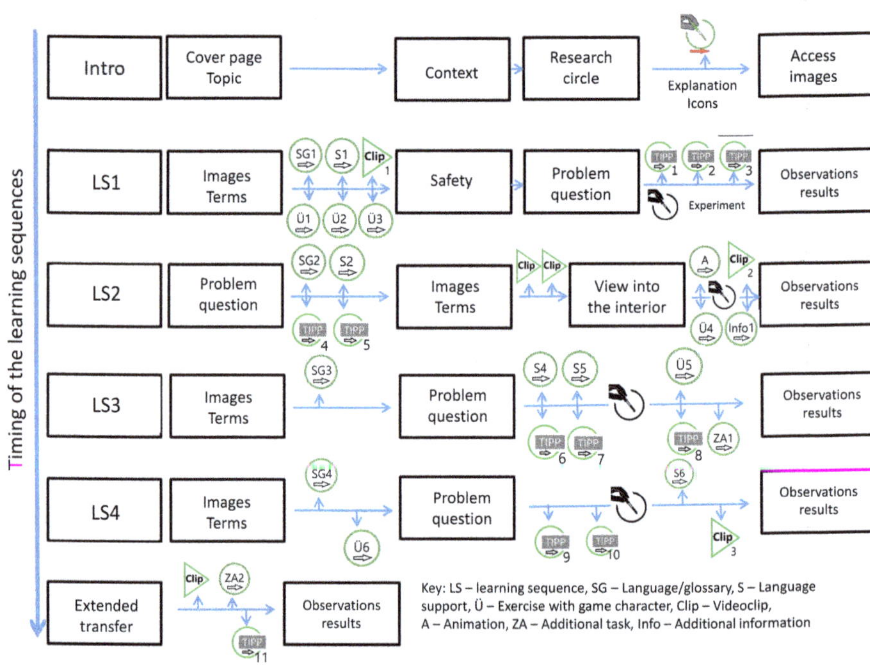

Figure 12.4 Structure with interactive parts of the "fire and flame" multi-touch learning book.

Figure 12.4 illustrates the pronounced interactivity of the digital multi-touch learning book. The vertical arrows pointing up and down show the clickable icons that provide individual learning pathways. Navigation runs *via* arrows and icons, the clickable icons lead to different information, animations, video clips and tutorials. They serve to explain and deepen problems, problem-solving strategies and contexts.

The widgets with playfully designed exercises for self-assessment contain feedback on the solution of the tasks with hints on the errors that occurred. The tips work either as individual learning aids or, in the case of more demanding experiments, as staged learning aids. In learning sequences (LS) 1–3, the level of difficulty of the problems increases, which is relieved by appropriate learning aids in terms of content and language in the multi-touch learning book.

In summary, the following can be said about the designed multimedia learning environment "fire and flame": the digital learning environment is interactive, but not adaptive, because the software does not react automatically to the learning path of the students. It meets essential educational requirements for interactive textbooks and Open Educational Resources (OER), it is freely accessible, can be used free of charge and it runs on all end devices (see www.tiemannlab.de).

12.3.2 Learning Analytics—Effective Use of Prompts in the Multi-touch Learning Book

Although learning with digital media has many advantages, learners in such learning environments often have difficulties in coping with the systemic characteristics of the learning environment.[21] In a comprehensive literature review, Devolder, van Braak and Tondeur[22] showed that the presentation or use of prompts is the most effective way to support self-regulated learning (SRL)[23] in digital learning environments. Prompts should be understood as procedures that induce and stimulate cognitive, metacognitive, volitional and cooperative activities during learning.[24] Prompting uses eye-catchers in the multi-touch learning book to direct the learner's attention to certain objects or behaviors and thus supports the learning process (see Figure 12.5).

12.4 Selected Results

The teaching unit was quantitatively tested with seven classes ($N = 165$, grades six and seven, ages 12–13). No control group was involved in the study. Situational questionnaires and fisheye 360° video recordings were used to record the perceived fit of the teaching offer with the individual performance of students, supplemented by questionnaires for teachers. The results regarding the pre-post surveys showed positive, significant changes for the students' self-assessment regarding fire and flame knowledge (high school ($N = 51$), primary school ($N = 43$), Integrated Secondary School (ISS) ($N = 71$), 23 items, ($t\ (164) = 32.66, p < 0.001$; $\alpha_{FWpre} = 0.927$; $\alpha_{FWpost} = 0.846$)). The Integrated Secondary School offers qualifications ranging from basic school-leaving certificates to university entrance qualifications. We showed that even students with a low interest in natural sciences answered all tasks in the knowledge test at a very high level.

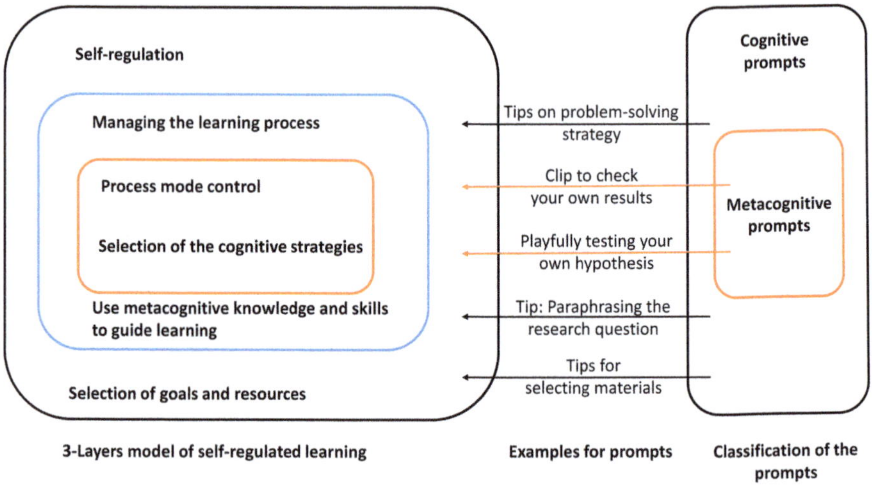

Figure 12.5 Prompting in the multi-touch learning book "fire and flame" with regard to the connection of the type of prompts with the model of self-regulated learning.[23]

154 out of 165 students passed the competence test. Of the 10 competency test tasks, an average of 8.5 were solved ($M = 8.51$, S.D. = 1.57), so the pass rate can be classified as very high. This rate decreases from high school students to primary school students to ISS students. The differences are significant ($F(2, 87) = 20.51, p = 0.004$). The effect size shows a medium effect $\eta^2 = 0.0679$.

On average, the students performed 8.42 clicks out of 37 possible clickables in the multi-touch learning book. ISS students reached the highest score for the number of clicks. They realized a mean of 8.95 clicks out of a possible 37 ($M = 8.95$, S.D. = 3.53). The use of the clickables did not differ significantly for the different types of schools (high school, ISS, primary school), *i.e.*, the interactive offer is accepted comparably well in the sum of the clicks.

The multi-touch learning book appeals equally to girls and boys. Both genders show no significant differences in the self-assessment of domain specific knowledge, the use of the interactive digital medium and in the competence test.

Video tutorials are used equally across all school types. In contrast, the other interactive offers are used in a school-specific way: learners from integrated secondary schools make greater use of the learning and language aids, high school students are interested in the additional offers and information, and primary school students increasingly click on the exercises with gaming character. The respective school-specific differences are significant. In an extreme group analysis, it could be demonstrated that students who frequently used the tips—especially those from the integrated secondary school—were able to produce glass products and a stirring rod made of glass of comparable quality to students who produced the product without using the tips.

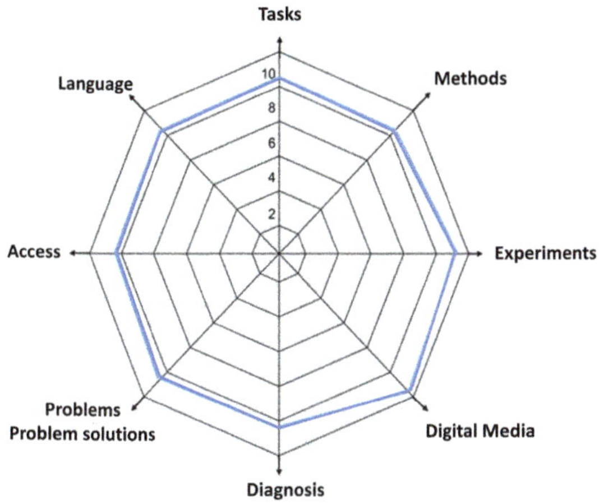

Figure 12.6 "Analysis spider" for analyzing the inclusive aspects of the multi-touch learning book (Legend: 10—full inclusive characteristics, 0—non-inclusive characteristics).

In an expert rating with master students of chemistry education ($N = 21$), the inclusive aspects of the multi-touch learning book tasks, methods, experiments, digital media, diagnosis, problems/problem-solutions, access and language were evaluated (see Figure 12.6).

The mean values of the assessment range between 8.20 (S.D. 1.28) and 9.40 (S.D. 1.095) on a scale of 0–10, where 10 means complete fulfilment of the inclusive aspect. The results for the different inclusive aspects are not significant. Overall, the expert rating (see Figure 12.6) shows good coverage of the inclusive aspects. The intended focus with the use of real experiments and the selected digital application was positively connoted. It was also noted positively that the students worked in the learning environment on a platform in which all other digital applications were embedded. This avoided time-consuming changes between digital media.

Based on the successful transfer of the MiC approach into an inclusive, analogue–digital learning environment, it could be shown that the three steps of the MiC approach are manageable and practically implementable. At the same time, the suitability and potential of the multi-touch learning book for inclusive chemistry teaching could be demonstrated.

12.5 Conclusion

With regard to the teachers' assessment of the suitability of the interactive textbook for inclusion, it can be stated that the entirety of the investigated inclusive aspects of the learning environment and especially of the multi-touch learning book was positively assessed. The results as a whole show the potential of the multi-touch learning book as a platform for inclusive learning environments, and at the same time the successful transfer of MiC into a practical, inclusive learning environment could be demonstrated. It could be shown that the three steps of the MiC approach are manageable and practically implementable. The exploration of the research object leads *via* the individual results of the evaluation and *via* their consideration from several perspectives (pentangulation) to the indirect verification of the suitability of the model. Considering the largely identical strategies of knowledge acquisition in the subjects of biology and physics, an extension of the model for inclusive chemistry teaching (MiC) into a **m**odel of **i**nclusive **s**cience **t**eaching (MiST) could be envisaged.

References

1. S. Becker, T. Bruckermann, A. Finger, J. Huwer, E. Kremser, M. Meier, L.-J. Thoms, C. Thyssen and L. von Kotzebue, *Digit. Basiskompetenzen–Orientierungshilfe und Praxisbeispiele für die Univ. Lehr-amtsausbildung den Naturwissenschaften*, 2020, pp. 14–43.
2. See Chapter 10: M. Usher and M. Barak, in *Digital Learning and Teaching in Chemistry*, ed. Y. J. Dori, C. Ngai and G. Szteinberg, Royal Society of Chemistry, United Kingdom, 2023.
3. J. Muuß-Merholz, *Individ. fördern mit Digit. Medien. Risiken, Erfolgsfaktoren. Gütersloh*, 2015, pp. 224–307.
4. J. Kranz and R. Tiemann, in *Naturwissenschaftliche Kompetenzen in der Gesellschaft von morgen. Gesellschaft für Didaktik der Chemie und Physik Jahrestagung in Wien 2019*, ed. S. Habig, Universität Duisburg-Essen, 2020, pp. 796–799.
5. J. Kahlert and U. Heimlich, in *Inklusion in Schule und Unterricht*, ed. U. Heimlich and J. Kahlert, Kohlhammer, Stuttgart, 2014, pp. 153–190.
6. Y. Anders, I. Hardy, S. Pauen, J. Ramseger, B. Sodian and M. Steffensky, *Wissenschaftliche Untersuchungen zur Arbeit der Stiftung „Haus der kleinen Forscher“*, Schaffhausen, 2013.
7. S. Prediger and C. von Aufschnaiter, in *Umgang mit Heterogenität in Schule und Unterricht*, ed. T. Bohl, J. Budde and M. Rieger-Ladich, Klinkhardt, Bad Heilbrunn, 2017, pp. 291–307.
8. K. Vogt, in *bwp@ Spezial 5 – Hochschultage Berufliche Bildung 2011, Fachtagung 11*, ed. I. Kettschau and K. Gemballa, 2011, pp. 1–9.
9. J. Hattie, *Visible Learning: A Synthesis of over 800 Meta-analyses Relating to Achievement*, Routledge, NY, 2009.
10. See Chapter 11: D. Marchak, M. Kesner and M. Frailich, in *Digital Learning and Teaching in Chemistry*, ed. Y. J. Dori, C. Ngai and G. Szteinberg, Royal Society of Chemistry, United Kingdom, 2023.
11. See Chapter 14: R. Lavi, in *Digital Learning and Teaching in Chemistry*, ed. Y. J. Dori, C. Ngai and G. Szteinberg, Royal Society of Chemistry, United Kingdom, 2023.
12. See Chapter 9: R. Blonder, in *Digital Learning and Teaching in Chemistry*, ed. Y. J. Dori, C. Ngai and G. Szteinberg, Royal Society of Chemistry, United Kingdom, 2023.

13. J. Musch, *Z. fur Padagog. Psychol.*, 2006, **13**, 148–160.
14. H. M. Niegemann, *Neue Lernmedien: konzipieren, entwickeln, einsetzen*, Verlag Hans Huber, Göttingen, 2001.
15. J. Koppelt and R. Tiemann, in *Entwicklung naturwissenschaftlichen Denkens zwischen Phänomenen und Systematik*, ed. D. Höttecke, Lit Verlag, Münster, 2009, pp. 173–175.
16. K. Groß and C. S. Reiners, *CHEMKON*, 2012, **19**, 13–20.
17. J. Huwer, A. Bock and J. Seibert, *Am. J. Educ. Res.*, 2018, **6**, 763–772.
18. G. Reinmann-Rothmeier and H. Mandl, in *Pädagogische Psychologie. Ein Lehrbuch*, ed. A. Krapp and B. Weidenmann, Beltz Psychologie Verlags Union, Weinheim, Basel, 2006, vol. 5, pp. 601–646.
19. J. Huwer and I. Eilks, in *Lernprozesse mit digitalen Werkzeugen unterstützen. Perspektiven aus der Didaktik naturwissenschaftlicher Fächer*, ed. J. Meßinger-Koppelt, S. Schanze and J. Groß, Joachim Herz Stiftung, Hamburg, 2017, pp. 81–94.
20. F. W. Hesse and H. Mandl, in *Studium Online – Hochschulentwicklung durch neue Medien*, ed. B. Stiftung and H. N. Stiftung, Verlag Bertelsmann Stiftung, Gütersloh, 2000, pp. 31–50.
21. J. A. Greene and R. Azevedo, *Educ. Psychol.*, 2010, **45**, 203–209.
22. A. Devolder, J. van Braak and J. Tondeur, *J. Comput. Assist. Learn.*, 2012, **28**, 557–573.
23. M. Boekaerts, *Learn. Instr.*, 1997, **7**, 161–186.
24. M. Bannert and C. Mengelkamp, *Metacogn. Learn.*, 2008, **3**, 39–58.

CHAPTER 13

Can YouTubers Provide Powerful Tools for Addressing Heterogeneity in the Classroom? An Analysis of Videos About the Periodic Table Using the TPACK Framework

MARÍA JOSELEVICH*[a], PABLO D. MORO[a] AND M. AGUSTINA MARTÍNEZ[a]

[a]Programa de Tecnologías de la Información y la Comunicación en Aplicaciones de Interés Social (TICAPPS), Universidad Nacional Arturo Jauretche, Provincia de Buenos Aires, Argentina
*E-mail: mjoselevich@unaj.edu.ar

13.1 Introduction

Scientists and educators around the world are concerned about low motivation levels and interest in learning science and following science study courses among students.[1] The causes of this phenomenon are manifold, but one of the most significant seems to be the use of traditional teaching methods in school that are not sufficiently contextualized with respect to new cultural formats,[1] showing the obsolescence of traditional teaching methods,

Advances in Chemistry Education Series No. 11
Digital Learning and Teaching in Chemistry
Edited by Yehudit Judy Dori, Courtney Ngai and Gabriela Szteinberg

Published by the Royal Society of Chemistry, www.rsc.org

which prioritize the written text. In contrast, many students prefer images to text.[2] This has led to an increase in the presence of visual images in textbooks,[3] in line with Mayer's multimedia learning principles[4] that argue that students can achieve a deeper understanding when explanations are presented in the form of words and pictures rather than words alone.

Digital technologies are fully integrated with the way people communicate. In the educational sphere, they start to become part of the processes of constructing both individual and collective meanings and are becoming informal learning environments.[5] Incorporating these environments is a relevant educational challenge, as they add up and combine with legitimate experiences of social life and different ways of learning that shape, nurture and sustain knowledge networks.[6]

In the field of chemistry education, digital resources have been useful for different purposes, such as replacing laboratory work,[7] and building mental models of abstract chemical concepts. There are numerous examples[8] of multimedia presentations and electronic tools that appear to improve student learning outcomes when compared to textbooks.[9]

In this chapter, we are particularly interested in one of the most important and abstract topics in chemistry: the organization of the chemical elements within the Periodic Table. We found of particular interest the work of Blonder *et al.*, (2013),[6] in which they indicate that the use of digital tools, such as videos, makes it possible to work on the relationship between different levels of organization. YouTube videos can be effective in enhancing student conception about the structure of the developing atomic models and promote a positive attitude towards learning chemistry.[10] Even more, in the case of crowded classes with students who have no experience in laboratory work, oral and auditory engagement with YouTube videos has been reported to lead to meaningful learning about the periodic properties of the elements even in cases out of context.[6]

YouTube is the largest open platform for audio-visual content and was the most visited site worldwide in 2020.[11] This platform presents a challenge for teaching and learning in closed school contexts as all kinds of content, knowledge and common knowledge are freely available outside formal educational settings, sometimes presenting more access, engagement and participation than formal educational actions and activities offered by conventional institutions and methods. YouTube is also a social networking environment. It functions as an effective tool for shaping the psychosocial identity of adolescents[12] who interact there both with their peers and with YouTubers.[12] Science YouTuber followers can add up to the millions. As a wonderful example, we mention the Spanish YouTuber Javier Santaolalla (*Date Un Vlog*, https://www.youtube.com/c/dateunvlog). YouTubers share their world with young people and teenagers and can spread a great passion and curiosity for understanding scientific topics that should undoubtedly lead to deeper and more meaningful learning. According to Tapia-Jara *et al.*, (2020)[13] many students replace school textbooks with videos produced by YouTubers.

YouTubers are close to their followers in the sense that they share similar features (age, language, culture, social context, *etc.*). This facilitates rapid identification as they are perceived by young people as "peers". Further, they also add qualities (creativity or talent) that they admire.[14] Their communicative styles are key in determining their success or failure in terms of number of views and interaction with users. Teenagers interact with YouTubers by "liking" or commenting on their videos, developing a sense of closeness and rapport that is fundamental to the success of the YouTuber.[15]

YouTubers' videos deal with certain aspects that we found analogous with those studied under the Technological Pedagogical Content Knowledge (TPACK) framework[16] that was developed to analyze different types of knowledge that teachers have for teaching using technological resources. In this chapter, we have decided to study these characteristics by making an analogy with the TPACK model. To narrow the scope of this work, given the scarce literature found on this particular topic, our approach was to carry out a qualitative and descriptive analysis of 20 videos hosted on YouTube. We chose videos considered successful by the YouTube platform itself that dealt with the Periodic Table of the Elements. Those videos were studied using categories analogous to those of TPACK (aspects related to content, use of technology and pedagogical choices). Finally, we analyzed the information found to answer the following questions:

(1) Which students' learning styles are stimulated when approaching scientific knowledge from the videos studied?
(2) How do these videos contribute either to the resolution or to the deepening of misconceptions related to the periodic organization of the elements?

This chapter is divided into five sections. Following the Introduction (Section 1), we detail in Section 2 the analogy with the TPACK framework that we built for the analysis of the chosen videos. Section 3 details the characteristics we decided to observe in the videos and analyzes them accordingly. The findings of the analysis are presented in Section 4. Finally, Section 5 presents the conclusions of the study together with recommendations for teachers using this type of video in the classroom.

13.2 Adaptation of the TPACK Framework

The TPACK framework describes teachers' knowledge by analyzing the intersections of three fields: content, pedagogy and technology, and the development within each particular context. Although the definition of TPACK does not include the precision of theoretical frameworks for its components, they necessarily appear in their application. For the purposes of this chapter, we defined new categories related to TPACK fields, which are presented below.

13.2.1 The Pedagogical Field

Digital technologies applied to science education seem to contribute to the creation of meaningful environments that lead to the construction of mental models and favour and foster student engagement.[8] The development of such mental schemas requires active cognitive engagement on the part of learners, which is triggered by resources that require a trade-off between experience and existing knowledge schemas. According to Mayer,[4] active construction mediated by multimedia messages complements attention to the characteristics of how information is processed by its recipients to achieve meaningful learning.

In line with the above, constructivist pedagogy places the teacher in the role of generator and designer of an environment of interactions suitable for making the learning of a given content viable in a specific context.[17] Since teaching should be situated within the specific contexts of the learners, learning styles must be considered central to the construction of teaching materials suitable for the majority of students.

Learning styles have been studied by various authors. We agree with those who take a multidimensional approach built from cognitive, affective, and physiological aspects.[18] According to these authors, the *cognitive* aspects are associated with how individuals interact with information, their level of structuring, contextual or analytical strategies, and of reflexivity or impulsivity as well as differences in sensory perception (visual, auditory and kinesthetic). The *affective* dimension is directly related to the motivation to learn certain subjects in certain situations and can be analyzed, from the students' perspective, from three dimensions:[19] ability to succeed (self-efficacy); sense of belonging; and perception of the relevance of their study (intrinsic motivation). *Physiological* aspects are addressed through neurophysiological studies and were left outside the scope of this study.

Thus, it is of interest to study which characteristics do the videos promote for each type of learning style to build teaching materials that lead to more meaningful learning.

13.2.2 The Technological Field

Audio-visual resources, such as videos and animations, have been used for multiple pedagogical purposes related to both the production of knowledge[20] and the increase of dynamism in the class.[21] The appropriate use of videos[22] allows the presentation of operating processes in sequential ways, as well as the analysis of the relationship between the parts and the whole in a model or process. In addition, they can help to develop skills and attitudes because they require global processing of the information they contain. Furthermore, they are viewed as means of representing scientific knowledge through modelling.

In this context, YouTube is used as a pedagogical tool to facilitate the provision of content and in which it is possible to build learning communities.

It is also used to teach the skills necessary to search, select, save and evaluate audio-visual representations, both of their own creation and of others.[5,21] However, we must warn against trivializing the pedagogical uses of these resources, as only genuine inclusion can lead to an improvement in the quality of education.[23]

13.2.3 The Content Field

The periodicity of the physical and chemical properties as shown in the Periodic Table is a fundamental law of chemistry. However, with it comes many misconceptions and difficulties for teaching and learning.

In relation to this topic,[24] strategies based on active participation that foster understanding of the periodic properties law, increase motivation to deal with this topic and promote meaningful learning in relation to this law. Franco-Mariscal & Oliva-Martínez (2012)[25] published an interesting paper in which they propose the existence of several misconceptions that hinder the learning of this content. They indicate as a first obstacle the requirement for students to learn merely by remembering a huge amount of information with empty and abstract meaning, organized by the notion of periodicity. In fact, periodicity itself is pointed out as a complex concept, since it is built from a criterion of organization, from which another criterion, regularity, is derived. Another obstacle mentioned is the attempt to interpret the contents simultaneously from a macroscopic and a submicroscopic point of view, leading to possible confusion in the levels of organization and the properties emerging from them. An additional problem is the use of intrinsically complex notions, such as the double meaning of the term "chemical element", which can be interpreted as referring to a real substance or to an ideal constituent of the chemical compounds of which it is a part.[25] In this sense, for example, when talking about the conduction of electricity, one should refer to the properties of the *substances* built from metallic elements and not to the properties of the metallic elements themselves. Finally, the above authors refer to deficiencies in students' prior learning of key concepts such as the atom, atomic mass, chemical change, chemical formulation and nomenclature. These latter obstacles must be reviewed in the context of the individual student and will not be studied here.

13.3 Methods

For this study, we examined the top 20 videos as determined by YouTube. They were found by searching on YouTube with the phrase "*Tabla Periódica*" (Periodic Table). The search in YouTube was in Spanish, as it was explicitly configured to do so, and in such a way that it would not take into account the user's browsing history. We chose 20 videos marked as successful by the platform,[26] based on an algorithm that takes into account the number of views, "likes", and subscribers of the channel as well as the average viewing time.

13.3.1 Analysis of the Videos' Pedagogical Approach

Although YouTubers do not explicitly refer to a learning theory, they are necessarily built on implicit approaches that we categorized as constructivist or traditional.[27] We consider a video to be *constructivist* when: (a) both the type of presentation and the proposed interaction are meant to establish a direct connection with users; (b) it resorts to a variety of strategies for the construction of learning, and (c) it requires a high degree of cognitive activity by the viewer. This approach complements the concept-based strategies of the cognitive learning theory.[4]

13.3.1.1 Analysis of the Learning Styles Favoured by the Videos

We characterized learning styles according to the degree of structuring, the contextual or analytical nature of the strategies, their levels of reflexivity or impulsivity and the sensory perceptions.[18] Using this basis, we categorized the videos as follows.

- Contextual—videos that start dealing with rather general topics and gradually move towards more detailed or specific explanations.
- Analytical—videos that start from specific explanations and then move on to describe the whole phenomenon within its context.
- Structured—videos that are supported by an orderly script and follow orderly steps or linear discourses.
- Unstructured—videos that contain digressions and detours.
- Reflective—videos that are driven by exposition and require the viewer to recall prior knowledge.
- Active—videos that encourage interaction rather than passive observation.
- Visual—videos focusing on the use of visual aids (*e.g.*, videos, images, animations, editing, *etc.*).
- Verbal—videos focusing on speech.

Kinesthetic learning is related to those learning styles in which learners are meant to perform concrete actions involving the touch perception or body movement. Some videos were considered to be close to this learning style if they recall the memory of an experience or if they propose some experimentation.

13.3.2 Analysis of the Videos' Context

The videos' context analysis was performed by defining whether the videos are intended for scientific divulgation or educational purposes.[28,29] According to our criterion, *divulgative* videos include controversial topics, refer to everyday life or to a high social impact issue, in the context of public interest. These types of videos usually deal with content in a multidisciplinary and multidimensional way, are aimed at a general audience and the speakers are shown as science enthusiasts or scientists who may or may not play a teaching role. In contrast, *science education* video conductors only play the role of teachers, contain only curriculum content and confine their contexts to the field of the discipline. In some cases, these types of videos focus on offering algorithmic logic tools to solve exercises. Scholar videos deal with the products of science rather than its processes or nature.

13.3.3 Analysis of the Technological Level of the Videos

Regarding the analysis of the technological aspects, we consider that the *technological development* of a video depends on the inclusion of modern and diverse sound, visual and post-production resources.[30]

13.3.4 Analysis of the Videos' Content

With regard to the content, we consider that the aim of a video is purely *rote learning*, when it consists merely of the presentation of mnemonic rules or similar tools to remember systematic information. We analyzed if the criterion of *periodicity* is worked on and whether this is done on the basis of an analysis of *periodic properties*. We studied whether the videos showed a simultaneous and undifferentiated approach to the contents from a *macroscopic* or phenomenological point of view and from a *submicroscopic* or model point of view. In particular, we observed whether the double meaning of the term "chemical element" was used in an undifferentiated way.

13.4 Findings

The following section presents the findings of the analysis of the 20 selected YouTube videos about the Periodic Table. The viewing and categorization of these videos was carried out by the authors from their own disciplinary, pedagogical and didactic perspectives.

The section begins by explaining our findings regarding the pedagogical approaches and communication styles used by the YouTubers in the videos analyzed. Subsequently, Table 13.1 shows the categorization of the videos according to the technological resources, the learning styles that we consider "stimulative" and the misconceptions related to the Periodic Table that we consider can be reinforced.

When analyzing the pedagogical approaches underlying the videos, we observed that only four of the videos analyzed (5, 7, 18 and 20) may suggest a constructivist view. In the other videos, we found very little variety in the strategies, procedures and didactic resources, showing a great predominance of traditional teaching models.

In terms of communication styles, we noticed that instructive videos prevail over divulgative videos. Videos are mostly directed at secondary school

Table 13.1 Analysis of videos related to the Periodic Table.

	Technological approach	Pedagogical approaches			Approaches to content			
Video	Technological resources (*)	Learning styles stimulated (**)	Constructivist (C)/ Traditional instructional (TI)	Scientific Divulgation (SD)/Scientific Education (SE)	Memorizing (M)/ Comprehensive Learning (CL)	Macro and submicroscopic interpretation	Confuses element whit elemental substance	Periodic Properties (P)/Periodicity (PP)
1	A, C, D: simple but funny.	C, S, Ac, Vi-Ve	TI	SE	M	—	—	—
2	B: excellent; D: very modern.	C, S-U, R, Vi	TI	SE	M	Yes	Yes	P and PP
3	B: good; C, D: very modern.	A, S, R, Vi	C	SE	—	—	—	P
4	D: poor; B: excellent.	C, S, Ac, Ve	TI	SE	M	Yes	Yes	—
5	B: excellent; C, D: simple.	A, U, R-Ac, Vi-Ve	C	SD	CL	Yes	Yes	P and PP
6	A; C.	A, S, Ac, Ve	TI	SE	M + CL	Yes	Yes	P and PP
7	B: excellent; C; D: simple.	A, U, R-Ac, Vi-Ve	C	SD	M + CL	Yes	Yes	P
8	Animated video with synthesized voice; B: good.	C, S, Ac, Vi	TI	SE	M	—	Yes	P and PP
9	Recording of a TV program.	C, S, Ac, Ve	TI	SD	—	—	Yes	P and PP
10	Voice-over. B: very good; D: very modern.	C-A, S-U, R-Ac, Vi-Ve	C/TI	SE	M	—	Yes	PP

(*continued*)

Table 13.1 *(continued)*

	Technological approach	Pedagogical approaches			Approaches to content			
Video	Technological resources (*)	Learning styles stimulated (**)	Constructivist (C)/ Traditional instructional (TI)	Scientific Divulgation (SD)/Scientific Education (SE)	Memorizing (M)/ Comprehensive Learning (CL)	Macro and submicroscopic interpretation	Confuses element whit elemental substance	Periodic Properties (P)/Periodicity (PP)
11	B: very good; D: very good.	A, S, R, Vi-Ve	C/TI	SE	M	—	Yes	—
12	A; C.	C, S, Ac, Vi-Ve	TI	SE	M	—	Yes	—
13	A; C.	C, S, Ac, Vi-Ve	TI	SE	M	—	—	—
14	A; C.	C-A, S, Ac, Ve	TI	SE	M	—	Yes	—
15	B: excellent; D: very modern.	C, S, Ac, Vi	TI	SE	M	—	Yes	P and PP
16	D: very modern; B: excellent. Voice-over	C, S, Ac, Vi-Ve	TI	SE	M	—	—	P and PP
17	Very simple video of a ppt presentation	C, U, Ac, Ve	TI	SE	M	—	—	—
18	B: very good; C; D: interesting	A, U, R, Vi-Ve	C	SD	—	—	Explains the difference but then reintro-duces confusion	—
19	A; C	C, S, Ac, Vi-Ve	TI	SE	M + CL	—	Yes	P
20	B: excellent; D: very modern	A, U, R, Vi-Ve	C	SD	Refers to the elements but not to the Periodic Table	—	—	—

(*) Technological resources: the YouTuber shows himself explaining (A), whiteboard (B), animation and images (C), Video editing (D). (**) Learning styles: contextual (C)/analytical (A); structured (S)/unstructured (U); reflective (Re)/active (Ac); visual (Vi)/verbal (Ve).

students, young people and the general public. We found a wide range of communication styles: formal, informal, colloquial, and traditional. Comical or humorous language appears alongside the use of youth slang, songs and body language. The relationship between complex topics and everyday situations appears in most of the videos analyzed. In addition, we found that direct interaction between YouTubers and users is very scarce. Indeed, interaction occurs mainly in the comments section as "likes" given to some comments. We have not found, within the context of any video, references to comments made by users over previous videos. However, there is a certain development of a YouTuber's discourse that employs active communicative and dialogic formats, which increase interactivity by encouraging the construction of concepts by the audience.[27]

As per the analysis result in Table 13.1, YouTubers use a wide variety of technological resources, with a predominance of animations and images with very good quality and the inclusion of whiteboards, videos, voice-overs, presentations and close-ups of YouTubers.

Regarding the stimulation of learning styles, we observed a predominance of videos with a general and contextual perspective over those with only analytical views. Most of the videos promote structured styles, with an ordered and sequenced discourse. In addition, the simultaneous stimulation of the verbal and visual channels is observed in half of the videos, including those with a more constructivist vision.

As a last item, we analyzed how a YouTuber's approach to content can influence the most common misconceptions in the learning of periodic properties and the Periodic Table. We found that the vast majority of the videos analyzed are oriented towards purely rote learning data and values regarding the properties and location of the elements in the Periodic Table without a conceptual approach to the topics. The inclusion of micro and macroscopic approaches was found in most of the videos but only two of them differentiated explicitly the properties of elementary substances, built from the elements, with those of the elements themselves. In this sense, videos 5 and 18 refer to the double meaning of the term *chemical element*. Unfortunately, video 18 confuses the concepts by adding a "Periodic Table" of elementary substances. Only three of the videos analyzed (5, 7 and 19) involve students' active thinking in any way. Finally, just over half of them mention concepts related to the periodicity of properties.

13.5 Conclusions

Based on the results of the analysis in the previous section, we can infer that YouTuber videos considered successful by the platform itself, have heterogeneous characteristics in relation to our adapted TPACK dimensions.

In terms of the technological dimension, the videos analyzed present a variety of resources and most of them show post-production quality. In the pedagogical dimension, the number of different strategies is notably reduced, while most of the videos have traditional teaching formats that do not tend to mobilize active mechanisms for the construction of knowledge in learners. The learning styles stimulated are mostly contextual, general and structured. In terms of perceptual styles, there is a focus on stimulating both verbal and visual channels.

Regarding the content dimension, we found that none of the videos help to overcome the misconceptions mentioned previously when learning the Periodic Table. We even consider that, if viewed without the guidance of a teacher, most of these videos would contribute to their misconceptions becoming entrenched and aggravated.

However, despite the aforementioned problems in the use of these videos, the statistics show that the videos analyzed are successful and that a large number of young people actively and autonomously use them. This is in complete contrast with the lack of interest that students show in science in formal settings.[1] We consider that one of the causes of this phenomenon is that the diversity of resources offered by some videos makes them attractive to students with diverse learning styles, including some that are often neglected in traditional school settings. This diversity of audio-visual resources simultaneously stimulates different channels of perception. In addition, the videos' varied discursive formats alternate between the specific languages of the discipline and those of everyday life, which are closer to young people. This conveys to young people a sense of proximity to the subjects studied, together with the development of a learning environment that is perceived as familiar. Such environments enhance belonging and a sense of self-efficacy and relevance. This is especially true when the topics are addressed by the sender through expressive resources that peak interest, and when they include examples from everyday life, related to socio-scientific problems, *etc.* All these components can contribute to the development of motivation[19] and increase learners' engagement with the topics to be studied.

The aim of this chapter was to contribute to teachers' understanding in establishing criteria for the careful selection of YouTubers' videos that lead to the genuine inclusion of technology in the classroom.[23] These selected videos would be characterized as "genuine", as mentioned in Section 13.2.2. We consider that the generation of didactic purposes that include YouTube videos within an appropriate TPACK framework turns these videos into a powerful teaching tool. In line with Mitra *et al.*, (2010),[21] we consider that such a framework should include a critical analysis of videos by teachers and students. Another factor to address in terms of diversity is the digital gap, which requires considering not only access but also support in the use of technology.[31]

We invite teachers to take the curation of YouTubers' videos as a teaching and learning activity. Thus, videos previously chosen by teachers could be analyzed together with students according to the categories that have been proposed in this work and others that may arise in the context of the group.

Acknowledgements

The authors would like to thank Marcos Ravaschino for assistance with methodology and Pedro Joselevich for his help with the translation of this paper. We would also like to show our gratitude to Gabriela Szteinberg for comments that greatly improved the text.

References

1. P.-Y. Liou, C.-L. Wang, J. J. H. Lin and S. Areepattamannil, *J. Exp. Educ.*, 2021, **89**(4), 605–624.
2. D. A. Morales and F. J. P. Palacios, *Enseñanza de las ciencias: revista de investigación y experiencias didácticas*, 2018, **36**(3), 41–58.
3. A. Raviolo and A. S. Farré, *Educ. Quim.*, 2020, **31**(3), 119–133.
4. R. E. Mayer, *New Dir. Teach. Learn.*, 2002, **89**, 55–71.
5. J. García Martínez and E. Serrano Torregrosa, in *X Jornades de Xarxes d'Investigació en Docència Universitària: la participació i el compromís de la comunitat universitària*, Universidad de Alicante: Instituto de Ciencias de la Educación, Alicante, España, 2012, pp. 2918–2931.
6. R. Blonder, M. Jonatan, Z. Bar-Dov, N. Benny, S. Rap and S. Sakhnini, *Chem. Educ. Res. Pract.*, 2013, **14**, 269–285.
7. T. J. Bohloko, M. Makatjane, T. Mokuku and M. J. George, *Afr. J. Res. Math. Sci. Technol. Educ.*, 2019, **23**(1), 75–85.
8. S. Kohler and T. C. Dietrich, *Front. Commun.*, 2021, 581302.
9. See Chapter 12: J. Kranz and R. Tiemann, in *Digital Learning and Teaching in Chemistry*, ed. Y. J. Dori, C. Ngai and G. Szteinberg, Royal Society of Chemistry, United Kingdom, 2023.
10. R. N. Liwanag and H. J. M. Ramirez, *Asian J. Educ. Res.*, 2019, **7**, 38–49.
11. 84 YouTube Statistics You Can't Ignore in 2022, https://invideo.io/blog/youtube-statistics/, accessed on February 2022.
12. V. Pérez-Torres, Y. Pastor-Ruiz and S. Abarrou-Ben-Boubaker, *Comunicar*, 2018, **55**, 61–70.
13. J. Tapia-Jara, A. Sánchez Ortíz and C. Vidal-Silva, *Form. Univ.*, 2020, **13**(1), 3–12.
14. W. Westenberg, *The Influence of YouTubers on Teenagers. A Descriptive Research about the Role YouTubers Play in the Life of Their Teenage Viewers*, University of Twente, Países Bajos, 2016, ch. 1, pp. 1–35.
15. M. Berzosa, *Youtuber y otras especies: El fenómeno que ha cambiado la manera de entender los contenidos audiovisuales*, Editorial Ariel, Madrid, 2017.
16. M. Koehler and P. Mishra, *Contemp. Issues Technol. Teach. Educ.*, 2009, **9**(1), 60–70.
17. R. Vera Velazquez, C. Castro Piguave, I. Estévez Valdés and K. Maldonado Zúñiga, *Revista Científica Sinapsis*, 2020, **3**(18), DOI: 10.37117/s.v3i18.399.
18. C. M. Alonso,D. J. Gallego and P. Honey, *Los Estilos de Aprendizaje: Procedimientos de diagnóstico y mejora. Edición: 7ª*, Editorial Mensajero, España, 2007.

19. V. Tinto, *presented in part at the Congreso Internacional "Ingresos e ingresantes a la universidad" Conferencia inaugural*. Universidad de Río Negro, Argentina, February 2021.
20. M. E. Barredo Hernández, H. Martínez Martín, V. Barbero Aparicio, T. Rilova García and A. Segura-Marrero, in *Prácticas Innovadoras inclusivas: retos y oportunidades*, ed. A. Rodríguez-Martín, Universidad de Oviedo, Servicio de Publicaciones Editorial: Oviedo: EDIUNO, España, 2017, vol. 4, pp. 1997–2004.
21. B. Mitra, J. Lewin-Jones, H. Barrett and S. Williamson, *Res. Post-Compuls. Educ.*, 2010, **15**(4), 405–414.
22. M. A. Martinez, A. Graieb, V. Fantini and M. Joselevich, in *Memoria Académica: Actas. Ensenada: Universidad Nacional de La Plata. Facultad de Humanidades y Ciencias de la Educación. Departamento de Ciencias Exactas y Naturales*, 2015.
23. I. Dussel and L. A. Quevedo, *Educación y nuevas tecnologías: los desafíos pedagógicos ante el mundo digital*, Santillana, Buenos Aires, 2012.
24. P. Carabelli, A. S. Farré and A. Raviolo, *Educación en la Química en Línea*, 2020, **26**(2), 212–225.
25. A. J. Franco-Mariscal and J. M. Oliva-Martínez, *Rev. Cient.*, 2012, **16**(2), 53–71.
26. P. Covington, J. Adams and E. Sargin. *Proceedings of the 10th ACM Conference on Recommender Systems (RecSys '16)*. Association for Computing Machinery, 2016, pp. 191–198.
27. I. Bellou, N. M. Papachristos and T. Mikropoulos, *Digital Learning Technologies in Chemistry Education: A Review*, 2018, DOI: 10.1007/978-3-319-73417-0_4.
28. A. Blanco López, *Revista Eureka sobre Enseñanza y Divulgación de las Ciencias*, 2004, **1**(2), 70–86.
29. J. C. Olmedo Estrada, *Revista Eureka sobre Enseñanza y Divulgación de las Ciencias*, 2011, **8**(2), 137–148.
30. R. Córdova-Tapia, S. Celly-Alvarado and D. Rivera-Rogel, in *Metodologías Activas Con Tic en la educación del siglo XXI*, ed. O. Buzón García and C. Romero García (coords.) Editorial: S. L. Dykinson, España, 2021, vol. 50, pp. 1036–1055.
31. P. Attewell, Comment: The First and Second Digital Divides, *Sociol. Educ.*, 2001, **74**(3), 252–259.

CHAPTER 14

A Formalised Conceptual Model-based Approach for Fostering and Assessing Students' Systems Thinking in Undergraduate Chemistry Education

R. LAVI*

School of Engineering, Massachusetts Institute of Technology, 77 Massachusetts Avenue, Cambridge, MA, USA
*E-mail: realavi@mit.edu

14.1 Introduction

As systems become more complex, industry demand for employees who possess systems thinking is increasing.[1] Systems thinking is about comprehending systems, as well as about describing and analyzing questions, circumstances, or problems as if they were systems.[2–4] Within K–12 science education, the National Research Council (NRC)[5] identified systems and system models as crosscutting concepts which should be implemented when teaching science to students. The Next Generation Science Standards by the NRC further solidified these guidelines.[6]

Advances in Chemistry Education Series No. 11
Digital Learning and Teaching in Chemistry
Edited by Yehudit Judy Dori, Courtney Ngai and Gabriela Szteinberg

Published by the Royal Society of Chemistry, www.rsc.org

14.1.1 Systems Thinking in Chemistry

Chemistry is both a natural science and an applied one. It involves both the creation of fundamental scientific knowledge and applications for improving human life and society.[7] Chemistry deals with complex systems,[8,9] and the need for infusing systems thinking into chemistry education has been argued for previously.[8] Systems thinking is important to chemistry education because chemical systems can be complex.[9] In addition, since chemistry is taught in other STEM disciplines, the understanding of chemical systems is also important to STEM subjects other than chemistry.[10]

A system in science or in engineering can be described as an entity composed of interacting parts. The whole entity exhibits properties distinct from those of its various parts. The system's structure and behaviour together provide its function, which delivers a specified purpose, pre-intended by and for humans.[3,11] While natural phenomena can be described and analyzed as systems, scientifically, they cannot be said to have a purpose; instead, they have a natural outcome.[11,12]

Chemical systems can be natural and/or artificial. Miljanić[9] describes three behaviours that can be observed in complex chemical systems: (1) those of individual components apart from observing the large system; (2) those that can be observed only in the whole system yet can be predicted based on understanding of individual components; and (3) those that can be observed only in the system as a whole and cannot be predicted with understanding of individual components. To understand a complex chemical system, one needs to understand these three behaviours within it. However, due to the complexity of real-world chemistry, instructors often take a reductionist, rather than a holistic approach to teaching chemistry.[10]

14.1.2 Teaching Systems Thinking in Chemistry

Approaches to teaching complex chemical systems include visualization and conceptualisation. While visualization techniques make invisible objects and behaviours of the chemical system observable and interactive,[13] conceptualisation allows students to make sense of system complexity by organizing their thoughts about the system.[14] The method and tool presented in this chapter both concern the conceptualisation of chemical systems.

No approach for teaching chemistry systems thinking in the undergraduate level provides both a standardised tool for conceptual modelling of knowledge and tested assessment rubric, which together would allow for the fostering and assessment of systems thinking in chemistry education. In this chapter, such an approach is presented.

Chemistry understanding can be described at four distinct levels: macroscopic (sensory), microscopic (molecular), symbolic (formulae), and process (reactions).[15] To make sense of each level and especially of the relations between them, systems thinking is required.

The purpose of this chapter is to demonstrate how systems thinking about chemical phenomena and chemical products can be taught at the undergraduate level using a formalised approach for conceptualisation of chemistry systems.

14.2 Suggested Instructional Approach

The instructional approach outlined in this chapter is suitable for teaching in any sub-discipline of chemistry, or interdisciplinary teaching together with disciplines outside of chemistry. In addition, the tool described herein uses natural language and allows for using multiple languages. The transdisciplinarity and the multiple languages afforded by this approach both enable the inclusion of students in every culture and discipline (or subdiscipline) in the chemistry learning process. The approach relies on existing resources tools which are freely available to anyone with a computer and a connection to the World Wide Web.

This section begins with a description of a *model-based systems engineering* (MBSE) language for conceptual modelling of systems. Next, follows a description of an assignment given to small teams of teacher–students from various STEM disciplines, including chemistry, who were tasked with creating conceptual models of a real-world biochemical phenomenon. This assignment, the background for the study, and the findings based on the entire research sample ($N = 42$) are all covered in detail elsewhere.[16] Then, based on a conceptual model submitted by one of the teacher–student teams, how the MBSE language allows for expressing the four levels of chemistry understanding is described.[15,17] Next, an assessment rubric for systems thinking in science and engineering education, which has previously been developed and validated,[16,18] is described and the application of this rubric on the same teacher team's conceptual model is demonstrated. Finally, two freely available *massive open online courses* (MOOCs) for learning about and how to use the aforementioned MBSE language, and its software implementation is described.

14.2.1 Conceptual Modelling Language

This chapter introduces a formalised language for conceptual modelling of chemical systems called Object-Process Methodology (OPM). Lavi and Dori[16] have already demonstrated OPM can be used for modelling chemical systems by chemistry teachers and teachers from other STEM disciplines. OPM is an MBSE methodology and language which comprises of *objects*—things that either exist or do not exist, and *processes*—things that are either occurring or not occurring. Additionally, *links* connect between objects and process (procedural, cause-and-effect links) and between objects or between processes (structural, static links). An OPM conceptual model (system model) is modelled hierarchically, starting with the top, holistic view and refining objects and process in this level down to sub-objects and sub-processes. There is no theoretical limit to the number of levels which can be represented in an OPM model—it is the decision of the modeler. For all these reasons, OPM is an ideal language for modelling chemical phenomena in all four chemistry understanding levels. Finally, an OPM model comprises of graphical (Object-Process Diagrams, or OPDs) and textual (Object-Process Language, or OPL) modalities which mirror each other.[11]

OPM has been adapted for use in various domains of *science, technology, engineering, and mathematics* (STEM) and STEM education, including science,[19] engineering,[20] science education,[16] and engineering education.[21]

As expounded by Aubrecht *et al.*,[14] OPM is particularly suitable for chemical systems conceptualisation: (a) the language elements are multiple and distinct; (b) unlike simpler techniques such as concept maps, *systems-oriented concept map extension* (SOCME), or systemigrams, OPM's hierarchical modelling affords the ability to present the chemical system or phenomenon across various levels of detail, relating to levels of chemistry understanding;[15,17] (c) OPM can be used for representing both chemical phenomena (natural systems) and/or chemical products (artificial systems). Figure 14.1 shows an example of how each chemistry understanding level can be represented in a single OPD. The OPM elements corresponding to each level of chemistry understanding are as follows:

- Macroscopic: air; heat; wood; fire; smoke.
- Microscopic: dioxide; carbon dioxide; dihydrogen oxide (water); carbohydrate; thermal energy; light energy.
- Symbolic: $2CH_4 + 4O_2 \rightarrow 2CO_2 + 4H_2O$.
- Process: combusting.

There are two MOOCs available for learning how to use OPM for MBSE, and for model-based system design in particular: MBSE 101: Foundations† and MBSE 102: Advanced Approaches to OPM.‡ The instructional design of both courses was carried out by the author of this chapter. All the videos from both MOOCs are freely available on YouTube, in English, Arabic, and Hebrew.§ As an indication of the need for and the quality of these courses, since March 2019, about 7800 learners enrolled in either course, out of whom about 1200 paid to receive a certificate of completion for both courses.

The cloud-based software implementation of OPM¶ is also freely available online in multiple languages. This software allows for remote conceptual modelling, both individually and collaboratively, which means both the students and the instructor can join together in modelling, reviewing, and discussing conceptual models.

†https://www.edx.org/course/model-based-systems-engineering-foundations.
‡https://www.edx.org/course/model-based-systems-engineering-advanced-approache.
§https://www.youtube.com/channel/UCMJTbOJq-3RCM7E1mfHCF5Q/playlists.
¶https://www.opcloud.tech/.

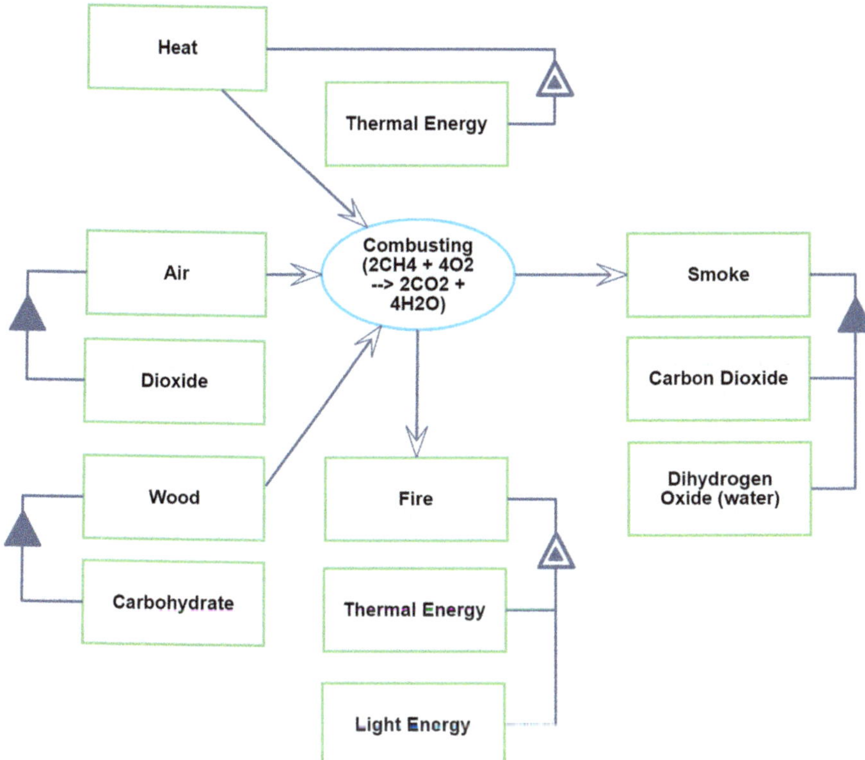

Figure 14.1 Representations of chemistry understanding levels in object-process methodology. Smoke consists of carbon dioxide and dihydrogen oxide. Wood consists of carbohydrate. Air consists of dioxide. Heat exhibits thermal energy. Fire exhibits light energy and thermal energy. Combusting consumes heat, air, and wood, and yields smoke and fire.

14.2.2 Conceptual Modelling Assignment

Lavi and Dori[16] adapted an article published in a peer-reviewed journal[24] to one page of text suitable for reading and understanding by students with secondary school science education but without background in biochemistry. The authors of the article describe how the *Alphanizomenon Ovalisporum* bacterium dominates the population of bacteria in a lake through a biochemical process called *enslavement*, which occurs as follows: (1) *Alphanizomenon Ovalisporum* bacteria release a toxin named Cylindrospermopsin into the water; (2) the toxin simulates a shortage of phosphates in the water, which causes other species of bacteria to release an enzyme named alkaline phosphatase; (3) the enzyme releases phosphate molecules from organic molecules in the water; and (4) *Alphanizomenon Ovalisporum*, adept at consuming phosphate molecules, consume these molecules and multiply. This process ultimately leads to an increase in the concentration of Cylindrospermopsin toxin in the water.

Following four hours of training in OPM software usage, which included the basic elements of OPM explained under 'The Exemplary OPM System Model' above, as well as a two-hour assignment on another chemistry phenomenon, teachers ($N = 42$), none of whom had prior knowledge or training in the conceptual modelling approach described herein, organised into self-selected pairs or triads, 16 in total. We then provided each team with the adapted scientific text and asked teams to select a phenomenon from the text and model it in two hours using OPM software.[‖] Since the assignment was a small part of a semester-long course, we provided a time limit of two hours for the assignment, and set some constraints for it, to help teachers produce adequate models on schedule: the conceptual model had to have (a) exactly two levels of complexity—one top view diagram (SD) and another detailed diagram, (b) four to six objects for the top view diagram, (c) two or three processes in the detailed diagram, and (d) one or two extra objects in the detailed diagram.

14.2.3 Identifying Chemistry Understanding Levels in the Conceptual Model

The four chemistry understanding levels[15,17] can clearly be distinguished in OPM models. Figures 14.2 and 14.3 show one teacher team's model based on the conceptual modelling assignment. The top view OPD (see Figure 14.2)

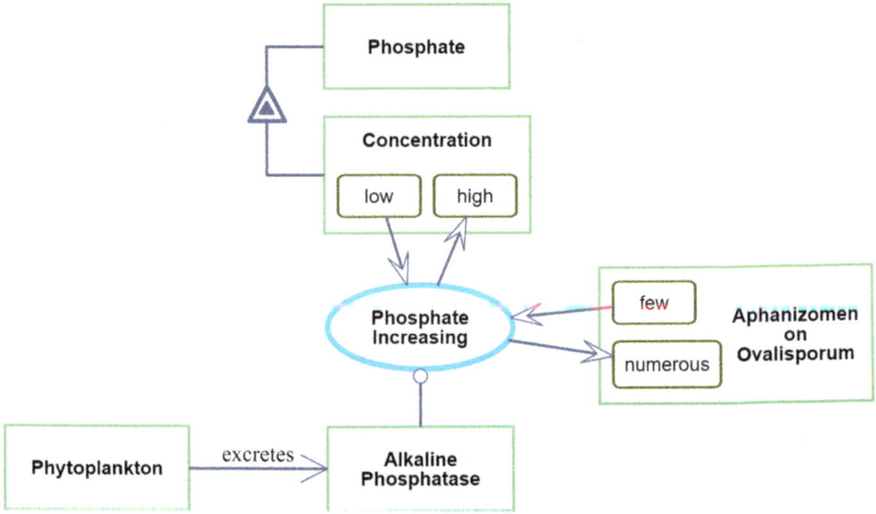

Figure 14.2 Conceptual model created by chemistry teacher-student team (top view). Phytoplankton excretes Alkaline Phosphatase. Phosphate Increasing requires Alkaline Phosphatase. Phosphate Increasing changes Concentration of Phosphatase from low to high. Phosphate Increasing changes *Aphanizomenon ovalisporum* from few to numerous. Only essential OPL sentences are shown.

‖Teams used an old version of the OPM software called OPCAT. See here: http://esml.iem.technion.ac.il/opcat-installation/.

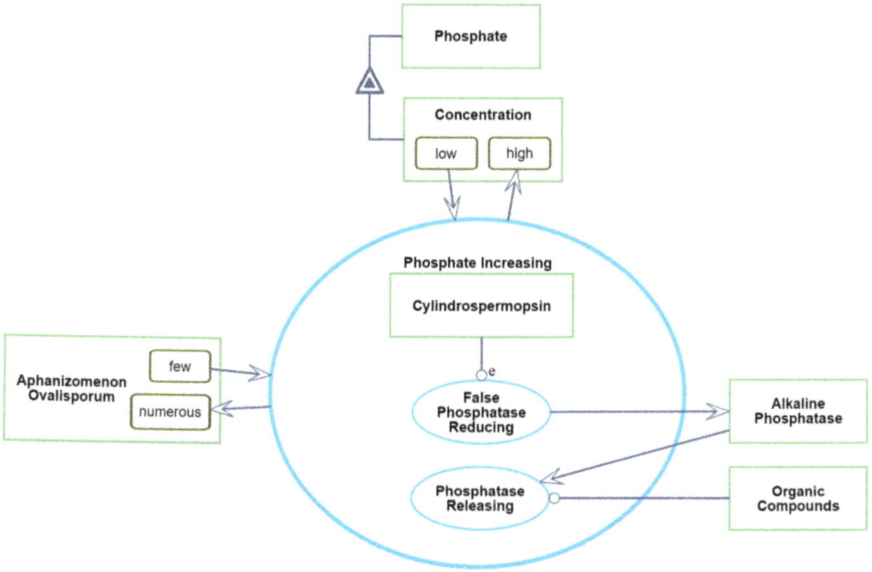

Figure 14.3 Conceptual model created by chemistry teacher-student team (detailed view). Phosphate Increasing from SD zooms in SD1 into False Phosphatase Reducing, and Phosphatase Releasing, as well as Cylindrospermopsin. Phosphate Increasing requires Alkaline Phosphatase. Phosphatase Releasing requires Organic Compounds. Cylindrospermopsin initiates False Phosphatase Reducing. Phosphatase Releasing consumes Alkaline Phosphatase. Phosphate Increasing changes Concentration of Phosphatase from low or high. Phosphate Increasing changes *Aphanizomenon ovalisporum* from few to numerous. False Phosphatase Reducing yields Alkaline Phosphatase. Only essential OPL sentences are shown.

shows the macroscopic level of the phenomenon being conceptualised: the object *Aphanizomenon Ovalisporum* is shown to change from state *few* to state *numerous*, which is a phenomenon observable with the human eye alone. As we refine the main process of the model—*phosphate increasing*—the detailed view OPD (see Figure 14.3) and its own, further refined OPDs, begin revealing the microscopic level of the phenomenon, as shown by the presence of the object Cylindrospermopsin. The process level of the phenomenon is shown by the main process shown in the top view OPD (see Figure 14.2) and its sub-processes, shown in the detailed view OPD (see Figure 14.3). While the symbolic level is missing from Figure 14.2, given more time and training, the teacher team could have refined the model enough for chemical formulae to be included in one or more of the lower-level OPDs.

14.2.4 Assessing Systems Thinking in the Conceptual Model

The *systems thinking assessment rubric* (STAR) for science and engineering education was designed by synthesizing papers on systems thinking assessment[18] and based on an established framework from systems

Table 14.1 Conceptual framework for systems thinking assessment rubric.

Aspect	Category	Attribute
Function	Emergent properties	A_1 – Intended purpose (synthesised system)/ Expected outcome (natural system)
		A_2 – Complexity levels
	General properties	A_3 – Main function
		A_4 – System boundary
Structure	Structural components	A_5 – Main object and its parts & attributes
	Structural relationships	A_6 – Structural relations
Behaviour	Procedural relationships	A_7 – Procedural relations
		A_8 – Procedural sequence
	System dynamics and temporal relationships	A_9 – Temporary objects and decision nodes

engineering—function, structure, and behaviour.[3,11] STAR has been applied successfully to OPM models in science and engineering education studies.[16,21–23] Table 14.1 shows the STAR conceptual framework. For an example of how the STAR is applied for scoring a specific conceptual model, see York, Lavi, Dori, and Orgill.[18]

Figures 14.2 and 14.3, and Table 14.2 show an example of applying the STAR for assessing an OPM system model of the chemistry phenomenon described above. We scored each attribute of systems thinking in every model with either 0 (no expression of attribute), 1, 2, or 3 (full expression of attribute). The model received a total score of 14 out of potential 27 (52%). The model presented herein and the explanation for how it was scored can elucidate the concepts of systems thinking and system function, structure, and behaviour.[3,11]

14.3 Summary and Suggestions

With chemistry's role as a central science for STEM,[10] and as a science that deals with exploring and designing complex systems,[9] it is not surprising that calls have been made to reorient chemistry education toward systems thinking.[8] Teaching systems thinking in chemistry requires an approach which enables the integration of (a) sub-disciplines of chemistry (and even of disciplines outside of chemistry), (b) the different understanding levels of chemistry, and (c) learning and assessment.

In this chapter, a conceptual model-based approach for fostering and assessing chemistry students' systems thinking at the undergraduate level was presented. This approach comprises of (a) a standardised conceptual modelling language—OPM, (b) its freely available software implementation OPCloud, (c) a previously tested rubric for assessing the systems thinking expressed in OPM models—STAR, (d) an example of a conceptual modelling assignment—how the *Alphanizomenon Ovalisporum* bacterium dominates the population of bacteria in a lake through a biochemical process called

Table 14.2 Scoring of a chemistry teacher team model.

Attribute and criteria for maximum score	Teacher team model OPD screenshot and scoring explanation
***A_1* – Expected outcome**	No example
Attribute score	0
Beneficiary: at least one required	None
Expected outcome: at least one required	None
***A_2* – Complexity levels**	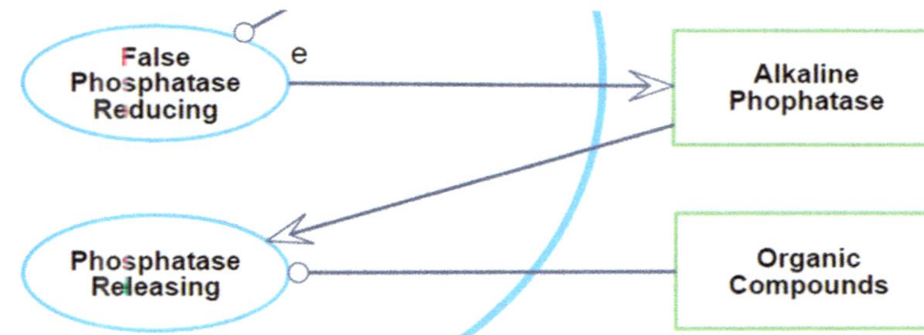
Attribute score	3
N complexity levels: at least two required	Two
N transformative sub-processes: at least two required	Two: (1) False Phosphatase Reducing yields Alkaline Phosphatase; (2) Phosphate Releasing consumes Alkaline Phosphatase

(*continued*)

Table 14.2 (*continued*)

Attribute and criteria for maximum score	Teacher team model OPD screenshot and scoring explanation
A_3 – Main function	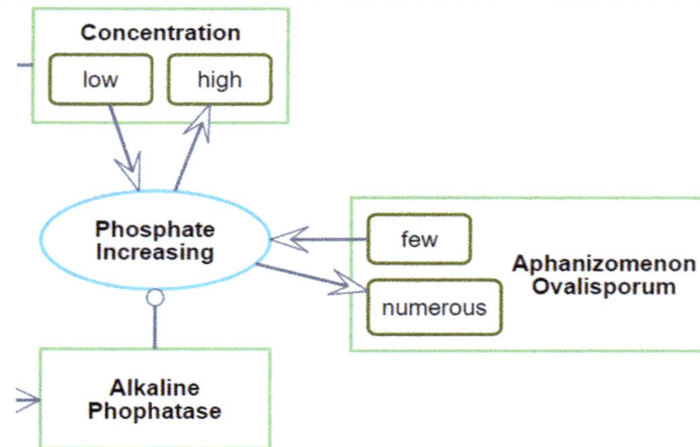
Attribute score	3
Systemic process: exactly one required	One: Phosphate Increasing
Transformed objects: at least one specific transformation required	One: *Aphanizomenon ovalisporum*, Concentration
Transformations: at least two required	Two: Phosphate Increasing changes (1) *Aphanizomenon ovalisporum* from few to numerous, and (2) Concentration from low to high
Enabler/s: at least one required	One: Alkaline Phosphatase
A_4 – System boundary	No example
Attribute score	0
N Environmental things: at least one	None
N system-boundary links: at least one	None

A_5 – Main object and its parts & attributes

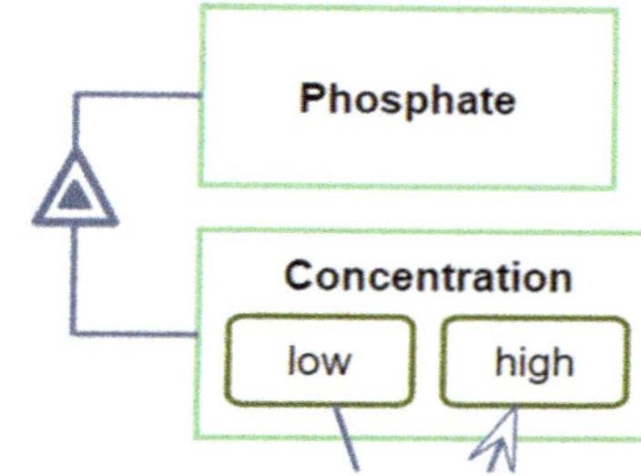

Attribute score	2
Main object: at least one required	One: Phosphate
N Sub-levels: at least two required	One

A_6 – Structural relations

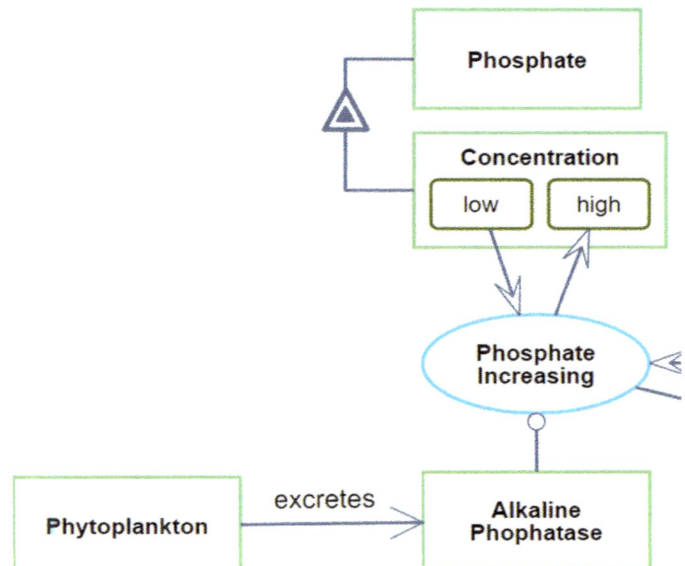

(continued)

Table 14.2 (continued)

Attribute and criteria for maximum score	Teacher team model OPD screenshot and scoring explanation
Attribute score *N link types: at least three required*	2 Two: (1) Exhibition-Characterization; (2) Unidirectional
A_7 – Procedural relations	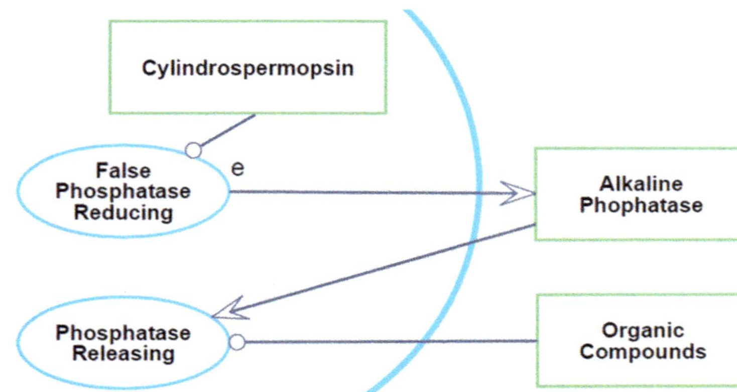
Attribute score *N basic link types: at least three required* *N advanced link types: at least two required*	2 Three: (1) Result; (2) Consumption; (3) Instrument One: Instrument event

A_8 – Procedural sequence		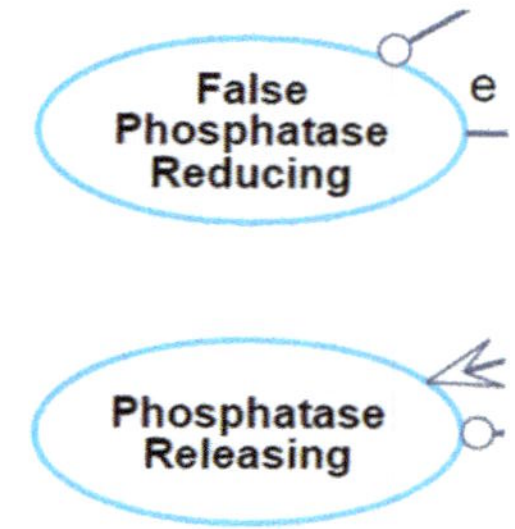
Attribute score	1	
Linear sequence: must be correct and complete	Correct but incomplete	
N non-linear sequences: at least one required	None	
A_9 – Temporary objects and decision nodes		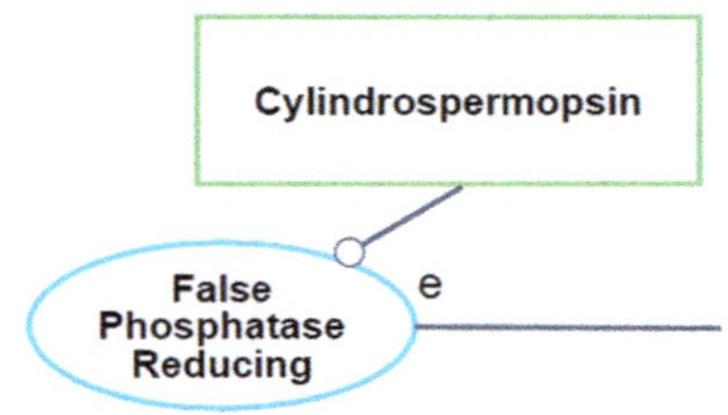
Attribute score	1	
N temporary objects: at least one required	One: Cylindrospermopsin	
N decision nodes: at least one required	None	

enslavement, and (e) two MOOCs for learning about and how to use OPM, which are freely available in English, Arabic, and Hebrew.

The approach presented herein can enable chemistry instruction and learning that pertain to real-world problems and to chemistry understanding across different disciplines and levels of chemistry understanding. The use of conceptual modelling using formalised language allows instructors to base their pedagogical decisions on a clear understanding of their students' own understanding of the chemistry concepts being taught. The approach described herein can be adapted, if needed, to remote or hybrid teaching of undergraduate chemistry, in line with best practices.[25] The STAR can be used by chemistry educators to conduct formative assessment of their students' understanding of the system being taught and provide those students with detailed and focused directions for improving their learning. The importance of formative assessment for chemistry learning has been expounded on by others (Wilkinson, *in preparation*). Using Table 14.2 as an example, the instructor can quickly surmise the modelers had trouble understanding the expected outcome of the system they had modeled, but not its main function. This means the instructor knows what students should focus on learning more about (the expected outcome of the system) and how to explain to those students what they did not understand well (the expected outcome of the system) using what they clearly understood well (the main function of the system).

While we suggest making use of all the aspects of the instructional approach we outlined in this paper, at a minimum, we prescribe students take the free version of Course 101.** This would take about 20 hours of students' and instructors' time, to learn the basics of OPM—enough to conceptually model two levels of detail of a chemical system. While becoming proficient in this approach requires time and effort from both instructors and students, it is a suitable candidate for achieving the reorientation towards systems thinking in chemistry education that is so required for solving humanity's biggest problems in the 21st century.

References

1. H. Jang, *J. Sci. Educ. Technol.*, 2016, **25**, 284.
2. O. B. Z. Assaraf and N. Orion, *J. Res. Sci. Teach.*, 2010, **47**, 1253.
3. E. Crawley, B. Cameron and D. Selva, *Systems Architecture: Strategy and Product Development for Complex Systems*, Pearson Higher Education: Hoboken, NJ, 2016.
4. M. Frank, *Syst. Eng.*, 2000, **3**, 163.
5. National Research Council, *A Framework for K-12 Science Education: Practices, Crosscutting Concepts, and Core Ideas*, National Academies Press, Washington, DC, 2012.

**https://www.edx.org/course/model-based-systems-engineering-foundations.

6. National Research Council, Next Generation Science Standards: For States, By States, National Academies Press, Washington, DC, 2013.
7. S. A. Matlin, G. Mehta, H. Hopf and A. Krief, *Nat. Chem.*, 2016, **8**, 393.
8. P. G. Mahaffy, A. Krief, H. Hopf, G. Mehta and S. A. Matlin, *Nat. Rev. Chem.*, 2018, **2**, 1.
9. O. Š. Miljanić, *Chem*, 2017, **2**, 502.
10. S. Pazicni and A. B. Flynn, *J. Chem. Educ.*, 2019, **96**, 2752.
11. D. Dori, *Model-Based Systems Engineering with OPM and SysML*, Springer, New York, NY, 2016.
12. O. Batzri, O. B. Z. Assaraf, C. Cohen and N. Orion, *J. Sci. Educ. Technol.*, 2015, **24**, 761.
13. M. Stieff, *J. Chem. Educ.*, 2019, **96**, 1300.
14. K. Aubrecht, Y. J. Dori, T. A. Holmes, R. Lavi, S. A. Martin, M. Orgill and H. Skaza-Acosta, *J. Chem. Educ.*, 2019, **96**, 2888.
15. Y. J. Dori and M. Hameiri, *J. Res. Sci. Teach.*, 2003, **40**, 278.
16. R. Lavi and Y. J. Dori, *Int. J. Sci. Educ.*, 2019, **41**, 248.
17. Z. Kaberman and Y. J. Dori, *Instr. Sci.*, 2009, **37**, 403.
18. S. York, R. Lavi, Y. J. Dori and M. Orgill, *J. Chem. Educ.*, 2019, **96**, 2742.
19. Y. Somekh, G. Haimovich, A. Guterman, D. Dori and M. Choder, *PLoS One*, 2014, **9**, e107085.
20. D. Dori and S. Thipphayathetthana, *Int. J. Satell. Commun. Netw.*, 2016, **34**, 295.
21. N. Wengrowicz, Y. J. Dori and D. Dori, *Assess. Eval. High. Educ.*, 2016, **42**, 607.
22. R. Lavi, Y. J. Dori and D. Dori, *IEEE Trans. Educ.*, 2020, **64**, 155.
23. R. Lavi, Y. J. Dori, N. Wengrowicz and D. Dori, *IEEE Trans. Educ.*, 2019, **63**, 39.
24. Y. Bar-Yosef, A. Sukenik, O. Hadas, Y. Viner-Mozzini and A. Kaplan, *Curr. Biol.*, 2010, **20**, 1557.
25. See Chapter 4: R. Iha, J. Luo, B. Lutes and G. Szteinberg, in *Digital Learning and Teaching in Chemistry*, ed. Y. J. Dori, C. Ngai and G. Szteinberg, Royal Society of Chemistry, United Kingdom, 2023.

CHAPTER 15

Chemistry Teachers' Awareness of Sustainability Through Social Media:Cultural Differences

MARINA TAL*[a,b], DANA ZREKE[c], MUHAMAD HUGERAT[c,d] AND AVI HOFSTEIN[c,e]

[a]MOFET Institute, Tel Aviv, Israel; [b]Faculty of Education in Science and Technology, Technion-Israel Institute of Technology, Haifa 3200003, Israel; [c]The Academic Arab College for Education in Israel-Haifa, Israel; [d]The Center of Science and Environmental Education, The Galilee Society—Shefaram, Israel; [e]Weizmann Institute of Science, Rehovot, Israel
*E-mail: marina.t@technion.ac.il

15.1 Introduction

In the modern era, great efforts have been made to help raise awareness of the potential damage humans can cause to the environment and the ways in which such damage can be eliminated or reduced. In 2015, the United Nations General Assembly proposed a set of 17 Sustainable Development Goals (SDGs) in the global agenda, Agenda 2030.[1] Most of the SDGs are directly related to the environmental crisis and require the development of environmental and sustainability education as a key tool for teaching and learning how to deal with this crisis.[2,3]

Advances in Chemistry Education Series No. 11
Digital Learning and Teaching in Chemistry
Edited by Yehudit Judy Dori, Courtney Ngai and Gabriela Szteinberg

Published by the Royal Society of Chemistry, www.rsc.org

Scientists as a whole, and chemists in particular, play a key role in sustainability, a topic that has gained increased relevance in recent years. Scientists have been working to solve current and future environmental problems in an effort to preserve and recycle natural resources such as drinking water, gas, oil, and minerals (all issues that are closely related to chemistry or chemistry-related sciences such as environmental issues). Chemists can contribute to the Affordable and Clean Energy goal by developing new materials for renewable energy, increasing energy efficiency in chemical processing, and advancing cleaner fuel technologies. For example, solar power can be converted into hydrogen fuel using natural and man-made photovoltaic cells.[4,5] Environmental chemistry provides methods and procedures to reduce pollutants and chemicals in the environment, consequently improving the air quality. Cleaner air with fewer chemicals leads to less damage to our health.[6,7] Chemistry teachers are in a unique position to help increase environmental awareness as well as incorporate educational materials related to environmental behavior, such as recycling, reducing consumption, and transitioning to the use of renewable energy sources.[8]

Chemistry that focuses on sustainability is both interesting and provides students with the opportunity to gain knowledge and skills as the future generation of scientists and leaders. Education for sustainability can be used as a tool to increase the relevance of science education as well as the scientific literacy of learners.[9,10] For chemistry teachers, training and workshops focusing on sustainability are important, enabling them to impart the knowledge needed to create sustainable, scientifically literate societies for future scientists and leaders.[11,12] Many studies around the world have investigated the sustainability awareness of pre-service teachers.[13,14] Burmeister and Eilks[14] described German pre-service and trainee chemistry teachers' understanding of sustainability. They indicated that the teachers showed positive attitudes towards education for sustainable development. However, in chemistry education, approximately two-thirds of the prospective teachers working in Germany during their probationary period had never heard of the concepts of green chemistry or sustainability.

Nowadays, social media applications are quite popular among students and have a significant impact on education regarding sustainability issues.[15–17] The use of social media has increased dramatically over the past decade, and the next generation cannot describe their lives without social media.[18,19] Although many teachers and educators find this phenomenon negative and dangerous, others are actively trying to utilize it for educational purposes.[20,21] Studies have shown the possibility of using social media to promote a new era of social learning. Through social media, high-school students can learn about chemistry through environmental contexts. This use of social media may be particularly useful for educators during the COVID-19 pandemic, since it is compatible with distance learning.[22,23] Technological and societal issues can provide easier contexts for learning chemistry,[9] since these issues are more likely to be directly perceived as meaningful and relevant to their own lives. Consequently, social media has the potential to further benefit the environment and elevate the level of environmental awareness among science students and teachers as well as among other social media users.[16] In addition, many teachers are likely to frequently use social media and can use it to achieve their educational goals and objectives related to sustainability education.

According to Goldman and colleagues,[24] society believes that the formal educational system is responsible for environmental education efforts. In Israel, as in many other countries, there is growing interest in environmental education, but its position in the formal education system remains unclear, since its effectiveness is questionable.[25] The issue of environmental awareness in Israel is particularly interesting because, although it is a developed country in many aspects, environmental concerns are not sufficiently addressed in Israeli government policy or in the education system.[26] In addition, Israel is a country facing considerable challenges due to the diversity of the cultural orientations of its inhabitants.[27] Teachers in different fields and sectors are addressing sustainability issues according to their cultural orientations, perceptions, and preferred teaching methods.[28–30] According to the literature, worldviews and educational perceptions differ from sector to sector.[27,31] As part of education for sustainability, the importance of the socio-cultural parameters of environmental problems and solutions has been increasingly recognized. Sustainability is understood in a variety of ways, reflecting the importance of culture when studying environmental problems.[32] Thus, there is a need to be aware of the cultural context in which an environmental education program or environmental research takes place.[33,34] The importance of multicultural education for environmental education has been discussed in several studies.[28,35] Owing to different worldviews, the Jewish and Arab sectors perceive sustainability in different ways.[36] In spite of the fact that any culture values its unique traditions, beliefs, and norms, social media has become a tool that promotes diversity and inclusivity by connecting people around the world, despite differences in culture and geography.[37,38] There is no inherent conflict between tradition and innovation regarding social media.[26] Moreover, some researchers believe that social media is more than just a technology—it is a new form of culture in itself.

Considering that we would like to promote sustainability in Israeli society through social media, since social media is a great tool for fostering a diverse and inclusive population, it is important first to identify the differences between the different sectors' sustainability awareness and then their social media usage strategies. In this chapter we report the impact of social media culture on the lives of science teachers and their environmental awareness as well as its potential to promote environmental education in Israel. The purpose of our study was to examine the environmental knowledge, awareness, and behavior of science teachers, particularly chemistry teachers in Israel as well as their perceptions of the impact of social media on environmental education and sustainability.

15.2 Research Plan and Methodology

In this section we present a study that examines how social media contributes to pre- and in-service science and chemistry teachers' environmental sustainability awareness. The participants were asked to discuss the most important sustainability topics according to their view of the world. The focus of this study was on the cultural and ethnic differences among the participants (based on the demographic information provided by the participants). The main goals of this study were to: (a) compare the sustainability awareness of pre- and in-service chemistry teachers from both the Jewish and Arab sectors; and (b) determine the role that social media plays in attracting chemistry teachers to environmental sustainability issues.

The study was guided by two research questions:

Research Question 1: What is the effect (if any) of cultural differences among pre- and in-service chemistry teachers on their environmental awareness?

Research Question 2: What social media sources are popular with chemistry teachers in Israel and what role do they play in creating sustainability awareness?

15.2.1 Research Participants

The study consisted of 128 participants, 85% of whom were female. All the participants in this study were pre- or in-service chemistry teachers. The participants were divided into two groups, based on whether they belonged to either a Jewish or an Arab sector (see Table 15.1).

15.2.2 Research Tool

To address the research questions, we designed an initial version of the questionnaire in Google Forms and administered it to five pre-service and five in-service chemistry teachers. Based on their feedback, we developed the final version of the Hebrew questionnaire. The questionnaire was translated from Hebrew to Arabic by an expert in both languages who has a PhD in

Table 15.1 An overview of the research participants.

Demographic variables		Jewish sector ($N = 49$)	Arab sector ($N = 79$)	All ($N = 128$)
Gender	Male	13	5	17
	Female	36	73	109
Teaching status	Pre-service	22	31	53
	In-service	27	48	75
Age	Under 35	24	49	73
	36–45	11	21	32
	46–55	9	6	15
	Over 56	4	3	7

Arabic. Thereafter, the translation was sent to two experts from the Arabic sector, the first with a PhD in Hebrew and the second with a PhD in Arabic. The two experts were asked by the researchers to review all the questions and statements in the questionnaire and determine whether the source of the questionnaire matched the translation. There were minor comments from the two experts, which were taken into account in the next version of the questionnaire. Then, two experts in chemistry and environmental education, one from the Jewish sector and one from the Arab sector, carefully examined all the statements and queries in the questionnaire in its latest version, to check whether there was a complete match between the source and the translation. Here, too, a number of minor corrections were taken into account and included in the next version of the questionnaire.

The questionnaire included both close-ended and open-ended questions to assess chemistry teachers' sustainability knowledge, attitudes, and awareness regarding specific environmental issues as well as their behaviour. In addition, items about the role of social media in raising awareness of sustainability were added to the questionnaire. The first three parts of this questionnaire examined teachers' sustainability awareness using previous questionnaires developed by researchers to assess knowledge, attitudes, and behaviour related to environmental sustainability issues.[24,39,40] The questionnaire included 62 items based on a Likert scale, aiming to better understand the teachers' perspectives on sustainable development. These items' statistical significance was determined by calculating Cronbach's alpha, which gave a value of 0.89, indicating high reliability. Mean values were calculated and any resulting differences were tested for their significance.

In addition to the close-ended items, the questionnaire also contained a number of open-ended items. In one of the open-ended questions, the teachers were asked to select two important environmental issues (based on a PISA 2006 questionnaire published in OECD) and to describe which important facts they thought students should know about.[41] In the last part, we added close-ended and open-ended questions about social media as a whole and specifically regarding its role in raising sustainability awareness. For example, in the open-ended question the participants were asked to describe ways in which they usually express their support for sustainable development on social media.

15.2.3 Data Analysis

We conducted the study using a mixed-methods methodology, including both quantitative and qualitative elements. We used statistical analyses[42] to compare close-ended item data and content analysis[43] to identify categories with open-ended item data.

Comparisons between the mean questionnaire scores among the subgroups were made using the one-way ANOVA test. If more than two subgroups were compared, the ANOVA test was followed by the Tukey post-hoc test for multiple comparisons. For the quantitative variables, the mean and standard deviation (S.D.) were calculated. Categorical variables were presented as percentages. The percentages were compared using the *chi*-square test or the Fisher exact test. Bonferroni adjustment for multiple comparisons was applied when needed. SPSS version 24 was used for statistical analysis. $P < 0.05$ was considered statistically significant. To triangulate the data, we analyzed the responses to the open-ended items related to the most important environmental issues and the effectiveness of social media as a platform for conveying sustainability messages. This research was approved by the MOFET Institute's Review Board and conducted according to the approved protocol. Approval number RB2103PD was received on March 7, 2021. All teachers in the study gave their consent to be part of the research and filled in the questionnaires. To protect confidentiality, participants were given codes and any identifying details were removed from the data.

15.3 Results and Discussion

The findings presented in this section address the two research questions. First, we refer to the effect of cultural aspects on environmental sustainability among pre- and in-service chemistry teachers in Israel. Then, we describe the role of social media among chemistry teachers in Israel and discuss the role the social media sources might play in creating sustainability awareness.

15.3.1 Sustainability Awareness

Table 15.2 presents the average score of the chemistry teachers' awareness, knowledge, attitudes, and behaviour concerning environmental sustainability issues. We found significant differences between the average score of the Jewish and Arab teachers (see Table 15.2). According to the teachers' self-reported data, the level of awareness, attitudes, and behaviour of chemistry pre- and in-service teachers from the Arab sector is significantly higher than that of their Jewish counterparts. In contrast, other studies have argued that the Arab sector has a lower level of environmental awareness.[28]

Within the questionnaire, we asked participants to identify from the nine options (based on PISA 2006)[41] the most important issue related to the

Table 15.2 Sustainability awareness average scores and differences between Jewish and Arab teachers.

Behaviour	Jewish $\bar{X}$(SD)	Arabic $\bar{X}$(SD)	
Variables	$N = 49$	$N = 78$	Statistical value
Awareness	3.59 (0.39)	3.87 (0.43)	$t(125) = -3.58^a$
Knowledge	3.31 (0.81)	3.25 (1.05)	$F(1,122) = 1.37$
Attitudes	3.86 (0.43)	4.16 (0.49)	$t(122) = -3.52^a$
Behaviour	3.07 (0.61)	3.48 (0.60)	$F(1,122) = 18.16^a$

$^a p < 0.001$.

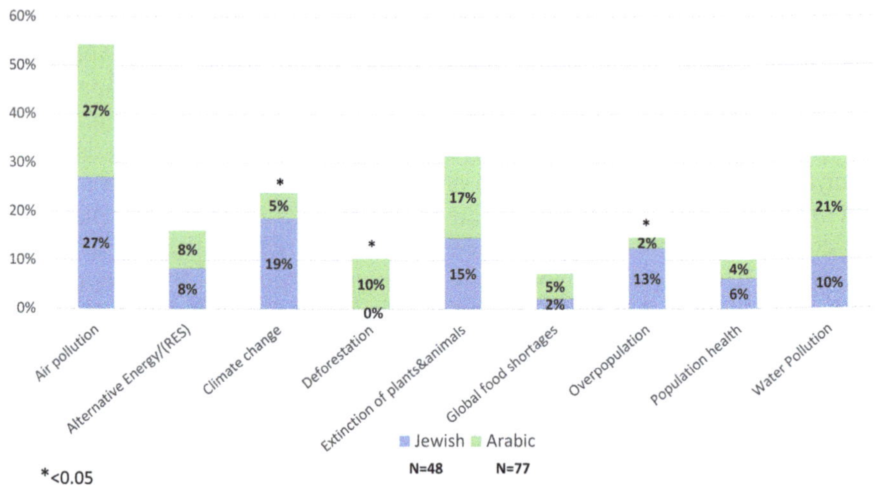

Figure 15.1 The most important environmental issues among the Jewish and Arab sectors.

environment. Figure 15.1 presents the teachers' perceptions of the most important sustainable environmental issue.

In terms of environmental importance, air pollution was rated as the top issue. Jewish and Arab teachers are equally concerned about this issue, perhaps because it is something that they experience every day and hear about in the media all the time. Additionally, teachers are also concerned with water pollution and the possible future extinction of plants and animals. We have seen differences between Jewish and Arab teachers on a number of environmental issues. The issue of water pollution, for example, seems to be of particular concern to the Arab sector. This topic was selected by 21% of the Arab teachers, but by only 10% of the Jewish teachers. Moreover, significant differences were found between both sectors regarding overpopulation, climate change, and deforestation issues. The following is a description of the statistical tests conducted and the differences discovered between Jewish and Arab teachers. Fisher's exact test revealed only a marginally significant difference between 12% of Jews and 2.5% of Arabs who chose overpopulation, $p = 0.053$. Pearson's *chi*-squared test revealed a significant difference between 18% of Jews and 5% of Arabs who chose climate change, $\chi^2(1) = 5.87$, $p = 0.015$. Fisher's exact test revealed only a marginally significant difference between 10.1% of Arabs and 0% of Jews who chose deforestation, $p = 0.023$.

Even though Arab teachers are very concerned about issues such as air pollution and deforestation, almost 30% of them mentioned that they commonly use wood to heat their houses. The Wilcoxon–Mann–Whitney test revealed a significant difference between the two sectors, ($U = 1090$, $p < 0.001$); only 14% of the Jewish teachers use this type of heating. This difference may be because most people in the Arab sector live in private houses in villages. This is in contrast to the majority of Jewish teachers, most of whom live in high-rise apartment buildings in cities.

We evaluated participants' declared knowledge regarding the nine environmental issues according to their knowledge about these nine issues. The mean value of the declared knowledge score was medium (ranging from 3.05 to 3.41) for all nine environmental issues. A summary of the close-ended question concerning the importance of nine different environmental issues raised in the questionnaire indicated that the three environmental issues they rated as the most important were air pollution, plant and animal extinctions, and water pollution. Nevertheless, even for these three environmental issues, the respondents indicated that their level of knowledge was incomplete (only moderate). Following these close-ended items, the participants were requested to select two of the nine environmental issues listed in the previous two questions and to describe which important facts they think students should know about in detail.

The following examples represent some of the responses to this question:

<u>#2004—Jewish pre-service teacher</u>
Air pollution, in my opinion, is the main and most important issue, since it is the main risk factor affecting human health. Air pollution affects both man and animals as well as the environment. Many serious problems are associated with air pollution, including cancer and breathing difficulties. Moreover, the ozone layer is also a concern. Furthermore, I selected the extinction of plants and animals, since most people currently do not attach any importance to it and do not realize that animals also have a right to live. Often people kill them and use them for their own gain.

<u>#4028—Arabic pre-service teacher</u>
It is very important to address the problem of air and water pollution because these two pollutants cause numerous environmental and health problems. Both of these factors determine the course of one's life, so it is necessary for students to be aware of these problems and reduce them, if possible.

<u>#1003—Jewish in-service teacher</u>
Regarding overpopulation—it is very important for students to understand that population increases are out of balance when a person consumes more resources than he replenishes or returns. In terms of climate change—whether it is caused by human actions or not (for example, due to recycling processes in our Solar System/the Earth), students should understand that with reasonable behaviour, wise consumption habits, wise resource use, science and technology, man will have to effectively deal with climate change and find solutions.

<u>#3048—Arabic in-service teacher</u>
The forests and trees are considered the foundation of life on Earth, and their importance is due to the fact that trees provide us with oxygen as well as remove greenhouse gases that are harmful to the ozone layer from the atmosphere. Trees, plants, and tropical forests provide many endangered animals with a natural environment, and man must preserve and protect biodiversity on Earth.

15.3.2 The Role of Social Media

The last part of the questionnaire provides information on how science and chemistry teachers utilize social media and how they view social media in relation to the sustainability issue.

We asked teachers where they discovered the concept of sustainable development for the first time.

Table 15.3 shows teachers' responses to the question "*Name the sources where you were first exposed to the concept of sustainable development*" (respondents could select more than one source). Table 15.3 shows that there were no meaningful differences between Jewish and Arab chemistry teachers regarding the source where they were first exposed to sustainable development. The main difference is between the pre- and in-service teachers. This is probably due to the generation gap between them. Almost 95% of pre-service teachers are under 35-years-old.

Figure 15.2 compares pre-service and in-service teachers in the two sectors.

Overall, Figure 15.2 provides us with a sense of the popularity of different information sources. When it comes to types of information sources, the division between in-service teachers and pre-service teachers allows us to observe how things are rapidly changing in our world. These results indicate that academic lectures (43%) are the most common exposure source for in-service teachers for the topic of sustainable development, whereas social media (42%) is the most common exposure source for pre-service teachers. Historically, academic lectures, the press, and textbooks have been important sources of knowledge and information (they were even the only ones for a while), but now this role is increasingly dominated by social media. It seems that newspapers are only used as a source of information by the older generation, since only 2% of the pre-service teachers revealed that they learned about it from the press, compared with 18% of the in-service teachers.

These results indicate that 12% of pre-service teachers and 6% of in-service teachers, who actually teach in the education system, have never been exposed to the concept of sustainable development. Despite the fact that the

Table 15.3 The source of the initial exposure to the sustainable development concept (all groups).[a]

	Jewish		Arab		Statistical value	
Source	In N = 24 (%)	Pre N = 21 (%)	In N = 48 (%)	Pre N = 31 (%)	χ^2	p
Textbooks	22.2	13.6	33.3	32.3	3.68	n.s.
Academic lectures	37.0	4.5	50.0	25.8	15.10	< 0.01
Newspapers	22.2	0.0	18.8	3.2	9.63	< 0.05
TV	25.9	22.7	29.2	22.6	0.56	n.s.
Social media	18.5	27.3	27.1	51.6	8.57	< 0.05
First exposure to this issue	3.7	27.3	6.3	0.0	14.99	< 0.001

[a]n.s. = not significant.

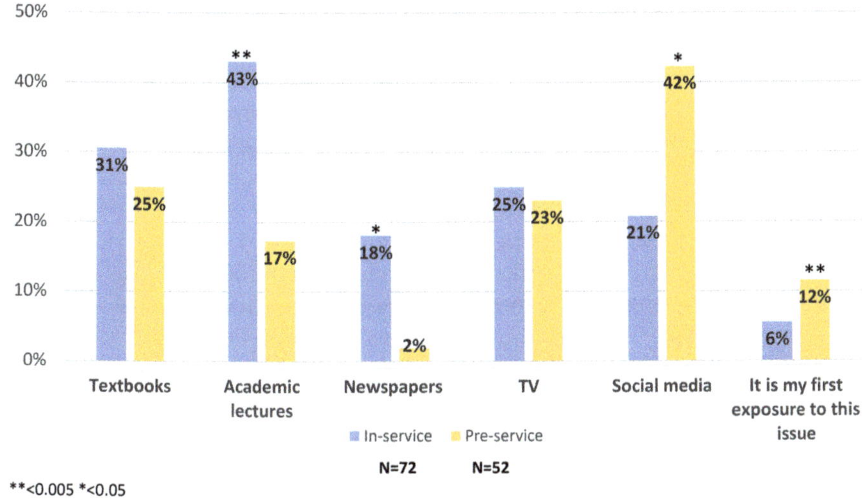

Figure 15.2 The source of the initial exposure to sustainability issues.

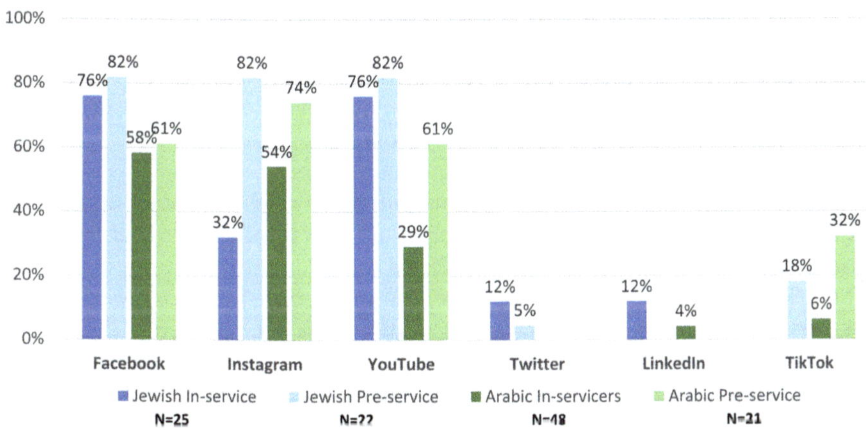

Figure 15.3 Popular social media platforms.

percentage of in-service teachers is significantly lower than the percentage of pre-service teachers, this finding is still very concerning. Regarding the information sources discussed above (academic lectures, newspapers, and social media), significant differences were found between the proportion of pre- and in-service teachers who were first exposed to the sustainable development concept through these information sources (see Figure 15.2). The increasing use of social media places it as a major player among the various sources of information.

Next, we asked the teachers about the most popular social media platforms. Figure 15.3 shows teachers' responses to the question "*Mark three social media sources you frequently use*". Facebook, Instagram, and YouTube are the three most popular social media platforms among the participants. The younger generation clearly access more social media sites and from a wider variety of sources than the older generation. Pearson's *chi*-squared test revealed a significant difference between 77.4% of the pre-service teachers and 45.3% of the in-service teachers who frequently use Instagram, $\chi^2(1) = 13.13$, $p < 0.001$ as well as between 26.4% of the pre-service teachers and 4.0% of the in-service teachers who frequently use TikTok $\chi^2(1) = 13.54$, $p < 0.001$. Regarding the cultural differences, apparently the use of social media is higher among pre- and in-service teachers from the Jewish sector than among those from the Arab sector, as a whole. Pearson's *chi*-squared test revealed a significant difference between the 73.5% of the Jewish sector and 41.5% of the Arab sector who frequently use YouTube, $\chi^2(1) = 12.21$, $p < 0.001$ as well as between 8.2% of the Jewish sector and 0% of the Arab sector who frequently use Twitter, $\chi^2(1) = 6.66$, $p < 0.05$. When compared to the Jewish in-service teachers, the Arab in-service teachers used fewer social media tools. However, based on Al-Kandari and colleagues,[44] social media, specifically Instagram, is increasingly popular in the Arab world today. Currently, Instagram is one of the most popular apps among young people, probably due to its wide range of options, ranging from sharing pictures of everyday life to informing the public of scientific developments.[44] The results indicate that Facebook is still a very important application across all sectors, because others with similar interests can be found (such as environmental movements) by following the messages they post and share with others, and by the events they organize. In addition, the Arab sector is less seen on Twitter or LinkedIn since most of this population instead uses Facebook and Instagram (see Figure 15.3). TikTok is a relatively new application that has only emerged in the last two years, so its scope of use is limited, compared with other platforms, and its users are largely young people (about 18% of Jewish pre-service teachers and about 32% of Arab pre-service teachers).

Finally, we asked the participants about the effectiveness of social media as a platform for conveying sustainability messages. Figure 15.4 shows teachers' responses to the question "*Mark the sources of social media that you believe are accessible and convenient for conveying sustainability messages*". Significant differences were found across sectors regarding perceptions of the effectiveness of the YouTube and Twitter apps in promoting sustainable development. Pearson's *chi*-squared test revealed a significant difference between 61.2% of the Jewish sector and 38.0% of the Arab sector who perceived YouTube as an effective platform for promoting sustainability issues, $\chi^2(1) = 6.50$, $p < 0.05$ as well as between 18.4% of the Jewish sector and 5.1% of the Arab sector who perceived Twitter as an effective platform for promoting sustainability issues, $\chi^2(1) = 5.82$, $p < 0.05$.

According to Figure 15.4, the majority of participants (65%) perceive Instagram as a very useful tool for communicating sustainability issues. The younger generation especially supports using Instagram for such purposes. Pearson's *chi*-squared test revealed a significant difference between 81.1% of

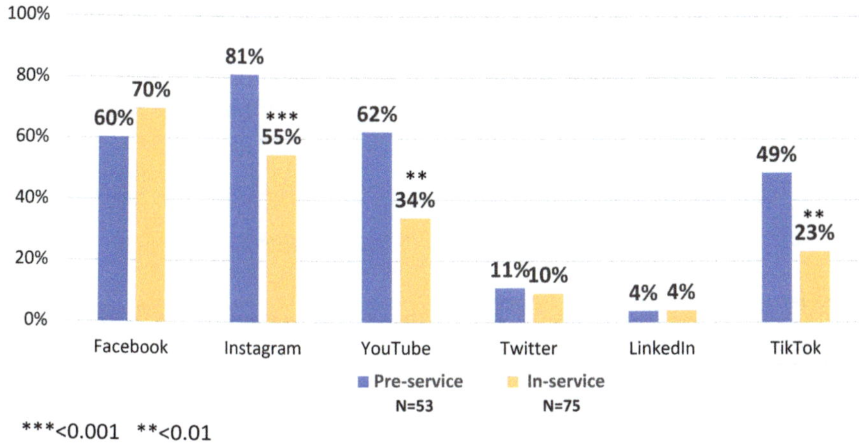

Figure 15.4 The effectiveness of social media as a platform for conveying sustainability messages.

the pre-service teachers and 55.3% of the in-service teachers who perceive Instagram as an effective platform for promoting sustainability issues, $\chi^2(1) = 10.53$, $p = 0.001$. Although pre-service teachers can use different social media platforms to promote sustainability, in-service teachers cling to more familiar apps such as Facebook and YouTube. The participants evidently perceive video clips as a very effective way to convey messages, probably due to their visualization capability regarding environmental issues such as climate-related natural disasters. We assume that this might explain why teachers used YouTube and TikTok at such high rates. This is especially surprising since TikTok is not their usual application of choice (see Figure 15.3). Pearson's *chi*-squared test revealed a significant difference between 49.1% of the pre-service teachers and 22.7% of the in-service teachers who perceive TikTok as an effective platform for promoting sustainability issues, $\chi^2(1) = 9.70$, $p < 0.05$ as well as between 62.3% of the pre-service teachers and 36.0% of the in-service teachers who perceive YouTube as an effective platform for promoting sustainability issues, $\chi^2(1) = 8.60$, $p < 0.05$.

Perhaps due to their unfamiliarity with these platforms, some in-service teachers perceive them as less effective. The older participants may not realize that they can also be used for beneficial purposes like promoting sustainability issues because they have heard negative reviews about them. Generally, the results indicate that teachers support teaching about sustainability using social media, such as YouTube, Instagram, and TikTok. Differences between pre-service and in-service teachers regarding the effectiveness of different social media sources might indicate generational differences. Following the close-ended items regarding the use of social media, the participants were requested to describe how they express their support for sustainable development through social media.

The following examples represent some of the responses to this question:

#2016—Jewish pre-service teacher
I support a page on Instagram that encourages green thinking. I send videos related to sustainability.

#1001—Jewish in-service teacher
I use social media passively—I don't upload or share posts.

#3044—Arabic in-service teacher
Through my Facebook page, I promote clean events and conservation of the environment.

#4008—Arabic in-service teacher
I share short video clips that explain sustainability issues.

15.4 Summary and Conclusions

Chemistry plays a key role in solving sustainability problems, which have become increasingly relevant in recent years as the world grapples with environmental crises, such as climate change and other environmental hazards. Chemistry teachers are in a unique position to help increase environmental awareness as well as incorporate educational materials related to environmental behaviour.[8,10] Sustainability awareness could be developed not only through traditional teaching methods such as academic lectures and textbooks, but also *via* a wide range of sources such as TV and social media.[19] Israeli young people, both Jewish and Arab, are increasingly using social media application tools, which may have a major impact on chemical education including sustainability issues. In this chapter, we examined the impact of social media on chemistry teachers from various cultures, their environmental awareness, and how social media might be used to promote environmental education in Israel.

The first research question was: *What is the effect (if any) of cultural differences among pre- and in-service chemistry teachers on their environmental awareness?*

The Arab sector is regarded as one that is less aware of environmental issues.[28] However, we found that Arab teachers exhibited greater sustainability awareness, attitudes, and behaviours than Jewish teachers. Our recommendation is to conduct a more comprehensive study using a more diverse set of science disciplines and a larger sample of teachers. Based on our present research, although Arab teachers are much more concerned about water pollution, they do not seem to be bothered by the issue of population growth. This is probably because the water and sewage infrastructure in Arab localities is in worse shape than that in the Jewish sector.[36] Israel is a small country, characterized by very high population growth, in contrast to other young countries where population growth is much slower. The problem of overpopulation is one of the major environmental issues that Israel will face in the coming years, and we are concerned that so few teachers (especially in the Arab sector) have viewed this issue as a significant environmental problem. The fact that almost 10% of the chemistry teachers who participated in our study have never heard of the concept of sustainable development is alarming and unacceptable according to Western world standards. This increasing problem must be addressed as soon as possible.

The second research question was: *What social media sources are popular with chemistry teachers in Israel and what role do they play in creating sustainability awareness?*

There is no doubt that social media sources have become a major player among the sources of information. Today, during the COVID-19 crisis, we are witnessing an unprecedented impact of social media. Often without any scientific basis, it is a source of exclusive information for its users. Although many teachers and educators find this phenomenon negative and dangerous, it is a fact that younger generations cannot adequately describe their lives without social media.[18] The significant differences between pre- and in-service teachers demonstrate that the world is rapidly changing. Unfortunately, the modern curriculum ignores the social and pedagogical effects of the changes related to social media that have taken place in the world.

The fact that we found no significant difference between Jewish and Arab pre-service teachers in terms of the variety and extent of social media use indicated that future teachers may be very motivated to introduce social media as a novel and effective teaching method. Because Arab pre-service teachers use social media frequently, the Arab education system should be encouraged to use advanced learning and teaching strategies. This is especially important because previous studies indicated that Arab educational systems tend to use traditional teaching methods and have made slow progress in implementing student-centered approaches.[27] The teachers themselves support the idea of teaching about sustainability through social media and point to apps such as YouTube, Instagram, and TikTok as suitable and effective platforms. In conclusion, regarding the use of social media, we agree with Senator James E. Watson who said "*If you can't beat them, join them*". The use of social media should begin to permeate the current teaching methods and strategies as part of the modernization process.

References

1. U. Nations, *70/1. Transforming Our World: the 2030 Agenda for Sustainable Development Transforming Our World: the 2030 Agenda for Sustainable Development Preamble*, 2030.
2. M. Rieckmann, *Education for Sustainable Development Goals: Learning Objectives*. Unesco Publishing, 2017.
3. D. Olsson, N. Gericke and J. Boeve-de Pauw, *Environ. Educ. Res.*, 2022, **28**, 405.

4. R. I. Pinhassi, D. Kallmann, G. Saper, H. Dotan, A. Linkov, A. Kay, V. Liveanu, G. Schuster, N. Adir and A. Rothschild, *Nat. Commun.*, 2016, **7**, 1.
5. J. Z. Zhang and E. Reisner, *Nat. Rev. Chem.*, 2020, **4**, 6.
6. A. Gopinath and K. Kadirvelu, *Environ. Chem. Lett.*, 2018, **16**, 1137.
7. S. Yu, *Environ. Chem. Lett.*, 2019, **17**, 565.
8. M. Burmeister, F. Rauch and I. Eilks, *R. Soc. Chem.*, 2012, **13**, 59.
9. I. Eilks, J. Sjöström and A. Hofstein, *Relevant Chem. Educ. Sustainability*, 2017, **44**, 18.
10. See Chapter 7: J. Huwer, C. Barth, A. Siol and I. Eilks, in *Digital Learning and Teaching in Chemistry*, ed. Y. J. Dori, C. Ngai and G. Szteinberg, Royal Society of Chemistry, United Kingdom, 2023.
11. J. Álvarez-García, C. P. Maldonado-Erazo, M. de la C. del Río-Rama and F. J. Castellano-Álvarez, *Sustainability*, 2019, **11**, 6034.
12. J. M. Marcos-Merino, I. Corbacho-Cuello and M. Hernández-Barco, *Sustainability*, 2020, **12**, 7445.
13. P. Bezeljak, M. Scheuch and G. Torkar, *Sustainability*, 2020, **12**, 6892.
14. M. Burmeister and I. Eilks, *Sci. Educ. Int.*, 2013, **24**, 167.
15. E. Gori, A. Romolini, S. Fissi and M. Contri, *Sustainability*, 2020, **12**, 4658.
16. S. Hamid, M. T. Ijab, H. Sulaiman, R. Md. Anwar and A. A. Norman, *Int. J. Sustainability High. Educ.*, 2017, **18**, 474.
17. C. G. Knight and L. K. Kaye, *Innov. Educ. Teach. Int.*, 2016, **53**, 145.
18. M. M. Alamri, M. A. Almaiah and W. M. Al-Rahmi, *Sustainability*, 2020, **12**, 6471.
19. See Chapter 8: N. Belova, A. Heckenthaler, J. M. Tietjen and C. Zowada, in *Digital Learning and Teaching in Chemistry*, ed. Y. J. Dori, C. Ngai and G. Szteinberg, Royal Society of Chemistry, United Kingdom, 2023.
20. Q. Wang, W. Chen and Y. Liang, *The Effects of Social Media on College Students, Alan Shawn Feinstein Graduate School*, Providence: Johnson & Wales University, 2011.
21. J. Knight-McCord, D. Cleary, N. Grant, A. Herron, J. Success, T. Lacey, T. Livingston, S. Robinson, R. Smith and R. Emanuel, *J. Undergrad. Ethn. Minority Psychol.*, 2016, **2**, 21.
22. P. E. Danjou, *J. Chem. Educ.*, 2020, **97**, 3168.
23. M. O. Hight, N. Q. Nguyen and T. A. Su, *J. Chem. Educ.*, 2021, **98**, 1283.
24. D. Goldman, B. Yavetz and S. Pe'er, *J. Environ. Educ.*, 2006, **38**, 3.
25. N. Zaradez, R. Sela-Sheffy and T. Tal, *Environ. Educ. Res.*, 2020, **26**, 812.
26. D. Goldman, R. Hansmann, J. Činčera, V. Radović, A. Telešienė, A. Balžekienė and J. Vávra, *Conceptualizing Environmental Citizenship for 21st Century Education*, Springer Cham, London, 2020, ch. 8, pp. 115–137.
27. S. Markic, I. Eilks, R. Mamlok-Naaman, M. Hugerat, N. Kortam, I. Dkeidek and A. Hofstein, *Teach. Teach.*, 2016, **22**, 131.
28. M. Negev and Y. Garb, *J. Environ. Educ.*, 2014, **45**, 143.
29. T. Tal and I. Alkaher, *Cult. Stud. Sci. Educ.*, 2010, **5**, 325.
30. T. Tal and E. Peled, *Environ. Educ. Res.*, 2017, **23**, 1032.
31. S. Asli, N. Kortam, S. Algamal, N. Sheme and M. Hugerat, *Eurasia J. Math. Sci. Technol. Educ.*, 2021, **17**, em1948.

32. C. A. Bowers, *Education, Cultural Myths, and the Ecological Crisis: Toward Deep Changes*, State University Of New York Press, Albany, 1993.
33. V. Dkhar, *Int. J. Res. Soc. Sci.*, 2019, **9**, 634.
34. C. Marouli, *Can. J. Environ. Educ.*, 2002, **7**, 26.
35. I. Alkaher, D. Goldman and G. Sagy, *Sustainability*, 2018, **10**, 3721.
36. Knesset Research and Information Center, *Environmental Issues in Arab Sector Settlements*, Knesset, Jerusalem, Israel, 2016, http://fs.knesset.gov.il/globaldocs/MMM/e0c7a649-29b8-e511-80d0-00155d0acb9e/2_e0c7a649-29b8-e511-80d0-00155d0acb9e_11_9920.pdf.
37. N. Ahmad, S. Hussain and N. Munir, Social Networking and Depression among University Students, *Pak. J. Med. Res.*, 2018, **57**, 77.
38. R. Sawyer and G. Chen, *Intercult. Commun. Stud.*, 2012, **21**, 151.
39. A. C. Michalos, H. Creech, N. Swayze, P. M. Kahlke, C. Buckler and K. Rempel, *Soc. Indic. Res.*, 2012, **106**, 213.
40. T. Tal, *Int. Res. Geogr. Environ. Educ.*, 2010, **19**, 263.
41. OECD, *PISA 2006 Technical Report*, 2009.
42. F. Erickson, in *Second International Handbook of Science Education*, ed. B. J. Fraser, K. Tobin and C. J. McRobbie, Springer, New York, 2012, ch. 9, p. 1451.
43. J. W. Creswell, R. Shope, V. L. Plano Clark and D. O. Green, *Res. Sch.*, 2006, **13**, 1.
44. A. A. Al-Kandari, T. K. Gaither, M. M. Alfahad, A. A. Dashti and A. R. Alsaber, *Public Relat. Rev.*, 2019, **45**, 101774.

CHAPTER 16

Using Visualization and Laboratory to Promote Learning in Science

J. KRAJCIK*

CREATE for STEM, Michigan State University, USA
*E-mail: krajcik@msu.edu

16.1 Introduction

Using computer-based technology tools to support students in learning science has captured the imagination of teachers, designers, and researchers since they were first used in science classrooms in the early 1980s. Interactive and visual features along with the ability to capture, display, and analyze data have grown tremendously since then and have made computer technologies ideal for providing opportunities to learn for a wide range of students. Students can use computer technology to examine complex aspects of science including building and visualizing complex molecules, constructing models to show the mechanism of how various phenomena occur, and analyzing large and complex data sets. As visualization and interactive features along with various forms of technology become more affordable and more sophisticated, such as the ubiquitous use of smartphones, designers and teachers have developed scaffolds to support learners in complex tasks, problem solving, and model building. Visualization tools can help learners imagine and explore complex and large data sets and explore micro worlds too small to be seen, even including the quantum level.[1] However, advances in technology

Advances in Chemistry Education Series No. 11
Digital Learning and Teaching in Chemistry
Edited by Yehudit Judy Dori, Courtney Ngai and Gabriela Szteinberg

Published by the Royal Society of Chemistry, www.rsc.org

hardware are only part of the story. Researchers have now designed modeling tools that have more sophisticated visualization capabilities along with the capacity to generate data based on the model's behaviour and compare the simulated data to real-world data to judge how valid the model is.[2] Such tools have unprecedented opportunities to promote deep learning that allow students to use their knowledge to make sense of new phenomena or solve ill-structured problems that mirror real-world situations. With more portable technologies in schools owned by learners, including interactive tablets and smartphones, the use of technology tools to support student learning through visualizing complex ideas and the nano world of atoms and molecules as well as the astronomical world of solar systems and galaxies has grown tremendously. The portability of tablets and smartphones and their network capabilities allow students to analyze data, build models, visualize the unseen worlds of cells and molecules, and explore galaxies. The flexibility, interactive power and networking, customization, and multiple representation capabilities of these portable devices could help to change the structure and purpose of the science classroom. Classrooms now have the potential to become learning environments where students interactively construct their knowledge and provide evidence to support their claims, instead of environments where students receive and recall information dispensed by the teacher or an electronic medium. For instance, technology tools allow students to visualize the world of atoms and molecules and observe what happens when the learner manipulates the temperature of the system. The chapters in this section further discuss these potentials.

Even with the advancements and ubiquity of technologies, however, the potential to support student learning in science classrooms will only occur when technology tools are used to support learning that cannot be accomplished without their use. To promote usable knowledge more than just showing a visualization to learners must occur.[3] Learners need to interact with the visualizations so that it allows learners to make connections among ideas. Visualizations need to allow learners to see and interact with "unseen worlds" like the world of cells and molecules by manipulating variables and observing the outcomes. Interactive visualization and simulations give learners the ability to interact with the visualization by changing variables and then observing the consequence. Interactive visualizations allow learners to examine and explore phenomena that are too small, too large, too fast, or too dangerous to explore. As such, interactive visualizations invite students to unseen worlds of cells and molecules, or galaxies. They can also explore phenomena that are too dangerous to examine—like volcanoes, or too fast, like earthquakes; or too slow to observe—like the growth of an organism over time. But the power of interactive visualizations is that they can present underlying causal accounts of the phenomena that allow learners to explore what happens when variables change. Students can change a variable and observe the consequence. We see this form of students interacting with technology in the work presented by Bernholt and colleagues (this book).[4]

Technology tools can also help students learn how to participate in authentic disciplinary practices, such as modelling and constructing explanations. As such, technology tools have the potential to increase student interest and learning by extending what students can do in the classroom by seamlessly connecting in-class and out-of-class settings and allowing collaborations with others outside the classroom. Technologies serve as powerful tools that can support learning by using authentic practices that help students in building connections among disciplinary ideas, forming a deeper and richer understanding that students can use to make sense of the world they live in and solve complex real-world problems.

Learning with visualizations provides a powerful example of how technology tools are a critical aspect of science education and in learning chemistry. Multimedia instruction, including the use of visualizations, have the potential to promote students to develop understanding that they can use in different contexts. With many K-12 schools and colleges moving to virtual instructions during the COVID-19 pandemic, the use of visualization tools by learners to support learning challenging ideas and experiencing phenomena virtually became critical. The chapters in this section explore the use of visualization and simulated labs to promote student learning.

In Chapter 17, *Applications of Digital Technology in Chemical Education*,[5] Han and Fung describe how researchers at the National University of Singapore (NUS) developed several Educational Technologies (EdTechs) to enhance student learning in chemistry during online learning situations. Researchers developed several tools including an augmented reality (AR) mobile app, a 360° immersive laboratory, virtual reality (VR) excursions, and a VR crime scene experience. Han and Fung argue that these tools promote collaborative learning, encourage critical thinking, and support students in self-directed learning by leveraging available technologies to build immersive learning experiences.

Han and Fung outline the development of these various tools which are open-educational resources that operate on smartphones and computers that most students own or can access. For instance, the NUS Department of Chemistry offers an elective module, titled Environmental Chemistry, in which students take a first-hand excursion to make observations of the environment. However, because the excursion is not always possible, the team designed a simulated excursion using virtual reality that allows for student interactivity and immersion in virtual worlds.

Laboratory work is very important in chemistry, but prior to doing laboratory work, students experience pre-laboratory experiences. Often, instructors use some form of video to introduce students to laboratory work. But traditional videos have limitations. For instance, traditional videos do not allow students to see all angles of the experimental set-up. To overcome the limitation of traditional video, Han and Fang make use of 360° video for pre-laboratory instruction. Students report, *via* survey, that the edtechs supported their learning. Although first-hand experiences in chemistry are preferred, Han and Fung demonstrate that students can have powerful experiences using interactive technologies.

In Chapter 18, *Designing Virtual Chemistry Visualizations Featuring Environmental Dilemmas to Promote Equitable Knowledge Integration*,[6] Donnelly-Hermosillo, Gerard and Linn argue how visualizations designed and developed based on principles from the knowledge integration (KI) framework can enhance student learning. The authors focus on examples of environmental dilemmas to illustrate how the KI framework can guide the design of learning technologies featuring visualizations. The KI framework has four main components: (1) eliciting students' ideas based on their cultural and instructional experiences; (2) using instruction to support students in analyzing scientific evidence so that students discover new ideas relevant to their lives; (3) helping students distinguish among ideas by comparing the evidence for the various ideas; and (4) guiding students to form an explanation based on evidence for the phenomenon and its role in local environmental dilemmas. These are powerful design principles that if designers would incorporate them into the design of simulations could improve learning and participation.

The focus on environmental issues is a key aspect of the chapter because it brings into focus issues of social justice. Donnelly-Hermosillo, Gerard and Linn[6] discuss how one high-school chemistry program decided to study chemical contaminants in their city's water after reading about the Flint, Michigan water crisis. The context allowed the class to have a meaningful purpose for learning about electrons in the outermost energy level of atoms, a topic that is very hard to contextualize meaningfully, but electrons in the outermost level are important to explain why metals and non-metals react the way they do. The authors argue that using environmental situations provides meaning for studying chemistry as they often bring in issues of social justice which many high-school learners find important.

What is particularly valuable about this chapter is the importance of using theoretical ideas to guide the development of visualization software that can motivate students to learn challenging ideas presented in the high school chemistry curriculum. Often the principles used to design visualizations are not discussed by developers, this chapter shows the benefit of using theoretical ideas, in this case the KI framework, for designing and developing instructional visualization in chemistry.

In Chapter 19, *Designing Tutorial Videos to Support Students' Learning of Reaction Mechanisms in Organic Chemistry*,[4] Bernholt and colleagues explore the potential of using tutorial videos with embedded instructional supports to aid students in understanding reaction mechanisms in organic chemistry. Their chapter focuses on tutorial videos as a special multimedia instruction format and illustrates evidence-based principles for selecting and designing video learning resources to promote understanding of challenging scientific ideas, *i.e.*, reaction mechanisms, in organic chemistry. These designers took advantage of the unique aspects of technology by incorporating highlighting techniques that guide students' attention to relevant parts of a representation, helping them follow various components of the reaction mechanism. These techniques are important instructional scaffolds that can guide student learning. The researchers argue that designers of instructional videos need to follow what is known about developing instructional videos. They highlight techniques, including signaling, as a form of scaffolding. In signaling, the designer purposefully highlights essential elements of a scientific idea or task. The researchers also provide evidence that this visualization technique can promote learning for students with low prior-knowledge of chemistry. This is a critical finding as they report that instructional videos in chemistry are used extensively. The finding is also critical as it points to strategies in how designers can enhance the learning of underrepresented students in science.

The authors argue that just including videos will not support learning for a wide range of learners; rather instructional videos need to be purposefully designed using evidence-based principles. Educators need to purposefully design videos with specific features that can promote learning (*i.e.*, signaling is one such technique). The use of these research-based principles increases the likelihood that students, including students with low prior knowledge, will learn from using instructional videos. For instance, often explanations presented in videos are not designed using evidence-based principles. Bernholt and colleagues[4] present principles to guide the development of videos that show chemical mechanisms including specifically stating the learning goals and pointing out contrast between reactions. Although the results presented in this chapter are encouraging, particularly with respect to use of signaling, more large-scale research across different contexts needs to occur.

Chapter 20, *Digital Tools for Equitable In-person and Remote Chemistry Learning* by Pullen and colleagues,[7] provides an analysis resource to guide practitioners and researchers for classifying and judging the value of digital tools or remote laboratories. They present six forms of learning: (1) acquisition; (2) inquiry; (3) practice; (4) production; (5) discussion; and (6) collaboration. The researchers also provide digital examples of each category. For instance, acquisition learning might include accessing various digital documents to find information; whereas collaboration includes learners accessing various online tools that support students working together on various aspects of projects.

In their chapter, Pullen and colleagues[7] judge several forms of chemistry teaching laboratories. For instance, remote-activated labs allow learners to engage in laboratory experiences, assisted by technology, by manipulating and completing an experiment in a real laboratory and in real time from a distance. To do such work, learners log into a site to conduct an experiment and receive real time audio-visual feedback on the progress of their experiment. While firsthand experiences are certainly preferred because of tactile and other sensory inputs that support learning such a smell are missing, the authors argue that when students cannot take part in firsthand experiences, such as during lockdowns during COVID-19 or if the laboratory work is too dangerous to be done in person, remote laboratory experience provides opportunities for students to learn and experience phenomena that they otherwise would not experience. As currently configured, Pullen and colleagues[7] judge remote activated labs to promote acquisition of theoretical ideas, support students in inquiry and allow students to practice laboratory skills, but remote activity labs in their current conditions, do not provide opportunities for discussion, production, or collaboration. Pullen and colleagues[7] also discuss the limitation of classical science laboratories not allowing students to take part in authentic inquiry activities. They argue that when the problem, methodology and outcome of a laboratory are all known, these features limit the learners in engaging in inquiry. It should be added that it also limits the joy of discovery that can happen. As such, they argue for the use of citizen science where learners collect and take part in authentic experiences.

Instructors, designers, and researchers interested in making judgments about how various digital tools and experience can support laboratory work will find the analysis resource system of value. Examples provided throughout the chapter show ways in which instructors can incorporate different ways of using technology tools from acquisition of information to constructing products to collaborating with others to make sense of data.

Milner-Bolotin and Milner in Chapter 21, *Smartphone Applications as a Catalyst for Active Learning in Chemistry: Investigating the Ideal Gas Law*,[8] discuss how interactive smartphone applications can engage students in chemistry learning in school and at home. Moreover, they argue that technologies, such as smartphone apps, need incorporation into STEM teacher education programs to support future teachers in realizing the possibilities of smartphones for learning and to help beginning teachers experience science learning using smartphones and other portable technologies.

Milner-Bolotin and Milner[8] describe how chemistry has many abstract science ideas that challenge students in learning and argue that technologies can support students in learning these concepts. They consider smartphones to have unlimited possibilities in supporting student learning as smartphones are "omnipotent gadgets already found in students' pockets" that can provide numerous authentic learning opportunities.

This chapter is valuable in that the authors make use of a technology that most high school and college students have readily available to them, smartphones, and make use of open-source interactive simulations, such as the PhET simulations to promote learning, as well as data collection and analysis software, Phyphox (**phy**sics **pho**ne e**x**periments) app that can be downloaded from the web at no cost. The PhET simulations offer powerful interactive simulations that run on smartphones that can promote student learning of challenging ideas. In this chapter, the authors describe an interactive simulation that allows learners to make a connection between macroscopic variables, such as pressure and temperature, and the effects that occur in the microscopic world of atoms and molecules. Supporting students in making these connections is critical to promoting learning and using apps that run on smartphones that students can use just about any place at any time provides greater learning opportunities.

The other advantage of this chapter is the innovative use of open educational resources for students to collect data on experiments they can conduct using smartphones. For instance, students can explore the effects of temperature on pressure by placing their smartphone that is running the Phyphox app in a sealed plastic bag and heating the plastic bag using a hair dryer. These types of experiments provide students with a source of data to support claims. The Phyphox app allows a much great number of students to have first-hand experiences with phenomena and the ability to collect data to support claims. Allowing students to collect and analyze data to support their claims has never been more important than it is today as many individuals do not grasp that science is based on evidence that is reproducible. Learners need to realize that the claims in science are supported by empirical evidence.

The most powerful point Milner-Bolotin and Milner[8] make is the importance of using technologies in science teacher education. While many future teachers can use computing for word processing, search the web and email, very few know how to use technologies, including smartphones, for teaching. They emphasize that we need to use technologies, like smartphones, in teacher preparation programs to support future teachers on how to use these technologies for teaching.

16.2 Concluding Comments

The chapters in this section offer innovative uses of visualization and other technology-based tools to support student learning of challenging scientific ideas in chemistry and the potential to explore chemical phenomena virtually. Several of the chapters in this section call for the use of what we know about design of visualizations and technology tools to promote learning. Donnelly-Hermosillo and colleagues show how the knowledge integration framework can be used to design interactive visualization including guiding students to construct explanations using evidence to explain local phenomena important to the lives of learners. Bernholt and colleagues describe other evidence-based ideas, including the use of scaffolds, such as signaling to guide student observations. Hans and Fung use the idea that students need to experience science phenomena to build understanding of ideas by building immersive and interactive experiences so learners can experience science when first-hand experiences are not possible.

The chapters in this section, while emphasizing that technology tools such as visualizations need to be designed to support learning, also stress the importance of embedding the technology tools in learning environments. Technology tools embedded within a learning environment that can involve students in finding solutions to meaningful questions, that support students in meeting specified learning goals, and that contain scaffolds to support students in accomplishing challenging tasks are essential if the potential of technology tools will realize their potential to promote learning of challenging scientific ideas.[9] Unfortunately, often the use of various technology tools is short-lived because the necessary support for secondary teachers and college professors to use them appropriately to promote learning and engagement is not present.

Technologies and visualization tools allow students to participate in a variety of science practices that they would not otherwise be able to do, like building and testing dynamic models, using interactive visualizations of unseen phenomena, experiencing science principles that students cannot do so firsthand, and analyzing and visualizing complex data sets. Learning technologies allow students to ask and explore "What if…?" questions and provide them opportunities to engage in these "What if" questions multiple times because there is not as much cost from both an expense and time perspective in redoing the investigation. Additionally, unlike exploring some potentially dangerous real-world phenomena, simulations and interactive visualizations are low risk with respect to safety. These various technology tools allow students to use tools and engage in similar activities as scientists.

Visualizations and interactive technology tools will change the role of teacher at the secondary level and that of instructors at the college levels as these new tools require that teachers and instructors make use of new teaching practices that promote students in various active learning strategies, but this does not diminish the role of the teacher or instructors in science classrooms. Science teachers and instructors play a critical role in promoting learning and in building classroom cultures that promote scientific disposition and attitudes. Although designers can embed scaffolds into the technology, teachers and instructors provide in the moment scaffolding for students as they engage in challenging practices of building models, analyzing, and making sense of data, observing interaction at the micro levels to see the effects at the macro level.

Technology-based tools, by themselves, will not support learning. To use technology-tools effectively curriculum materials, professional learning and administrative financial support and buy in are critical. We cannot expect secondary teachers or college instructors to know how to teach with technology-based tools. Accompany curriculum materials and sustained professional learning that show how to use technology tools for learning are critical. As argued in the Milner-Bolotin and Milner chapter,[8] more needs to be done in science teacher education programs to seamlessly integrate the use of technologies to support future teachers in how to use these powerful learning technologies to support students in developing deep understanding of important scientific ideas. Instructors at the college level will also need professional learning.

What evidence supports the use of visualization and interactive technologies for the teaching and learning of science? Some of the chapters in this section provide evidence of the promise of using technologies, but more careful research is needed. Research is needed to explore and refine further the design principles to develop interactive visualization and accompany teaching practices. Not only should teacher educators and designers develop new technology tools and visualization that embody the design principles described in the chapters in this section, but teachers and instructors also need to develop professional networks to share opportunities to enrich their teaching practices and strategies for using visualization and technology effectively in science teaching. Research that provides further insights into the design principles and process needed for effective long-term professional learning and teacher education also needs to occur.

Visualization tools and interactive technologies can support the development of challenging scientific ideas in chemistry at macro, micro, quantum, symbol, and process levels. Technologies and visualization tools need to be embedded in a much larger framework that includes curriculum materials, teaching practices, long-term professional learning, and administrative support. When technology tools become integrated with teaching systems, learning technologies will become necessary tools for providing a wide range of learners the opportunity to develop usable knowledge of chemistry and science.

References

1. Y. J. Dori, V. Dangur, S. Avargil and U. Peskin, *J. Chem. Educ.*, 2014, **91**, 1306–1317.
2. J. S. Krajcik and K. Mun, in *Handbook of Research on Science Education*, Routledge, 2014, vol. II, pp. 351–374.
3. National Research Council, *Education for Life and Work: Developing Transferable Knowledge and Skills in the 21st Century*, ed. J. W. Pellegrino and M. L. Hilton, The National Academies Press, Washington, DC, 2012.
4. See Chapter 19: S. Bernholt, J. Eckhard, M. Rodemer, A. Langner, G. Asmussen and N. Graulich, in *Digital Learning and Teaching in Chemistry*, ed. Y. J. Dori, C. Ngai and G. Szteinberg, Royal Society of Chemistry, United Kingdom, 2023.
5. See Chapter 17: J. Y. Han and F. M. Fung, in *Digital Learning and Teaching in Chemistry*, ed. Y. J. Dori, C. Ngai and G. Szteinberg, Royal Society of Chemistry, United Kingdom, 2023.
6. See Chapter 18: D. D. Hermosillo, L. Gerard and M. C. Linn, in *Digital Learning and Teaching in Chemistry*, ed. Y. J. Dori, C. Ngai and G. Szteinberg, Royal Society of Chemistry, United Kingdom, 2023.
7. See Chapter 20: R. Pullen, A. Motion, S. Schmid, S. George-Williams, S. Wilkinson and S. Leach, in *Digital Learning and Teaching in Chemistry*, ed. Y. J. Dori, C. Ngai and G. Szteinberg, Royal Society of Chemistry, United Kingdom, 2023.
8. See Chapter 21: M. Milner-Bolotin and V. Milner, in *Digital Learning and Teaching in Chemistry*, ed. Y. J. Dori, C. Ngai and G. Szteinberg, Royal Society of Chemistry, United Kingdom, 2023.
9. J. D. Bransford, A. A. L. Brown and R. R. Cocking, *How People Learn Brain, Mind, Experience, and School Committee*, 2004, vol. 116.

CHAPTER 17

Applications of Digital Technology in Chemical Education

JIA YI HAN[a] AND FUN MAN FUNG*[a,b]

[a]Department of Chemistry, National University of Singapore, Singapore;
[b]Institute for Applied Learning Sciences and Educational Technology (ALSET), National University of Singapore, Singapore
*E-mail: fun.man@u.nus.edu

17.1 Introduction

Since the days of the SARS outbreak in 2003, flagship universities such as the National University of Singapore (NUS) have leveraged technology, such as an online learning management system, to ensure that learners continue to have access to high quality education from home.[1,2] The outbreak of COVID-19 in 2020, however, caused an unprecedented change in the way we teach by intensifying the adoption of technology in education. Classes had to abruptly move online in response to government mandated restrictions to contain the spread of the virus. As DeKorver and Herrington (*in publication*) noted, educators—particularly chemistry educators—find themselves navigating in unchartered waters in teaching classes online.[3] Tam and Fisher support this claim as they observed that institutes such as medical schools were reluctant to modify the learning materials for online teaching due to logistical limitations and fear that any modifications made would adversely affect the

Advances in Chemistry Education Series No. 11
Digital Learning and Teaching in Chemistry
Edited by Yehudit Judy Dori, Courtney Ngai and Gabriela Szteinberg

Published by the Royal Society of Chemistry, www.rsc.org

curriculum.[4] At NUS Chemistry, we created pedagogical methods to allow our students to foster collaborative learning and engagement. At the same time, we endeavoured to equip students with essential skills such as spatial visualisation, critical thinking, and laboratory techniques to prepare them for the working world. While there were tools available that target those skills,[5] most of them were less adaptable to our courses after considering the nature of our syllabi, assessment modes, and logistics. Thus, we developed our EdTechs organically in the hopes of altering the learning experience.

17.2 Chemist-curated Educational Technologies

The EdTechs that we curated make use of virtual reality (VR) technology or augmented reality (AR) technology to engage users in learning. VR technology allows users to be fully immersed with a human-made virtual space, where their perception is isolated from the physical world. Typically, an in-built VR headset or a pair of VR goggles is required to achieve immersivity. By being fully immersed with the digital environment, users feel as though they are part of the environment, and they can interact with the elements. In contrast, AR technologies do not isolate users from the physical world but adds to it. The use of wearables is not a compulsory requirement since most AR apps are compatible with smartphones.

The EdTechs discussed here are all freely available to students and the public and do not require specialised hardware. A smartphone or a computer, which most Singaporean students possess, is the minimum requirement to access these EdTech tools, and therefore available for most learners. For EdTechs using VR technology, a pair of VR goggles affixed to the smartphone can be used instead of the more costly VR headsets. These VR goggles cost around five US dollars, reflecting the equity in accessibility that was thoughtfully considered when we developed the EdTechs.

17.2.1 The AR Mobile App—NuPOV (Android, iOS)

17.2.1.1 Spatial Visualisation Ability in Learning Organic Chemistry

Organic chemistry is one of the complex branches of chemistry that relies heavily on spatial-visualisation skills to understand concepts such as reaction mechanisms. At NUS Chemistry, freshmen chemistry students enroll in the course Organic Chemistry I (code: CM1121) as a partial fulfilment of the Bachelor of Science degree.[6] The Learning Outcomes (LO) for CM1121 are shown in Table 17.1.

As members of the teaching team, we observed that our students often had difficulties translating two-dimensional (2D) molecular structures and reactions to three-dimensional (3D) space.

To help students in addressing this difficulty, educators frequently use the physical molecular model kits during lesson to demonstrate organic

Table 17.1 Learning outcomes for the freshmen organic chemistry course at NUS.

S/N	Learning outcomes (LO) for CM1121
1	Apply and evaluate simple organic reaction transformations, functional group interconversion and C–C bond formation reactions
2	Describe the mechanisms of organic reactions
3	Distinguish between *R*/*S* and *E*/*Z* configurations and differentiate the taxonomy in stereochemistry
4	Describe and account for the chemical reactivity of reaction

chemistry concepts to students. The use of these tangible kits, however, poses problems as well. If the lecture is conducted in a large lecture theatre, students may not be able to clearly see the molecular models that the educator is holding. Some educators approach this problem by passing the molecular models around the class after explaining the concepts, while carrying on with the lesson. In this case, students will have to focus on two things at once—maneuvering the molecular model and paying attention to the lecture.

17.2.1.2 Equity in Accessible Digital Technology to Support Spatial Visualisation Skills Training

Fortunately, smartphones are ubiquitous in Singapore. Therefore, we elected to develop an AR chemical application compatible with these devices so that students can visualise and interact with molecules in 3D with their own personal devices, thus training their spatial-visualisation skills. While there are many applications on the market that possess such features such as "Element 3D" by Yang, Mei, and Yue,[7] and "AR assisted visualisation" by Eriksen, Nielsen, and Pittelkow,[8] none of these applications specifically target organic chemistry reaction mechanisms. Furthermore, many of these apps lack user-interactivity, which is crucial in allowing students to understand and appreciate how molecules undergo changes during a reaction.[9] Thus, we partnered with the software engineers and educational technologists from NUS Information Technology to create a mobile app for our students.

The general premise of the app is to enable our freshmen organic chemistry students to visualise using AR how molecules undergo structural changes to give the product after a successful chemical reaction. For the choice of reaction, we first selected nucleophilic addition as this is one of the fundamental reactions in organic chemistry.[10] Thus, it is essential for our freshmen students to have a good understanding of this concept so that they can have the confidence to study more complex reaction mechanisms. To address the user interactivity of our app, we implemented a gaming element where users aim a green cannon on-screen containing nucleophiles and shoots them towards an AR generated electrophile. A reaction is determined successful when the nucleophile approaches the electrophile at the correct angle of attack from the right direction, and the user will be able to observe how the molecules change structurally to form the product molecule.[11,12] As this gaming element and observation of the molecule are taken from the perspective of the nucleophile, we named our AR app Nucleophile's Point of View (NuPOV).

NuPOV is accessible and affordable to users due to its availability on mobile app stores free of charge (Android, iOS). Users without background knowledge of Nucleophilic Addition reaction, such as non-majors, can still reap the benefits of using NuPOV by interacting and observing how molecules undergo changes during a successful reaction, thereby enhancing their spatial-visualisation skills. For users with some background knowledge of the reaction, they can enhance their understanding of the concept further by virtue of Mayer's Principle of Multimedia Learning[13] through interacting with the visual elements provided by NuPOV.

17.2.1.3 *Applying NuPOV in Our Class*

After developing NuPOV, we conducted a pre-intervention questionnaire to ascertain students' confidence in their perceived understanding of organic chemistry concepts. We then introduced the app to CM1121 students during the lesson. The NuPOV team guided students through the installation process, followed by a guide on how to use the app. Henceforth, students were given a few weeks to use the app in their own time. After which we conducted a post-intervention questionnaire to determine if there are any changes in students' perception of their understanding in organic chemistry. Overall, 65% of the 87 students surveyed indicated an increase in perceived confidence in solving harder questions related to nucleophilic addition. In addition, there were more students who responded that they felt more confident teaching and learning the concept through informal peer interactions after using NuPOV compared to before using it. A limitation in our study is that as students have other learning opportunities, it is difficult to establish that NuPOV is what actually enables students to have greater confidence in solving harder questions related to nucleophilic addition and not to other variables.

17.2.1.4 *Options for NuPOV*

NuPOV was curated to fulfil three out of the four LO for CM1121, namely LO1, LO2 and LO4. While these LO has yet to be assessed solely through the use of NuPOV, the app offers an opportunity for students to work towards these LO. Moreover, NuPOV may also support students in understanding the conditions required for nucleophilic addition to take place, thus students will be able to account for the chemical reactivity for the reaction to take place. NuPOV offers them the opportunity to distinguish the spatial arrangement of molecules, enabling the students to differentiate between the various stereoisomers of the molecule. Finally, the use of NuPOV can take place after attending the class on Nucleophilic addition to reinforce concepts learned afterwards. More details related to NuPOV can be found in the article by Aw and colleagues.[11] The project has also been featured in Higher Education Campus Conference hosted by NUS,[14] as a book chapter,[15] and on the Media & Learning website.[16]

17.2.2 A 360° Immersive Laboratory

Laboratory work is an essential part of chemistry as it helps to validate the theoretical frameworks put forth by scientists to explain natural phenomena. As such, laboratory work is one of the core components in many universities' undergraduate chemistry courses.[17] NUS Chemistry offers two compulsory laboratory courses for our third-year undergraduate chemistry majors—Laboratory Course 1, and Laboratory Course 2. The LOs for the two courses are shown in Table 17.2.

Typically at NUS, prior to each laboratory sessions (pre-lab), the instructor would teach the theoretical and safety aspects of the experiment to the students, followed by a general overview of the laboratory techniques used

Table 17.2 Learning outcomes for the third year laboratory courses.

Third year laboratory course	S/N	Learning outcomes (LO)
Laboratory course 1	1	Identify and apply safe laboratory practice by conducting comprehensive risk assessment of the experiments
	2	Conduct multi-step synthesis of organic and inorganic materials independently
	3	Isolate and purify organic and inorganic compounds and coordination complexes
	4	Use infrared and ^{1}H NMR spectra as diagnostic tools in the identification of synthesized compounds and assessment of their purities
	5	Solve problems in the practical act of chemical synthesis and connect these to their prior chemistry knowledge
Laboratory course 2	1	Bridge between academic and future professional development in their Final Year Project (FYP), Final Year Internship (FYI), and future career by providing an authentic (real-world) learning experience
	2	Apply and establish connections between theory and practice through hands-on practical experiences and making sense of experimental results
	3	Develop not only knowledge and technical skills but also analytical and communication skills in the field of Analytical and Physical Chemistry

in the experiment.[18] This could involve using multimedia such as video recordings[19,20] filmed using video recorder or smartphones to demonstrate the experimental setup and techniques, as well as key visual elements to look out for whilst conducting the experiment. The use of pre-recorded laboratory videos is especially salient during the pandemic as movement restrictions were placed in many parts of the world. More educators adopted this method to demonstrate laboratory techniques to students as a pre-laboratory preparation to supplementary tasks in place of face-to-face laboratories.[21]

17.2.2.1 Weakness of Traditional Laboratory Videos

However, there are several disadvantages of using a conventional video recorder to record the laboratory: (1) it is often difficult to find an optimal angle to film the experimental setup, especially when the setup is inside the fume hood; (2) the sash of the fume hood often diminishes the setup visibility; and (3) when filming a close-up shot, the videographer would need to bring the camera closer to the set up and adjust the camera's zoom settings to show the setup's point of interest. These maneuvers increase the shakiness and reduce the quality of the resulting footage.

An important implication stemming from these disadvantages is the students' frustration at not being able to make out the details of the setup due to the shakiness and poor quality of the video. Some of the students even commented that the excessive shakiness of the video induced motion sickness.

17.2.2.2 Producing 360° Videos

One way to overcome these challenges is to film 360° immersive videos for pre-lab instructional teaching. As part of the teaching team for the two laboratory courses at NUS, we applied this approach of filming for our pre-lab videos. Two types of filming equipment were used in our filming approach, utilising a Ricoh Theta S 360° camera and a tripod stand. As the 360° camera and the tripod stand were compact, they were able to fit inside a fume hood and be placed near the experimental setup during the filming without interfering with the setup. As the video recorder has a 360° panoramic view of the surroundings, we were able to concentrate solely on presenting the experimental techniques while recording the video.

During post-production, we used Camtasia® as our video editing software to edit our videos as we were familiar with it. We inserted texts and pictures to supplement the instructor's voice captured by the video recorder to catch viewers' attention and to make sure that they understood the key concepts required to perform the experiment. After editing was done, the video was uploaded to video sharing sites that supports 360° videos such as Facebook and YouTube for our students to watch.

When viewers watch the 360° video on their computer, they can navigate the 360° view by dragging the video using their mouse cursor, or their fingers if their computer supports a touchscreen. For those watching on mobile devices, they can either use their fingers or make use of their devices' built-in gyroscope to navigate the video. For added immersivity, viewers can pair their devices with a VR headset or a pair of VR goggles to provide the perception that they are viewing the experimental demonstration in person.

17.2.2.3 Opportunities for 360° Immersive Laboratory Videos

A decade ago, before we used pre-lab videos, we observed that students expressed some form of discomfort when they stepped into the laboratory for the first time. These observed feelings are understandable due to unfamiliarity of the premises and equipment with which students have no previous encounters. We believe that the application of 360° immersive laboratory videos has the ability to assuage students' anxiety before they enter the laboratory space. For Laboratory Course 1, apart from those LO that could only be achieved when students had their hands-on experience in conducting the experiment in the laboratory, our 360° videos demonstrated the various safety considerations in the experiment procedures that viewers should take note of through the supplementary texts and images in the video. Furthermore, the use of the supplementary texts and images enables viewers to note the theory behind the experiments, allowing them to connect the theoretical knowledge with the experiment, thus empowering them to solve problems that may arise during the experiment.

For Laboratory Course 2, in addition to allowing viewers to connect theoretical concepts with the practical aspects, the 360° videos potentially enhance the learning by looking at the blind spots that are not captured in traditional instructional videos. By viewing the educator's facial expression, it ostensibly provides more realistic learning experience. Thus, we opine that our students will have a better understanding of the technical aspects of the analytical instruments used in the lab, and better communicate these concepts during their in-person viva voce as part of their course assessments. For more details regarding 360° immersive video, readers can refer to the journal article published in *Journal of Chemical Education* by Ardisara and Fung.[22]

17.2.3 VR Excursion for Environmental Chemistry

While laboratory work is important in experimental chemistry, some of the more interdisciplinary sub-domains such as environmental chemistry also require one to embark on excursions for field work. Such excursions enable students to actively engage the environment through their senses, allowing them to connect ideas and concepts together and thereby promoting knowledge retention.[23]

NUS Chemistry offers an elective course titled Environmental Chemistry (code: CM3261) for third- and fourth-year chemistry students. This course covers how human activities impact the environment and *vice versa* from an interdisciplinary perspective. Although we understood the importance of having an excursion for our students to engage their senses, there were several roadblocks that prevented us from doing so. Our students' timetables at NUS are often complex to align, making it even more complex due to the teaching team's own work commitments in the faculty. As a result, we had to forgo the idea of having an excursion in favour of a traditional lecture. Students' feedback in the end of semester course revealed that they understood the contents taught but failed to see how these contents could be applied in society. Afterall, some of the problems caused by human activities on the environment may not be very obvious, especially given that the highly urbanised Singapore has strict environmental regulations to protect the wellbeing of their citizens and the environment. Thus, the effects of activities such as pesticide use or heavy metal poisoning from drinking water highlighted in the course, in our opinion, do not appear to be highly relevant to most of our students.

We considered using a camera-equipped drone to provide both a bird's eye view and a close up shot of the excursion site.[24] However, this method was not logistically feasible due to the authorities requiring permits for owning and operating a drone, and also due to many restricted areas in Singapore such as private properties and military installations where operating a drone is not allowed. Another idea similar to the one mentioned in the previous section was to use a 360° camera and film the excursion sites while the educator describes what he experienced to the viewers. Even though this method has an element of immersivity, it has its own limitation that is shared with the drone method in that viewers would not be able to interact with the elements found in the video, making both methods a glorified documentary show. As interactivity and immersivity are the two key factors that we feel are important in an excursion, we decided to conduct a simulated excursion through using VR.[25]

VR possesses not only the 360° panoramic view of the surrounding just like the 360° camera mentioned before, but it also provides an added advantage of having an on-demand interaction with the VR elements by viewers. These features increase immersivity, allowing the viewers to have a more realistic first-person experience with the environment. The development of a VR excursion is simple as minimum prior technological knowledge is required. This is made possible with many of the VR creation tools being straightforward to use.[26] Logistically, a VR excursion is as cost-effective as the 360° images, also known as photospheres, and can be reused again for future students.

As part of the CM3261 teaching team, the Singapore team partnered with educators from France in curating the VR excursion. As Singapore is mostly urban, we thought that showing a rural area for the VR excursion would provide an eye-opening experience for our students. Finally, we decided to have the VR excursion setting to be in Ungaran, Indonesia. The teaching team then travelled to the location and took 360° images of the excursion site using the same camera that was used in our 360° video mentioned before. The team then uploaded the images to edu2VR, a cloud platform for storing 360° VR images. Before showing the images to our students we added annotations to them and created teleportation features in the VR.

The VR excursion was introduced to our 74 CM3261 students over a span of two lecture sessions lasting 1.5 h each in the lecture theatre. During the first session, we got our students to become familiarised with the concept of VR excursion by showing them how to access the VR images using the mobile application called Binoo. We distributed VR goggles to our students which can be equipped to their mobile devices for added immersivity. Following which, we gave students a brief walkthrough on the features that were in place in the VR excursion before letting them explore the virtual excursions on their own. During the second session, the teaching team then guided our students in the VR excursion to the specific points of interests and encourage them to interact with the VR elements before letting the students to explore the excursion site further on their own. In both sessions, our students felt excited as they interacted with the VR elements. While it is positive that students expressed enthusiasm as they interacted with VR, this excitement might be temporary due to the Hawthorn effect.[27]

We conducted an anonymous questionnaire after the VR excursion session and found that 64% of our students agreed that the VR excursion experience was engaging. Although some students reported feeling mild dizziness when using the VR, students overall praised the VR excursion to be novel and that they did not think they would have an overseas excursion with their classmates right in the lecture theatre. Students also noted that the VR excursion enabled them to further consolidate their understanding of theoretical concepts and appreciate the far-reaching effects human activities have on the environment. More details about our VR excursion can be found in the article published by Fung and colleague.[25] This project is also published as a book chapter,[28] and was presented in the EDUCAUSE 2019 conference in Chicago, Illinois.[29]

17.2.4 VR Crime Scene

The application of VR is not limited to simulating outdoor teaching environment. Dunnagan *et al.* curated a VR organic chemistry laboratory for teaching infrared spectroscopy,[30] while Maksimenko *et al.* created simulated experience where users can interact with atomic and subatomic particles.[31] In our case, besides simulating an overseas excursion, we applied the VR technology to forensic science too. The forensic science course is for third year students minoring in forensic science and is jointly coordinated by educators from the Department of Chemistry and the Department of Biological Sciences. One of the important objectives of this course is to teach students how to analyse and collect evidence from a crime scene. This is achieved by having educators create a staged physical crime scene and having the students carefully collecting the evidence. Afterwards, students use the evidence to perform a crime scene reconstruction to propose the suspect and how the crime was committed. Through this practice, students develop their observational skills, and at the same time, enhance their critical and analytical thinking.

In response to the physical distancing requirements, an in-person staged crime scene was unfeasible. This lack of exposure to physical crime scenes posed a problem of losing the opportunity to practice the techniques that was being taught in class, and also missing the opportunity to enhance their soft skills such as critical and analytical thinking, and their observation skills which are essential to prepare them for graduate outcomes. Thus, armed with the previous experience in conducting virtual overseas excursions using VR, we collaborated with our counterparts from the Department of Biological Sciences to create a crime scene using VR.[32]

Using VR to simulate a crime scene is not a new idea. Educators have turned to VR to simulate crime scenes for their students to apply their theoretical knowledge into practice. Mayne and Green had developed their own VR crime scene using the Unity® Engine and open source 3D models to create the VR elements, and students visualised and interacted with those elements using Oculus VR headsets.[33] Wang *et al.* went a step further by creating a crime scene using a laser scanner to scan real world objects (such as physical rooms and household objects) before converting the images to VR elements using a powerful computer. The resultant elements are then visualised using a VR headset.[34] However, there are some limitations in the above-mentioned VR attempts. For Mayne's and Green's VR crime scene, the VR elements that were created do not look realistic enough as the models were built using Blender, a 3D modelling software for animation. Furthermore, the inclusion of a birds eye view mode does not present a real-life experience for students as crime scene investigators usually do not view crime scenes from this perspective. Although Wang and colleagues' VR elements look realistic, the equipment used are expensive and the process in creating the VR experience has a steep learning curve. These limitations can discourage people from using VR to simulate crime scene.

Our VR crime scene aims to be more affordable and accessible for educators and students by using inexpensive equipment. The tools we used in creating our VR crime scene are user friendly, allowing educators and students alike to create and use the VR without extensive programming knowledge.

Prior to generating the VR elements, we set up a simulated crime scene at one of the teaching team's house. We planted evidence such as handprints into the crime scene, and then used the Ricoh Theta S 360° camera to take panoramic photos of the scene. Finally, we uploaded the photos onto Uptale, a web-based VR software, for processing and creating the VR experience.

The completed crime scene can be accessed remotely using the web browser on the computer or on mobile. During the online activity, we directed our 37 students to visit Uptale's website to access the crime scene. Students were asked to complete a report by listing out the collected evidence, the method used to analyse the crime scene, and the conclusions they arrived at based on their analyzes.

The questionnaires results showed positive reception among our students: 79% of the 37 students enjoyed their VR crime scene experience, with 84% of them stating they would recommend their peers to try out the VR experience. More information can be found in the article by Kader and colleagues.[32]

17.3 Digitization of Chemical Education

In this chapter, we have illustrated that the digitization of chemical education can be shaped in various dimensions. From our empirical experiences, we have documented how we worked with technologists to implement four EdTechs in our courses ranging from laboratory and classroom setting across different sub-disciplines. The EdTechs curated may support the inclusivity and equity for our students as cost barriers were removed. With AR technology, students can visualise better using their smartphones without worrying about to pay for molecular models. VR excursion overcomes the limitation of financial and scheduling barriers so that students can still partake in the excursions in classroom with their friends. Through the VR crime scene, the pandemic-induced constraints were overcome, providing a possibility for more students to enroll in the course, and granting greater accessibility for students who are interested in forensic sciences.

Despite students' positive feedbacks regarding these EdTechs, longitudinal studies along with statistical analysis are necessary to ensure that these EdTechs substantially improve students' learning and understanding. Going forward, the digitization of chemical education provides opportunities for distance education courses for non-majors and offerings as on-demand resources for students to choose how to engage with the materials. Right now, students have these EdTechs at their disposals to use as pre-lab and pre-lecture materials, or post-lab and post-lecture as review materials. Further developed VR laboratory can be used to make-up for missed laboratory sessions—promoting equity and inclusivity.

Lastly, the tools required to curate each EdTech are summarised in Table 17.3. We hope that readers will be inspired by EdTechs and experiment based

Table 17.3 Summary of the EdTechs curated and the tools required.

	EdTechs			
What tools are needed to curate and to use?	NuPOV AR mobile app	360° immersive laboratory	VR excursion	VR crime scene
Mobile phone (Hardware)	Required	Optional	Required	Optional
360° camera (Hardware)	No need	Required	Required	Required
VR headset/goggles (Hardware)	No need	Optional	Required	Optional
Online platform for embedding 360° media (Software)	No need	Required	Required	Required
Is it an immersive EdTech?	No	Yes	Yes	Yes

on their needs. Our common lived experience during the COVID-19 pandemic has radically changed the educational landscape by causing various restrictions such as limiting travelling and face-to-face teaching. This dramatic disruption brings forth an opportunity for us to innovate and do something different. And even if the instructional modality shifts back to either full face-to-face or hybrid instruction, these innovations should and could continue to have a positive impact on learning and assessment processes, especially when they occurred online (see also Chapter 24 by Herscovitz *et al.*).[35] As we have explained, adopting digital technology is one possible way to solve teaching problems we face. Therefore, we believe that digital technology can engender equity and inclusivity so that students can continue to receive quality education they require to prepare them for the working world.

Acknowledgements

The authors would like to thank all co-workers and collaborators for the opportunities to work on the aforementioned EdTech projects. We are grateful for the support from our colleagues, students, and the department. We thank the financial support from the following grants: USPC-NUS grant for innovation in higher education (2017-03-T/USPC-NUS), (2018-02-EDU/USPC-NUS), (2019-01-EDU/USPC-NUS).

References

1. N. Cavus, Y. B. Mohammed and M. N. Yakubu, Determinants of Learning Management Systems during COVID-19 Pandemic for Sustainable Education, *Sustainability*, 2021, **13**(9), 5189.
2. M. L. Wong, L. G. Goh, D. Ng, C. N. Ong, K. H. Phua and P. Tambyah, *et al.*, Online Health Education on SARS to University Students During the SARS Outbreak, *Int. Electron. J. Health Educ.*, 2005, **8**, 205–217.
3. See Chapter 31: B. DeKorver and D. Herrington, in *Digital Learning and Teaching in Chemistry*, ed. Y. J. Dori, C. Ngai and G. Szteinberg, Royal Society of Chemistry, United Kingdom, 2023.
4. W. Tam and D. Fisher, COVID-19: a missed opportunity for medical education, Singapore Med. J. [Internet]. 3 October 2021 [cited 3 March 2022], available from: http://www.smj.org.sg/sites/default/files/CO-2021-074-epub.pdf.
5. W. K. Chiu, Pedagogy of Emerging Technologies in Chemical Education during the Era of Digitalization and Artificial Intelligence: A Systematic Review, *Educ. Sci.*, 2021, **11**(11), 709.
6. F. M. Fung and Y. Lam, How COVID-19 Disrupted Our "Flipped" Freshman Organic Chemistry Course: Insights Gained from Singapore, *J. Chem. Educ.*, 2020, **97**(9), 2573–2580.
7. S. Yang, B. Mei and X. Yue, Mobile Augmented Reality Assisted Chemical Education: Insights from Elements 4D, *J. Chem. Educ.*, 2018, **95**(6), 1060–1062.

8. K. Eriksen, B. E. Nielsen and M. Pittelkow, Visualizing 3D Molecular Structures Using an Augmented Reality App, *J. Chem. Educ.*, 2020, **97**(5), 1487–1490.
9. R. Tasker, ConfChem Conference on Interactive Visualizations for Chemistry Teaching and Learning: Research into Practice—Visualizing the Molecular World for a Deep Understanding of Chemistry, *J. Chem. Educ.*, 2016, **93**(6), 1152–1153.
10. B. J. Esselman and N. J. Hill, in *Integrating Computational Chemistry into an Organic Chemistry Laboratory Curriculum Using WebMO*, ed. Grushow A. and Reeves M. S., American Chemical Society, Washington, DC, ACS Symposium Series, 2019 [cited 13 June 2021], pp. 139–62, available from https://pubs.acs.org/doi/abs/10.1021/bk-2019-1312.ch011.
11. J. K. Aw, K. C. Boellaard, T. K. Tan, J. Yap, Y. P. Loh and B. Colasson, *et al.*, Interacting with Three-Dimensional Molecular Structures Using an Augmented Reality Mobile App, *J. Chem. Educ.*, 2020, **97**(10), 3877–3881.
12. F. M. Fung and C. Zimmermann, *Technology-enabled Blended Learning Experiences for Chemistry Education and Outreach*, Elsevier, 1st edn, 2021, p. 212.
13. R. E. Mayer, Multimedia learning: Are we asking the right questions? *Educ. Psychol.*, 1997, **32**(1), 1–19.
14. F. M. Fung, Y. Lam, J. Yap, T. K. Tan, B. Colasson and É. Blanc, in *Interacting with 3D Molecular Structures Using NuPOV to Teach Organic Chemistry*, Singapore, 2020, available from: https://nus.edu.sg/cdtl/docs/default-source/engagement-docs/conferences/e-hecc2020/ehecc2020-booklet.pdf.
15. J. Y. Han, Y. John, T. Teck Kiang, L. Yulin and F. Fun Man, Applying NuPOV to support students' three-dimensional visualization skills, in *Technology-Enabled Blended Learning Experiences for Chemistry Education and Outreach*, Elsevier, 1st edn, 2021. pp. 151–61.
16. F. M. Fung, *How NuPOV Revolutionises Organic Chemistry Education with the Help of AR [Internet]*, Media & Learning, 2020 [cited 27 May 2022], Available from: https://media-and-learning.eu/type/featured-articles/how-nupov-revolutionises-organic-chemistry-education-with-the-help-of-ar/.
17. S. L. Bretz, Evidence for the Importance of Laboratory Courses, *J. Chem. Educ.*, 2019, **96**(2), 193–195.
18. F. M. Fung and S. F. Watts, Bridges to the Future: Toward Future Ready Graduates in Chemistry Laboratories, *J. Chem. Educ.*, 2019, **96**(8), 1620–1629.
19. M. E. Anzovino, V. A. Mallia, M. S. Morton, J. E. Barker Paredes, R. Pennington and D. P. Pursell, et al., Insights and Initiatives While Teaching Organic Chemistry I and II with Laboratory Courses in the Time of COVID-19, *J. Chem. Educ.*, 2020, **97**(9), 3240–3245.
20. L. S. Nadelson, J. Scaggs, C. Sheffield and O. M. McDougal, Integration of Video-Based Demonstrations to Prepare Students for the Organic Chemistry Laboratory, *J. Sci. Educ. Technol.*, 2015, **24**(4), 476–483.

21. J. D. Aguirre and F. Selampinar, Teaching Chemistry in the Time of COVID-19: Memories and the Classroom, *J. Chem. Educ.*, 2020, **97**(9), 2909–2912.
22. A. Ardisara and F. M. Fung, Integrating 360° Videos in an Undergraduate Chemistry Laboratory Course, *J. Chem. Educ.*, 2018, **95**(10), 1881–1884.
23. A. A. Mackenzie and R. T. White, Fieldwork in Geography and Long-term Memory Structures, *Am. Educ. Res. J.*, 1982, **19**(4), 623–632.
24. F. M. Fung and S. F. Watts, Bird's-Eye View of Sampling Sites: Using Unmanned Aerial Vehicles To Make Chemistry Fieldwork Videos, *J. Chem. Educ.*, 2017, **94**(10), 1557–1561.
25. F. M. Fung, W. Y. Choo, A. Ardisara, C. D. Zimmermann, S. Watts and T. Koscielniak, *et al.*, Applying a Virtual Reality Platform in Environmental Chemistry Education To Conduct a Field Trip to an Overseas Site, *J. Chem. Educ.*, 2019, **96**(2), 382–386.
26. M. d. P. Carretero, S. García, A. Moreno, N. Alcain and I. Elorza, Methodology to create virtual reality assisted training courses within the Industry 4.0 vision, *Multimedia Tools Appl.*, 2021, **80**(19), 29699–29717.
27. S. A. Seixas-Mikelus, A. Adal, T. Kesavadas, A. Baheti, G. Srimathveeravalli and A. Hussain, *et al.*, Can Image-Based Virtual Reality Help Teach Anatomy? *J. Endourol.*, 2010, **24**(4), 629–634.
28. T. Koscielniak. A review of immersive learning technologies featured at EDUCAUSE annual conferences: Evolution since 2016, in *Technology-Enabled Blended Learning Experiences for Chemistry Education and Outreach*, Elsevier, 1st edn, 2021, pp. 163–84.
29. Curating a Bilateral Immersive Learning Experience: Our France-Singapore Story [Internet], EDUCAUSE, [cited 28 May 2022], available from: https://events.educause.edu/annual-conference/2019/agenda/curating-a-bilateral-immersive-learning-experience-our-francesingapore-story.
30. C. L. Dunnagan, D. A. Dannenberg, M. P. Cuales, A. D. Earnest, R. M. Gurnsey and M. T. Gallardo-Williams, Production and Evaluation of a Realistic Immersive Virtual Reality Organic Chemistry Laboratory Experience: Infrared Spectroscopy, *J. Chem. Educ.*, 2020, **97**(1), 258–262.
31. N. Maksimenko, A. Okolzina, A. Vlasova, C. Tracey and M. Kurushkin, Introducing Atomic Structure to First-Year Undergraduate Chemistry Students with an Immersive Virtual Reality Experience, *J. Chem. Educ.*, 2021, **98**(6), 2104–2108.
32. S. N. Kader, W. B. Ng, S. W. L. Tan and F. M. Fung, Building an Interactive Immersive Virtual Reality Crime Scene for Future Chemists to Learn Forensic Science Chemistry, *J. Chem. Educ.*, 2020, **97**(9), 2651–2656.
33. R. Mayne and H. Green, Virtual reality for teaching and learning in crime scene investigation, *Sci. Justice*, 2020, **60**(5), 466–472.
34. J. Wang, Z. Li, W. Hu, Y. Shao, L. Wang and R. Wu, *et al.*, Virtual reality and integrated crime scene scanning for immersive and heterogeneous crime scene reconstruction, *Forensic Sci. Int.*, 2019, **303**, 109943.
35. See Chapter 25: O. Herscovitz, M. Varsano, O. Faris and Y. J. Dori, in *Digital Learning and Teaching in Chemistry*, ed. Y. J. Dori, C. Ngai and G. Szteinberg, Royal Society of Chemistry, United Kingdom, 2023.

CHAPTER 18

Designing Virtual Chemistry Visualizations Featuring Environmental Dilemmas to Promote Equitable Knowledge Integration

DERMOT DONNELLY-HERMOSILLO*[a], LIBBY GERARD[b] AND MARCIA C. LINN[b]

[a]Department of Chemistry and Biochemistry, California State University, Fresno, USA; [b]Graduate School of Education, University of California, Berkeley, USA
*E-mail: ddonnelly@csufresno.edu

18.1 Introduction

Complex ideas in chemistry can be successfully captured in visualizations, simulations, and virtual laboratories.[1–3] Chemistry research using online learning environments has identified effective instructional features using design and comparison studies.[4–6] The constructivist Knowledge Integration (KI) framework, developed over 25 years, illustrates the value of designing instruction where students use visualizations to discover ideas and to distinguish among alternative ideas.[7] This chapter considers the potential of visualizations to not only support conceptual learning but also welcome

Advances in Chemistry Education Series No. 11
Digital Learning and Teaching in Chemistry
Edited by Yehudit Judy Dori, Courtney Ngai and Gabriela Szteinberg

Published by the Royal Society of Chemistry, www.rsc.org

each student in chemistry. We highlight chemistry activities that tackle local environmental dilemmas. They can illustrate issues that apply to the lives of students such as the production of pollutants. Furthermore, environmental dilemmas raise issues of social justice such as the distribution of pollutants across neighborhoods. We illustrate the potential of visualizations by discussing studies that integrate environmental and social justice issues to deepen secondary school students' interest in and learning of chemistry.

Although access to virtual visualizations remains an issue, most schools now provide 1 : 1 networked computers and students report having an accessible computer at home, particularly after the remote instruction demands during the COVID-19 pandemic.[8] Hence, the potential of technology, and in turn, visualizations to welcome each student in chemistry continues to expand.

18.2 Designing Visualizations to Welcome Each Student in Chemistry

To welcome each student in chemistry, KI instruction builds on the ideas students bring to class.[7] Research shows that students develop multiple, often incomplete, or contradictory ideas about scientific phenomena.[9] Informed by constructivist learning theories including KI, activity structures that build on these ideas result in robust understanding.[5,7] The KI process starts by *eliciting the ideas* students have based on their cultural and instructional experiences.[7] Next, instruction supports students to *discover new ideas* by analyzing scientific visualizations and evidence including personal experiences.[2,5] Students are then supported to *distinguish among ideas*, comparing the evidence that led to their different ideas, or comparing their ideas with those of a peer.[10] Finally, instruction guides students to *connect their evidence-supported ideas* to form an explanation for the phenomenon and its role in local environmental dilemmas.

To illustrate, in a unit focused on recycling processes, "What happens to old tires?" (WISE: wise.berkeley.edu, research-based, open-source, and customizable online learning units for secondary school students), students are asked about their existing ideas about recycling by explaining what happens to old tires.[10] They speculate about whether they can be recycled (elicit ideas). Students suggest pulverizing, burning, and other techniques. Next, students explore visualizations related to the molecular properties of tires (discover ideas; Figure 18.1). The visualizations of recycling help students discover new ideas and make sense of their own varied ideas.[5,7] Students then use a discussion tool called 'Idea Manager' to compare the properties of tires and other recyclable materials such as plastics.[10] They discuss and distinguish their ideas from those of their peers. Finally, students reflect on their efforts to distinguish among all the ideas and build an integrated explanation of recycling. They discuss the implications for what it means for their local

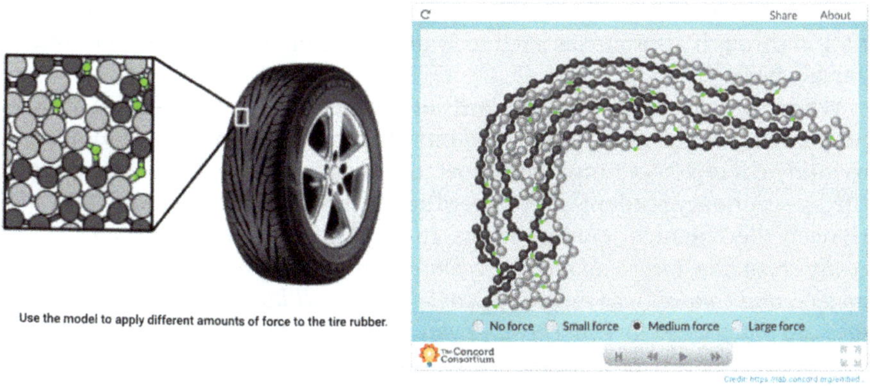

Figure 18.1 Visualization of the molecular properties of tires.

environment and beyond (connect ideas). In a design study, the research shows the value of student collaboration to distinguish ideas about how the molecular properties of materials impact the type and success of recycling processes.[10]

The KI perspective positions chemistry education as a means for respecting the ideas of each student, promoting each student's success, affirming students' cultural competencies, and fostering their identity as a chemistry learner. This approach stands in contrast to the typical view of chemistry education where settled facts are transmitted to learners.[11]

Chemistry visualizations enable learners to analyze and explain factors that are contributing to environmental inequities in their communities and generate effective strategies for change. This approach fosters student agency by supporting students to connect the new chemistry ideas with the well-formed ideas from their own experiences. Furthermore, applying chemistry to local environmental dilemmas helps students see the relevance of chemistry. What has often been characterized as students' lack of interest in pursuing chemistry may instead reflect students' resistance to typical chemistry curriculum materials that neglect links to real-world issues.[11,12]

To further illustrate, students often raise issues about chemical contaminants. After the water crisis in Flint, Michigan, a high-school chemistry class initiated a study of water quality in their neighborhood in Chicago. Students first investigated the reactivity of various metallic elements to learn that the low reactivity of lead is central to its use in water pipes. Students learned that while lead is relatively unreactive, it corrodes like other metals which causes water contamination. Understanding corrosion invited students to analyze the patterns of electrons in the outermost energy level of atoms to predict the formation of oxidized lead. They analyzed lead toxicity and identified contamination in a community garden and in a water fountain at a local park, both places situated in communities of people of color. Students explored how one community organized to use chemistry to decrease lead contamination through approaches such as water filtration and the design of raised garden beds.[11]

When learning complex, standards-aligned chemistry topics from a social justice lens, as in this example, students are more likely to become motivated to study chemistry—a topic often perceived as dry or irrelevant to everyday life by secondary students.[11] As one student from the above study who experienced the chemical contaminants curriculum reported, "*I thought it was really crazy how you could connect science, which before to me had been so technical. . .and connect it to everyday people and how it literally affects your lives a lot. . .So that's why I took AP chem...it was kind of like I want to pursue a career in science so that I can do some of these things.*"[11]

Teachers are making efforts to incorporate students' environmental interests into their instruction. However, they face obstacles such as limited curriculum resources/exemplars and assessment materials; a lack of alignment between teaching required standards and environmental dilemmas; and the substantial amount of time required to develop such materials.[13–15] We illustrate promising directions for leveraging visualizations about environmental dilemmas.

18.3 Research-practice Partnerships Localize Environmental Dilemmas in Chemistry

Grounding chemistry instruction using visualizations in local and community-based environmental dilemmas appeals to students by eliciting and building on students' personally relevant ideas.[2,7] Local dilemmas, such as the distribution of pollutants, give students opportunities to reflect on existing and new ideas from conceptual, ethical, and political perspectives. Two recent reviews of the literature, however, find that typical instructional approaches to environmental dilemmas such as climate change are potentially fragmentary and unsystematic, and largely ineffective in impacting students' attitudes and environmental stewardship, and thus further work on environmental dilemmas is needed.[14,15]

Including teachers in the research and design process is essential for connecting visualizations and constructivist pedagogy to address environmental dilemmas in chemistry. Research-practice partnerships (RPPs) are long-term, systematic efforts to improve pre-college science curriculum through research.[16] RPPs include science teachers along with other stakeholders such as science curriculum coordinators, students, researchers, computer scientists, and professional chemists.[17]

RPPs jointly develop chemistry visualizations and embed them in constructivist pedagogy to illustrate environmental dilemmas. Using web-based, open-source resources ensures that materials are free and customizable. This approach enables RPPs around the globe to modify the visualizations and surrounding curriculum to meet the needs of their learners. Often this involves refining visualizations built by experts to align with students' prior knowledge and interests. We provide several examples.

18.3.1 Chemical Reactions and Alternative Fuels

The WISE Chemical Reactions unit, for example, was built by an RPP of local teachers, education researchers, technologists, and scientists (25–30 participants each summer). In the unit, students explore visualizations to learn about the atomic properties of molecules, chemical reactions, formation of synthetic fuels, and the impact of fuels on global temperature. Students model chemical reactions themselves, contrasting combustion of hydrogen *versus* a hydrocarbon to gain insight into the reactants and products in gasoline use.[18] To enable students to explore the impact of varied fuel-types on climate, the RPP jointly designed an interactive model where students can vary the use of different fuel sources to power cars in a city, and analyze the impact on CO_2 emissions, global temperature, and cost over 100 years (Figure 18.2). The joint design of the model leverages each partner's expertise. Partners discussed, for instance: What are the most promising alternative fuel sources today that 8th graders could investigate? What calculation can you use to compare the CO_2 emissions of varied fuel types? What variables need

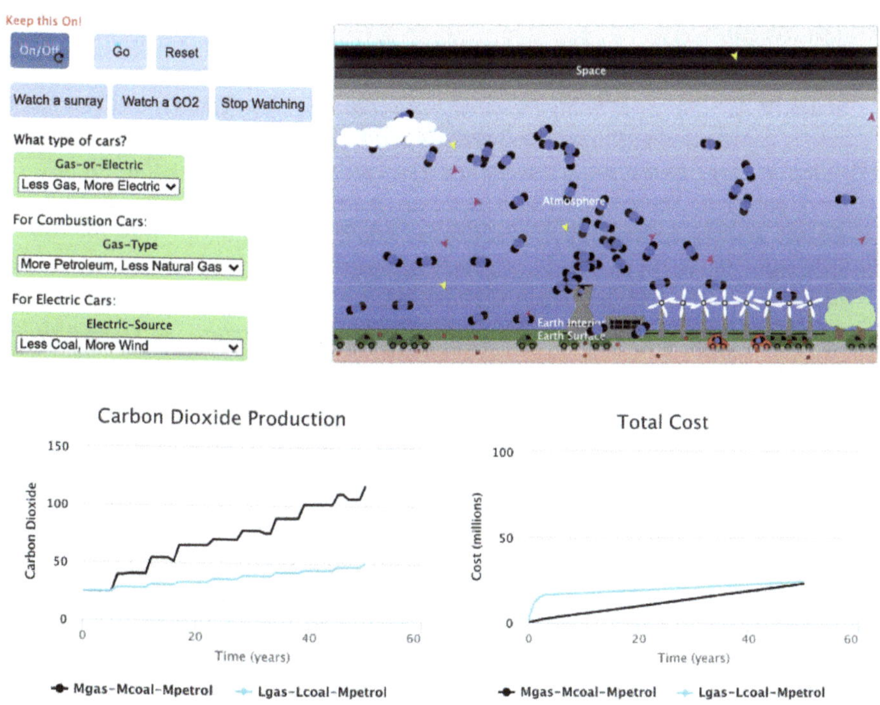

Figure 18.2 Visualization of CO_2 emissions by fuel types.

to be considered (*e.g.*, how far does one gallon of fuel move the car)? What is the efficiency with which each fuel is converted from chemical energy to the kinetic energy of the car? How much choice should we offer students in experimentation?

Students use evidence from the alternative fuels model, as well as climate visualizations in the unit, to write a letter to explain to a policymaker how gasoline combustion works, why using gasoline contributes to increases in global temperature, and how the policymaker can address this problem with alternative fuels or a combination of fuel types. Middle- and high-school chemistry teachers in the RPP further customized the unit to appeal to their students. Each customized a template letter within the unit to address a different policymaker who they thought was most relevant to their students (*e.g.*, local congressperson, CEO of a well-known car manufacturer). Each teacher also incorporated additional data into the unit for students to analyze because it was relevant to their selected policymaker (*e.g.*, the percent of each type of fuel used in their city for the local congressperson). Some teachers also modified the level of scaffolding within the letter template to help students integrate multiple data points from the visualization into their written argument. The customizations supported students to use their chemistry knowledge and evidence from the visualization to articulate the impacts of gasoline combustion on climate and propose a solution.

The following are two excerpts from 8th grade student letters written at the end of the unit:

> **Student A:** "*Energy is being released from combustion of gasoline. When gasoline is combusted with oxygen it releases carbon dioxide and water. From this reaction large amounts of energy are released, making it possible for you to work your car. The problem with gasoline fueled cars are that they disrupt the Earth's energy balance by producing too much carbon dioxide into the atmosphere for the earth to handle. A good alternative is driving petroleum (octane) fueled cars. Others might propose that we use natural gas. To convince the others in the city you need to tell them this. Using octane fuel less affects our temperature than using natural gas fuel by 1.7 degrees.*"

> **Student B:** "*...In a gasoline chemical reaction, H_2O (water) and CO_2 (carbon dioxide) are released into the atmosphere. This happens because when you burn gasoline the atoms rearrange themselves from octane and oxygen to water and carbon dioxide. So, when you burn/use gasoline, carbon dioxide is released into the atmosphere, contributing to climate change. Because of how much we drive gasoline-fueled cars it disrupts our Earth's energy balance. Burning gasoline produces lots of CO_2 into the atmosphere that the Earth doesn't have enough stuff to keep Earth's energy balanced. A good alternative to using gasoline would be wind. Using wind instead of gasoline would cut down the amount of CO_2 that is produced into the atmosphere. The only problem with using wind-powered cars is that it costs more. But wind is the best solution so far, I believe.*"

18.3.2 Guiding Drawing of Chemical Reactions for Alternative Fuels

One RPP using the alternative fuels unit collaborated with a computer scientist to design an activity where students draw the arrangement of atoms before and after methane and ethane combustion reactions. The RPP explored designing automated, personalized guidance on the drawings.[4] The RPP designed and tested automated guidance based on the KI framework, determining what type of guidance would help each student move one step further. Providing immediate, automated guidance on each student's work frees teachers to respond to individual students. This approach aligns with prior research demonstrating that using multiple representations of chemical reactions and providing students ways to visualize the particles involved in the reactions can build chemistry understanding.[18]

The automated guidance was designed based on an analysis of 98 student-generated drawings. It helps students build on ideas about conservation of matter, reactants, and products (see Figure 18.3). Each hint, assigned as soon as the drawing was completed, recognized a promising student idea (*e.g.*, conservation), posed a question targeting a challenge (*e.g.*, how many molecules in the products) to elicit their ideas, directed students to a relevant resource in the unit, and encouraged the student to revise their drawing.

Results showed that the automated guidance (AG) was as effective as teacher guidance (TG) in supporting students to revise and improve their drawings (Effect sizes: $d = 0.33$ ($n = 155$; AG) and $d = 0.39$ ($n = 108$; TG)).[4] Students generally appreciated the automated guidance. However, a few

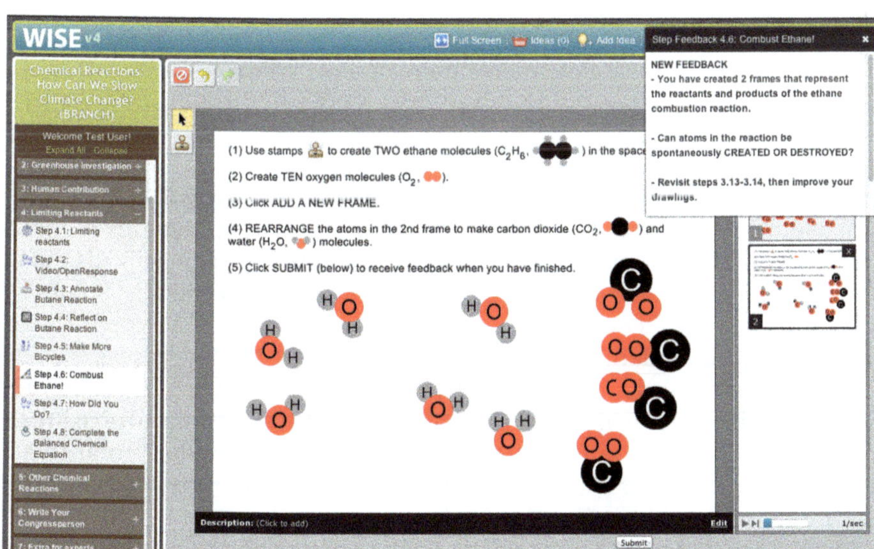

Figure 18.3 Visualization of the automated guidance step.

mentioned that they could often get the teacher to tell them the right answer while the guidance made them figure out the solution.

Using automated guidance can ensure that each student receives personalized guidance on their understanding of chemistry concepts such as chemical reactions in environmental dilemmas. Advances in learning analytics are creating new opportunities to auto-guide student-generated work in chemistry. These advances include algorithms to diagnose students' logged choices in virtual experimentation and natural language processing models to assess student-written explanations.

18.3.3 Promoting Collaboration During Virtual Experimentation

Students benefit from working collaboratively to explore dynamic chemistry visualizations. In a series of studies, an RPP combined the constructivist activity structure of collaboration with dynamic *versus* static visualizations for chemical reactions in photosynthesis and cellular respiration.[6] They focused on ways to support students as they distinguish among ideas. Both the static and dynamic visualizations provided prompts to help students distinguish key concepts and evaluate their ideas. The findings indicated that collaborators who explored the dynamic visualization formed significantly more accurate and integrated explanations of the energy and matter transformations in photosynthesis and cell respiration than the static visualization group. Drawing on analysis of classroom video data, the authors concluded that the dynamic visualizations were more effective than the static representations in helping students distinguish ideas about chemical reactions in photosynthesis and cellular respiration.

18.3.4 Designing Assessments that Make Student Ideas Visible for Teachers

To help teachers guide students as they explore visualizations, RPPs create embedded assessments that reveal the heterogeneity of student ideas about environmental dilemmas. These assessments can help teachers diagnose student challenges and identify student ideas for use as starting points in a class discussion.[19] For example, when studying chemical reactions, Zhang and Linn asked students to draw four distinct steps in the process.[18] They found that some students drew only two steps, expecting reactants to immediately turn into products; other students expected all the molecules to break into atoms and then reform. When embedded assessments reveal these views, teachers can plan a class discussion that guides students to analyze evidence and sort out their different views.

In another example, students studied a two-week project-based unit on desalination with a drought context.[20] In this KI-aligned unit, that was used within a California high-school, students' ideas about the California

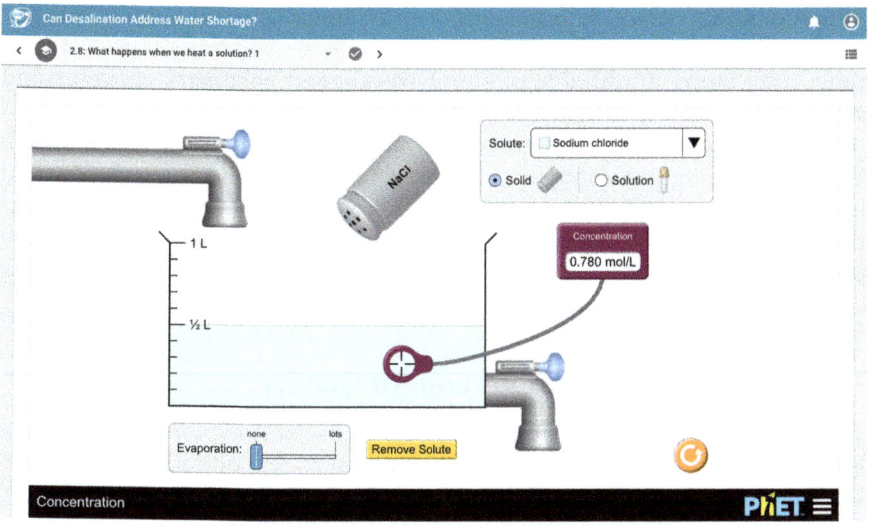

Figure 18.4 Visualization of evaporation impacting molarity.

drought, water accessibility, and associated concepts were elicited. Students explored chemical concepts related to desalination through visualizations of heat transfer involving solar radiation and a visualization illustrating how evaporation impacts the concentration of a saltwater solution (Figure 18.4). Students also had an opportunity to construct and test a desalination model in order to test their conceptual ideas for the unit. Embedded assessments across different stages of the unit asked students to explain heat transfer from the sun, solubility of salt in water, and molarity connected to evaporation. The teacher in this study commented that the ability to review and comment on embedded assessments within the unit supported her in providing better and more accurate guidance to students.[20]

In summary, RPPs are combining visualizations with KI constructivist pedagogy to address environmental dilemmas and welcome students in chemistry.[17] Activity sequences developed by RPPs take advantage of collaboration, drawings with personalized guidance, and assessments that make student ideas visible. These sequences follow the KI pattern, helping teachers to support students to connect scientific ideas to everyday life. Such opportunities can raise social justice issues around environmental dilemmas.

18.4 Addressing Social Justice in Chemistry

Incorporating environmental dilemmas into chemistry increases personal relevance and raises issues of social justice.[11] These dilemmas open up exploration of the political and ethical factors involved in understanding a scientific problem.[21] Social justice problems make chemistry meaningful to students and illustrate how their communities can benefit from understanding chemical processes and evidence-based argumentation in order to generate solutions. Rather than taking additional instructional time, integrating knowledge of standards-aligned chemistry concepts and social justice factors deepens student understanding.[11]

RPPs are drawing on expertise from organizations that serve communities experiencing environmental problems to respond to the lack of social justice chemistry lessons in most precollege and college-level chemistry curriculum materials.[16,17] They create visualizations that illustrate the chemical processes involved in the environmental problem as well as the social factors contributing to the inequitable impacts.

18.4.1 Chemical Reactions and Air Quality

This WISE middle school unit, designed by an RPP, explores the environmental impacts of hydrocarbon combustion in a local oil refinery.[22] The lesson examines the disparate impacts of the refinery, located near the school community, on asthma rates among Black Americans compared to people of other races. It engages students in gathering evidence from visualizations to identify the factors that contribute to this environmental justice issue. The RPP included two middle school chemistry teachers, a social justice expert who had lived in and served the focal community, a curriculum designer, education researchers, and scientists.

The RPP localized the unit by including imagery and datasets from within or nearby the school community, incorporated teaching of redlining as a political factor that has contributed to the inequalities and highlighted the factors that can be understood and rectified using chemistry. The curriculum begins by grounding the study of chemistry in the local community. The unit features images of the local oil refinery and videos of scientists and activists from the community who are using their understanding of air quality to advocate for change. It elicits students' ideas based on their experiences asking, *what do you think might be some of the impacts of living near the freeways, railroads and refineries in Refinery City?*

The unit connects the impact of the refinery to incomplete combustion. Students collaboratively interact with a visualization of a hydrocarbon combustion reaction and analyze a video about incomplete combustion. Both support students to analyze the products of a combustion reaction that occurs with insufficient oxygen. Embedded assessments prompt students to link chemistry ideas and social justice ideas. They are asked to use evidence from the visualizations to explain the products of incomplete combustion and its impacts on humans. Students are asked if they think everyone is impacted by the reactions the same amount, or, if some people are more impacted than others and why.

Next students explore interactive maps that display the quantity of particulate matter pollution produced by incomplete combustion by neighborhood, in Refinery city. They compare this map to another interactive map of asthma rates by neighborhood in the city. Students explore the chemical properties of the products of incomplete combustion including diesel particles and particulate matter. They examine how these affect the lungs and other body systems when inhaled. This activity is designed to help students link ideas about air pollution from the refinery with health impacts.

Students then analyze data represented in graphs illustrating the asthma hospitalization rates in the city by racial group. Students discover the unequal impacts of air pollution on Black communities who live near the refinery. To understand these inequities, the unit supports students to learn about the historical practice of redlining and to explore how redlining impacts who is impacted by the refinery caused air pollution. Students use maps showing redlining and rates of pollution to distinguish between the impact of redlining and the (lack of) impact of race on asthma rates.

Finally, students connect ideas in a reflection activity where they revisit (and revise if they choose) their explanation about who is impacted by air quality. They explain whether air quality impacts each person the same or differently. During the unit, students link their ideas about the chemical processes involved in an oil refinery, their ideas about air quality and asthma rates, and their ideas about how redlining has impacted who lives near the factory. For example, one 7th grader wrote, "*I think the impacts [of the refinery] are different depending on where a person lives because if they're near an area with a lot of incomplete combustion, they would probably inhale a lot of soot. Also, …the previously redlined areas are closer to highways and industrial & waste sites, so a lot of particulate matter pollute their air quality…Some of the groups or people affected by the impacts of air pollution include the workers in the refinery, the citizens of [city] and the African American/Black people living in the redlined areas closest to the refinery.*"

Last, to empower students to use their chemistry knowledge, students jointly generate ideas for policies or actions to ensure some communities do not experience more harmful environmental impacts than others. Students put forth ideas such as, "*First, have inspections that regulate how much pollutants that come out of refineries and if they exceed a certain point tell them they need a way to stop producing that much pollution.*". Research demonstrated that students made significant gains in their understanding of the differential impacts of pollution depending on proximity to pollutants.[22]

In summary, RPPs can design lessons that leverage visualizations to support students' evidence-based reasoning about chemistry in an environmental justice dilemma. These lessons welcome students' cultural, political, and ethical views as valuable ideas for learning chemistry. These lessons motivate students to construct integrated knowledge of chemistry concepts by linking their ideas from personal experience with ideas from instruction. Studies of how students reason about chemistry in environmental dilemmas are also broadening the boundaries of what constitutes chemistry knowledge. This change is driving designers to create new visualizations that support the investigation of social factors that illustrate how chemistry phenomena affect communities.

18.4.2 Affordances of Physical and Virtual Laboratories

Visualizations enable curriculum designers to create virtual laboratories to reinforce, replace, or enhance physical laboratories. Both virtual and physical laboratories have valuable affordances (see Table 18.1).[1,23–25] As this chapter illustrates, virtual laboratories can extend physical laboratories about chemical reactions to environmental dilemmas and to issues of social justice. Multiple studies combining virtual and physical laboratories show the benefit of connecting molecular visualizations to physical investigations in chemistry.[1,26] Comparisons of physical and virtual laboratories reveal the advantages of each format. A recent study found that students working with virtual models engaged in significantly more discussions about making predictions and understanding relationships between variables, than did students working with physical models.[24] Contrary to expectations, the virtual model prepared students as well as the physical model for subsequent physical experiments with physical materials.

Direct comparisons of virtual and physical laboratories often try to replicate the physical laboratory in the virtual format by equating conditions.[27] This approach eliminates the affordances of each format. Instead, research is needed that establishes the value of the unique affordances of each format. Further, research is needed to capture the opportunities that result from combining virtual and physical laboratories such as shown in research on environmental dilemmas and social justice.

To promote integrated understanding, analysis of studies comparing virtual and physical laboratories using the Knowledge Integration (KI) pedagogy reveals factors that differentiate investigations of virtual and physical laboratories. Most laboratories support students to discover new ideas. Virtual and

Table 18.1 Affordances of virtual and physical laboratories.

Virtual laboratory affordances	Physical laboratory affordances
• Can make visible the relationship between microscopic [molecular], vast [solar system], fast [motion], and complex [photosynthesis, force] phenomena. • Can give students agency to choose their investigation and even conduct "impossible" or dangerous experiments. • Can enable discussions where students distinguish among their ideas while planning a series of virtual experiments. • Makes testing multiple conditions efficient. • Can reduce student concerns about dangerous or toxic chemicals and about breaking glassware or apparatus. • Can enable students to manipulate the code behind the visualization and design alternative experiments.	• Can support students to test questions that go beyond those in a virtual laboratory. • Can build students' skill in using equipment and machines in physical laboratories. • Can support students to cooperate to conduct separate parts of the experiment. • Can embody the laboratory experience. • Can prepare students to deal with unanticipated consequences, experience "error" of measurement, and troubleshoot failed experiments.

physical laboratories differ in supporting students to distinguish among their prior and newly discovered ideas. Visualizations enable students to probe their own ideas by conducting multiple virtual experiments, even including experiments that would be impossible in a physical laboratory. In addition, as illustrated in this chapter, virtual laboratories can support students to investigate models of complex environmental dilemmas to deepen their understanding. While both physical and virtual laboratories can prompt students to reflect on their investigations, a key aspect of KI, reflections are often more powerful in virtual laboratories. In virtual laboratories, students have time to use interactive visualizations to intentionally seek evidence to investigate their conjectures; they can gather convincing evidence to support their reflections.

18.5 Conclusions

This chapter illustrates the value of visualizations to promote integrated understanding in chemistry. The examples show how visualizations, embedded in instruction featuring constructivist pedagogy, can support students to integrate complex science concepts in the context of compelling environmental dilemmas. By using open-source, customizable online units, designers including teachers can welcome each student to chemistry. Involving students in meaningful, glocal environmental challenges and social justice issues has the potential to motivate a lifelong fascination with issues that affect their lives. Virtual laboratory activities supported by interactive visualizations offer opportunities to enhance physical laboratories by extending instruction to broader, global issues.

Although many chemistry investigations integrate environmental challenges, few incorporate social justice issues.[28] Furthermore, when social justice is mentioned, it is often peripheral to the instructional goal.[15] In addition, there are limited studies that simultaneously integrate visualizations alongside environmental challenges and social justice issues, despite their potential to critically engage and invest students in chemistry topics.

Developing holistic curriculum, assessments, and technology that use visualizations for environmental dilemmas and other contemporary issues in chemistry are needed. The expansion of RPPs to include environmental stakeholders can strengthen the relevance and appeal of chemistry instruction.[16] Further, there are encouraging developments as RPPs are integrating engaging contexts such as air quality (asthma rates), desalination (water availability), and environmental destruction in precollege courses. The KI framework provides a highly-researched, pedagogical framework to guide and align the design of such curricula alongside their assessments and supporting visualizations.

References

1. T. de Jong, M. C. Linn and Z. C. Zacharia, *Science*, 2013, **340**, 305.
2. D. F. Donnelly-Hermosillo, L. F. Gerard and M. C. Linn, *Comput. Educ.*, 2020, **146**, 103748.

3. See Chapter 21: M. Milner-Bolotin and V. Milner, in *Digital Learning and Teaching in Chemistry*, ed. Y. J. Dori, C. Ngai and G. Szteinberg, Royal Society of Chemistry, United Kingdom, 2023.
4. A. N. Rafferty, L. Gerard, K. McElhaney and M. C. Linn, *Int. Conf. Learn. Sci.*, 2014, 386.
5. K. W. McElhaney, H. Y. Chang, J. L. Chiu and M. C. Linn, *Stud. Sci. Ed.*, 2015, **51**, 49.
6. K. Ryoo and K. Bedell, The effects of visualizations on linguistically diverse students' understanding of energy and matter in life science, *J. Res. Sci. Teach.*, 2017, **54**, 1274.
7. M. C. Linn and B. S. Eylon, *Science Learning and Instruction: Taking Advantage of Technology to Promote Knowledge Integration*, Routledge, New York, 2011.
8. Pew Research Center, *59% of U.S. Parents with Lower Incomes Say Their Child May Face Digital Obstacles in Schoolwork*, 2020, retrieved from https://www.pewresearch.org/fact-tank/2020/09/10/59-of-u-s-parents-with-lower-incomes-say-their-child-may-face-digital-obstacles-in-schoolwork/.
9. A. A. diSessa, in *A Friendly Introduction to "Knowledge in Pieces": Modeling Types of Knowledge and Their Roles in Learning - Invited Lectures from the 13th International Congress on Mathematical Education*, ed. G. Kaiser, H. Forgasz, M. Graven, A. Kuzniak, E. Simmt, and B. Xu, Springer International Publishing, Cham, 2018, pp. 65–84.
10. K. McElhaney, C. Matuk, D. Miller and M. Linn, *Int. Conf. Learn. Sci.*, 2012, 323.
11. D. Morales-Doyle, T. C. Price and M. J. Chappell, *Sci. Educ.*, 2019, **103**, 1347.
12. D. Morales-Doyle, *J. Res. Sci. Teach.*, 2018, **55**, 749.
13. L. Chen and S. Xiao, *Educ. Res. Rev.*, 2021, **32**, 100377.
14. D. Rousell and A. Cutter-Mackenzie-Knowles, *Child. Geogr.*, 2019, **18**, 191.
15. D. Bhattacharya, K. Carroll Steward and C. T. Forbes, *J. Geosci. Educ.*, 2020, **69**, 223.
16. C. E. Coburn and W. R. Penuel, *Educ. Res.*, 2016, **45**, 48.
17. R. Pea and M. C. Linn, *Front. Educ.*, 2020, **5**, 130.
18. Z. H. Zhang and M. C. Linn, *J. Res. Sci. Teach.*, 2011, **48**, 1177.
19. Z. Yaseen and P. Aubusson, *Res. Sci. Educ.*, 2020, **50**, 529.
20. M. Li, D. F. Donnelly-Hermosillo and J. Click, *J. Sci. Educ. Technol.*, 2022, **31**, 27.
21. T. D. Sadler and D. L. Zeidler, *J. Res. Sci. Teach.*, 2019, **46**, 909.
22. A. Bradford, K. Billings, L. Gerard and M. C. Linn, AERA meeting 2021.
23. C. Xie, R. Tinker, B. Tinker, A. Pallant, D. Damelin and B. Berenfeld, *Science*, 2011, **332**, 1516.
24. S. Puntambekar, D. Gnesdilow, C. D. Tissenbaum, N. H. Narayanan and N. S. Rebello, *J. Res. Sci. Teach.*, 2021, **58**, 392.

25. J. Chao, J. L. Chiu, C. J. DeJaegher and E. A. Pan, *J. Sci. Educ. Technol.*, 2016, **25**, 16.
26. Z. Kaberman and Y. J. Dori, *Int. J. Sci Math. Educ.*, 2009, **7**, 597–625.
27. H. O. Kapici, H. Akcay and T. de Jong, *J. Sci. Educ. Technol.*, 2019, **28**, 231.
28. G. A. Lasker, K. E. Mellor, M. L. Mullins, S. M. Nesmith and N. J. Simcox, *J. Chem. Educ.*, 2017, **94**, 983.

CHAPTER 19

Designing Tutorial Videos to Support Students' Learning of Reaction Mechanisms in Organic Chemistry

SASCHA BERNHOLT*[a], JULIA ECKHARD[b], MARC RODEMER[c], AXEL LANGNER[b], GYDE ASMUSSEN[a] AND NICOLE GRAULICH[b]

[a]IPN—Leibniz Institute for Science and Mathematics Education, Department of Chemistry Education, Olshausenstr. 62, 24118 Kiel, Germany; [b]Justus-Liebig-University Giessen, Institute of Chemistry Education, 35392 Giessen, Germany; [c]University of Duisburg-Essen, Department of Chemistry Education, Schützenbahn 70, 45127 Essen, Germany
*E-mail: bernholt@leibniz-ipn.de

19.1 Introduction

Learning with visualizations is an inherent part of science education. Multimedia instruction, in which visual and textual elements are combined, has been shown to bear high potential to promote student learning. In this context, videos that blend different instructional elements have been used intensively to deliver information and to support students' knowledge construction. This chapter focuses on tutorial videos as a multimedia instruction

Advances in Chemistry Education Series No. 11
Digital Learning and Teaching in Chemistry
Edited by Yehudit Judy Dori, Courtney Ngai and Gabriela Szteinberg

Published by the Royal Society of Chemistry, www.rsc.org

format and illustrates evidence-based principles for selecting and designing this type of learning resource referring to the context of organic chemistry.

Tutorial videos have become increasingly popular in recent years. Almost every possible topic in science is covered and available on platforms like YouTube. With regard to evaluating the quality of these videos, Kulgemeyer[1] summarized in his review that most research reports focus on best practice examples or on students' use and evaluation of tutorial videos. The design and quality of tutorial videos are also researched from the perspective of digital media or multimedia theories,[2] applying instructional design principles to judge the videos' quality (for a critique, see Schroeder and Traxler, 2017).[3] Also, cognitive load theory has been used extensively to derive guidelines that aim to optimize learning from tutorial videos.[4]

Substantially less literature is concerned with criteria derived from research in chemistry education to design explanations provided in these tutorial videos[5] as well as appropriately combining explanations with specific multimedia principles. Despite a tremendous body of science education literature on students' construction of scientific explanations and argumentation,[6] few have focused on the impact of instructional explanations on students' understanding, in general or specifically in tutorial videos.[4,5] However, this focus seems to be important because the explanations provided in the learning videos have a great impact on what students actually understand and learn from these resources.[5]

This chapter provides an overview about evidence-based principles for designing both videos (as a form of multimedia instruction) and instructional explanations (*i.e.*, explanations designed for the purpose of teaching and, in this context, provided to students). Implications from both lines of research are discussed with regard to the format of tutorial videos used in chemistry education. Building on this background, we present empirical findings from a case study in which tutorial videos were purposefully designed to support students' understanding of reaction mechanisms in organic chemistry.

19.2 Theoretical Background

19.2.1 Learning from Videos

Chemistry, as a natural science, is characterized by a great variety of representations, which are used to describe macroscopic phenomena with submicroscopic models, as well as structural and mathematical representations. All of these representations have their own special meaning and are used for different purposes. They represent chemical content in a well-defined and extremely compact way, making them an indispensable means of communication.[7] Accordingly, a great variety of resources are used in teaching to support students' learning of and with these representations, ranging from photos and illustrations to haptic molecular model construction kits to dynamic animations or interactive simulations (see Chapter 21 by

Milner-Bolotin and Milner),[8] to virtual or augmented reality applications (see Chapter 17 by Han and Fung).[9]

In this chapter, we focus on videos as a specific type of instructional resource. Several studies underline the effectiveness of videos as educational tools,[4] in terms of engaging students or to illustrate abstract or hard-to-visualize content. However, videos are not effective for learning *per se* as simply watching videos may only promote the illusion of knowledge or skill acquisition, while a person's actual abilities do not improve despite watching the video.[2,4,10]

With regard to principles that can guide instructors to design videos that are effective in supporting student learning, Brame summarized different key elements that relate either to cognitive load theory (CLT) or to student engagement and active learning.[4] The basic assumptions of CLT are that learning, as a cognitive process of schema construction, requires mental resources that are limited, and that these limited resources can be occupied by specific types of cognitive load.[11] While there is considerable debate regarding the typology of cognitive load, the original differentiation by Sweller postulated three types of cognitive load: intrinsic cognitive load (induced by the complexity of the task itself); extraneous cognitive load (induced by task-irrelevant aspects); and germane cognitive load (induced in form of allocating cognitive resources by the learner while solving the task).[12] In the context of multimedia learning, approaches to reduce extraneous cognitive load in instructional materials have received particular attention.[4]

Building upon CLT, the cognitive theory of multimedia learning (CTML) proposes several principles to reduce extraneous processing in learning with multimedia (Figure 19.1).[2] According to the coherence principle, students learn better when decorative graphics or task-irrelevant text and details are excluded. The redundancy principle emphasizes that learning is hindered when information is provided simultaneously in form of a verbal narration and printed text. Exceptions to this principle occur when the printed text provides a focusing function (*e.g.*, in form of headlines) or when the narration makes use of unfamiliar words or is provided in a second language. The spatial and temporal contiguity principle suggests that people learn better when corresponding words and pictures are presented near to each other or simultaneously rather than successively. Finally, the signaling principle states that learning from multimedia is supported when specific visual cues are added to highlight relevant parts.[2]

Both the CLT and CTML emphasize the relevance of students' active engagement when working with learning material.[2,4] For instance, the CTML also comprises several principles to foster students' generative processing of instructional material (mainly aiming at motivational and affective aspects) to minimize situations in which students have cognitive capacities available but do not choose to use them for engaging with the learning material. Generative processing may be fostered by engaging students in summarizing the learning content, by illustrating the learning content in form of drawings, or by explaining the content to a fellow student.[2]

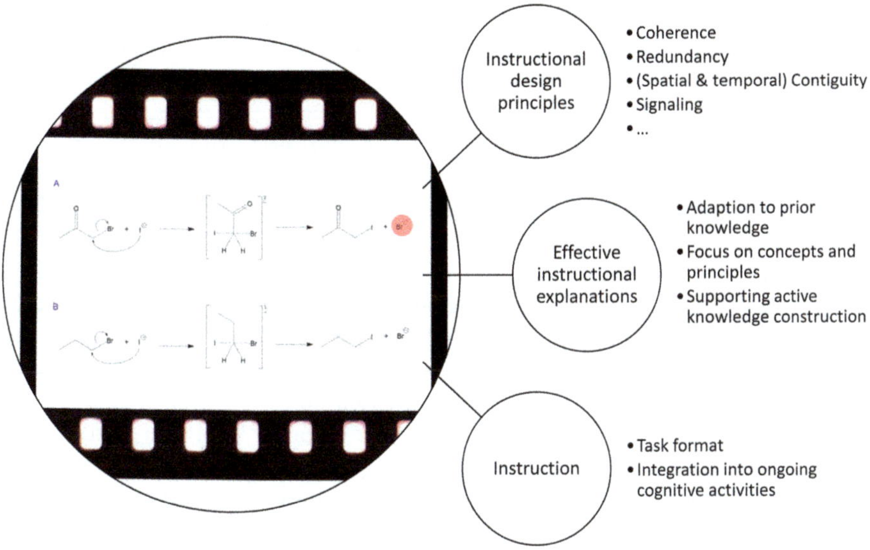

Figure 19.1 Overview of the theoretical and didactic methodological references that should be taken into account in the context of creating instructional videos.

Overall, CLT and the CTML propose specific principles to support instructors in selecting, designing, and arranging elements in multimedia instruction to make learning materials more accessible and easier to process. As these theories hardly make any statement about the selection and preparation of the learning content, the content is often only regarded in rather broad categories like declarative *vs.* procedural knowledge[1] or simple *vs.* complex material[2] instead of a relevant aspect to determine the design of instructional material. However, the learning content and how it is presented to students has major implications on the relevance and impact of the different multimedia instruction principles in supporting student learning.[5] For example, the specific explanation of a scientific topic in a tutorial video determines when students' attention should be directed to which specific relevant areas on the display, and thus the implementation of the signaling principle. Consequently, following multimedia design principles alone will not necessarily result in effective tutorial videos.

19.2.2 Instructional Explanations

Explanations in tutorial videos can be considered a specific instructional format that aims to make (science) content comprehensible to a specific group of learners.[5] In their review, Wittwer and Renkl proposed four essential factors that make instructional explanations effective. Instructional explanations should: (1) be adapted to the learner's prior knowledge; (2) focus on concepts and principles; (3) be integrated into the learner's ongoing cognitive activities; and (4) not replace the learners' knowledge-construction activities (Figure 19.1).[13] Partly overlapping frameworks have been proposed elsewhere.[1,14]

The adaptation of the explanation to the learners' prior knowledge and interests is a major point for effective instructional explanations.[1,13,14] Not only focusing on the content structure but taking the learners' needs into account increases the likelihood that the explanation is understood. While its importance is widely accepted in educational research and teaching, adaptation to an often larger and sometimes even unspecified target audience is probably one of the most challenging aspects in the production of explanation videos.

Ideally, the explanation explicitly addresses common misconceptions or mistakes with regard to the topic at hand. For instance, multiple studies have shown that students struggle in identifying nucleophiles in the context of substitution reactions, which indicates the relevancy to explain in more detail what properties characterize a nucleophile, instead of only assigning this label to a specific species among the given starting materials. Despite positive effects in many studies, the efficiency of this approach depends on whether the learners perceive this information (*i.e.*, the explicit contrasting of misunderstandings and intended understanding) to be relevant for their learning.

Within the context of adaptation, an important feature of successful explanations is to avoid digressions while supporting learners to focus on the core ideas. Accordingly, "minimal explanations" are often recommended, sometimes against the background of cognitive load theory.[5,11,13] Providing only the information needed to understand the problem at hand, in a precise and coherent form, aims at reducing any distraction from irrelevant details, while still providing sufficient details to make the explanation comprehensible (see the coherence principle in CTML).[2] In the context of identifying nucleophiles, it is important to address both structural features (charges, lone electron pairs) and the function resulting from these structures (capacity of donating electrons), but it may be rather distracting to mention further properties of the specific nucleophile (*e.g.*, boiling point) unrelated to its role in the reaction and task context. Generally, coherence is often attributed to naming the same object identically (even though experts might identify different terms as synonyms), to providing explicit connections between concepts and examples, and to making intensive use of connectives to illustrate the temporal and causal structure of the argument and, thus, to support the learner in constructing a mental representation of the explained content.[5] These features also support students who need to process the explanations in a second language.

In a similar vein, instructional explanations that focus on (often abstract) concepts can "provide learners with insightful ways of making sense of scientific or mathematical content" (p. 53).[13] As multiple studies have repeatedly evidenced that learners face severe difficulties in acquiring deeper levels of understanding of core concepts, instructional explanations provide means to support students in moving beyond concrete experiences provided by other learning tasks (*e.g.*, conducting an experiment or describing a phenomenon). This process is supported when the explanation also highlights the relevancy of underlying concepts to explain the observable effect(s), which requires students to engage in deeper processing of the provided information. Using the example of substitution reactions, explanations of the mechanism should not only describe the sequence of events but focus on explaining the underlying cause–effect relationships and principles that result in the pattern characteristic for the mechanism.

Here, specific prompts are often used to highlight the most important parts or essential ideas of an explanation. With regard to tutorial videos, the format allows to implement specific highlighting techniques that can guide the learner's attention to the relevant parts of the displayed information, *e.g.*, by means of spotlights, colouring, and zooming-in, or by use of on-screen text or symbols. For instance, highlighting specific parts of structural formulas that are addressed in the verbal explanation can guide the learner's attention to the relevant parts of the display, *e.g.*, reactive centres. In the context of multimedia learning, this approach refers to the signaling principle (see above).[2,4] Multiple studies have provided evidence for the benefit of this kind of attention guidance (*i.e.*, the signaling principle) on the acquisition of both conceptual as well as perceptual aspects of knowledge. The benefits of signaling are also often complemented by effects of segmenting or chunking, as the provided information is subdivided into smaller pieces of information that students can engage with sequentially.[4]

19.3 Research Context: Designing Tutorial Videos in Organic Chemistry

Building up on the theoretical background of multimedia learning and instructional explanations, we aim to illustrate how specific parts of the described principles can be implemented in designing explanation videos to support students' understanding of reaction mechanisms in organic chemistry. In the following sections, we briefly review research findings regarding students' understanding and difficulties in the context of reaction mechanisms (as a basis to adapt the explanation to the learner's prior knowledge[1,13]) to then describe specific assumptions we made about the task structure (which should align with the learners' ongoing cognitive activities and foster their knowledge-construction).[13]

19.3.1 Research on Students' Understanding of Reaction Mechanisms

While reaction mechanisms and the electron pushing formalism are common external representations and serve as explanatory and predictive tools for experts in the field,[1] previous research has demonstrated that college students struggle with understanding the underlying meaning of mechanistic representations used throughout these courses.[15,16] Rather, students often follow a surface approach by focusing on manipulating drawn structures and arrows to obtain the desired product or by memorizing specific patterns of reaction steps.[17]

A central reason for students' difficulties with reaction mechanisms is assumed to be the (missing) link between the representation of a chemical reaction (*i.e.*, symbols, Lewis structures) and the chemical concepts (*i.e.*, knowledge about bond polarity) that must be inferred from the representation.[18] To foster students in focusing on concepts and principles[13] and their ability in linking this conceptual knowledge to the representations at hand, there is evidence that an explicit cueing technique might be an appropriate strategy to guide students' attention while learning reaction mechanisms.[19,20]

19.3.2 Selecting a Task Format that Fosters Students' Engagement

In organic chemistry, students often tend to rely on memorizing mechanisms, which is additionally supported by common task formats, like "predict the product", that focus on *what* will happen and not on *why* it will happen.[21] While reaction mechanisms are commonly discussed with regard to specific examples, case comparison tasks display two similar reactions, which should be compared and discussed by the learner, such as determining the reactivity based on commonalities and differences between the two cases.[22] Case comparisons have been shown to more actively engage students in comparing explicit structural differences between reactants in two reactions, in weighing the implicit factors that result from these structural differences with regard to their influence on the course of the reaction, and thus, in eliciting students' reasoning of reaction mechanisms.[21,23] Consequently, case comparisons are assumed to provide means to foster students' active engagement with mechanism tasks.

19.4 Research Methodology

Based on the theoretical background and the specific assumptions we made about the task structure, as described above, we designed a series of tutorial videos to support students' understanding of substitution reactions. In designing the visualization and the explanation in these videos, a special emphasis was given to the redundancy and signaling principle.

19.4.1 Structuring the Verbal Explanation

Taking nucleophilic substitution reactions as a basic reaction mechanism in introductory organic chemistry courses, tutorial videos need to cover three main influential factors (leaving group ability, nucleophilicity, substrate effects) that determine the reactivity in this type of mechanism. All three factors are governed by implicit properties of the reactants, making it necessary for students to infer these properties from the explicit features of the given structural formulae and then, to make conclusions about the course of the reaction.

In the present study, case comparisons between specific reactions were visualized by Lewis structures of the reactions. Reactions differed slightly between cases to induce a comparison with regard to one of the three influential factors. As students in introductory courses have limited prior knowledge with this specific reaction mechanism and struggle with weighing multiple influential factors, only cases differing in one of the three factors (instead of varying multiple factors) were selected (Figure 19.2).

The verbal explanation aimed to provide a complete cause-effect description of each factor (*e.g.*, https://osf.io/r4sx3).[23,24] Also, the structure of the explanation took the task format of a case comparison into account. Consequently, a step-by-step structure was embedded in the explanation, covering seven aspects: (1) describing the chemical changes that occur in both reactions; (2) explicitly stating the overall goal of the case comparison task; (3) naming the commonalities and differences between the two reactions; (4) summarizing the role of the influential factors regarding reactivity; (5) explaining and contrasting the influences of the implicit properties; (6) summarizing the effects on the transition state to refer to the energetic account

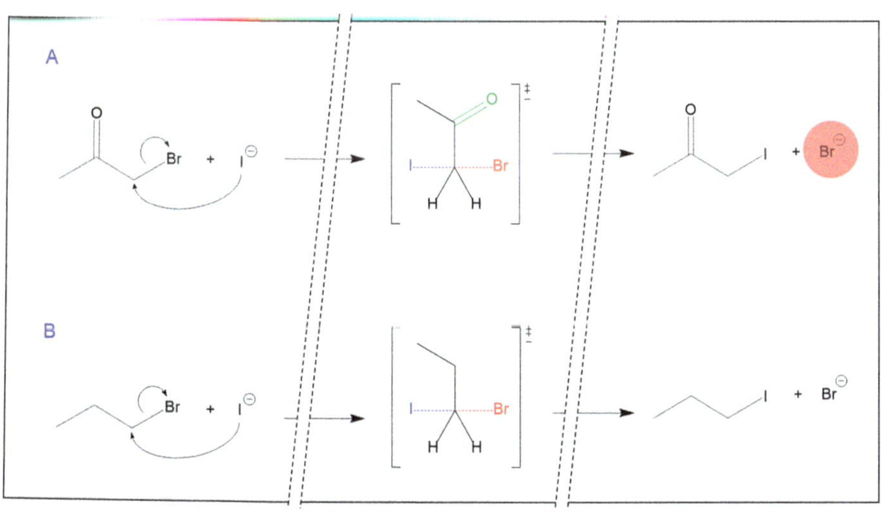

Figure 19.2 Image montage of an exemplary case comparison task as displayed in the videos. Students were asked to decide whether reaction A or B runs faster. No highlighting is shown in the control group (as shown on the left), functional groups are colour-coded in the static group (as shown in the middle), the translucent red dot moves synchronously to the explanation in the dynamic group (as shown on the right). Videos with each implemented highlighting technique can be found in German (original) and English (translated) at https://osf.io/r4sx3/.

of the reactions; and (7) concluding with a justified statement about the reactivity of both reactions.

Special attention was given to a consistent use of nomenclature and to explicitly address the relevant species, while avoiding unspecific terms or propositions (see the coherence principle in CTML). Assuming a generally low level of prior knowledge in the target audience, the fifth (implicit properties) and sixth aspects (transition state and energetic account) were described in detail to support students in linking these concepts to the representation at hand.

19.4.2 Implementing Different Highlighting Techniques

While the explanation was identical to all students, different highlighting techniques were implemented in the videos to further take into account different levels of student prior knowledge (see the signaling principle in CTML; videos with each implemented highlighting technique can be found in German (original) and English (translated) at https://osf.io/r4sx3/). While students with higher prior knowledge often do not require further cues, there is multiple evidence that students with lower prior knowledge benefit from this kind of support. Here, the implemented highlighting techniques predominantly aimed to support students in explicitly linking the key chemical concepts (described in the explanation) to the displayed representations of reaction mechanisms.

A moderate level of highlighting was achieved by making use of colour-coding of relevant functional groups, akin to common chemistry textbooks (Figure 19.2, centre). The permanent colour-coding (further referred to as 'static') aimed to highlight the relevant parts of the displayed Lewis structures and, thus, to support students in following specific groups of atoms across reaction steps. However, this technique still requires students to recognize the functional groups that are sequentially addressed in the verbal explanation.

A more specific implementation of highlighting was achieved in form of a moving translucent red dot that was synchronized with the verbal explanation (Figure 19.2, right). Accordingly, specific parts of the displayed Lewis structure were highlighted *in the moment* in which they were addressed in the verbal explanation. Compared to permanent colour-coding, this highlighting (hereafter referred to as 'dynamic') was sparser and more targeted, as only the relevant parts of the display were accentuated. The red dot moves on the display when the movement of specific entities (*e.g.*, flow of electrons, attack of a nucleophile) is addressed in the verbal explanation.

19.4.3 Study Design

Overall, we developed nine videos with a length of approximately 5 min each, with three videos for each of the three main influential factors on nucleophilic substitution reactions (leaving group ability, nucleophilicity, substrate effects) that differ in the implemented type of highlighting (static, dynamic, or no highlighting; see Section 19.3.4). The main research questions were:

1. To what extent do student learning gains differ between pretest and posttest as a function of the highlighting techniques used in the different videos?
2. To what extent do students with lower prior knowledge (LPK) benefit more from the different highlighting techniques than students with higher prior knowledge (HPK)?

Based on a randomized controlled trial, 171 students from beginner organic chemistry courses at two German universities completed a pre-test before watching the series of three videos. The test covered different essential aspects of nucleophilic substitution reactions (*e.g.*, information-retrieval from structural formulae, case comparisons of functional group properties or of reactivity between two comparable reactions). The three videos for each student incorporated the same level of highlighting (static, dynamic, or no highlighting). After each video, a survey consisting of four statements on the video's comprehensibility and demand was given to students. They were asked to rate how much they agree to each statement using a four-point Likert scale (1—"totally disagree" to 4—"totally agree"). Ratings were aggregated across the three videos (Cronbach's $\alpha > 0.90$). Finally, students completed the identical test. Reliabilities of pre- and post-test were sufficiently high (Cronbach's $\alpha > 0.68$).

19.5 Results

To analyze students' learning gains from pre- to post-test, a robust one-way repeated measurement analysis of variance was performed, including the pretest and dummy-coded group membership (*i.e.*, two separate grouping variables to contrast dynamic and static highlighting to no highlighting, respectively) as independent variables. The overall model explains a large proportion of the variance in students' posttest results ($R^2 = 0.53$) and indicates a significant increase in test performance from pre- to post-test ($F_t = 36.90$, $F_{crit} = 3.92$, $p < 0.05$, $\hat{\psi} = 0.06$ [0.04, 0.08], Yuen's $d = 0.32$), which can be considered a small to medium effect. Based on pairwise comparisons with Bonferroni correction, no statistically significant difference in the posttest score between the groups was found (for both main effects, $p_s > 0.14$), indicating that all three groups, on average, showed a comparable performance in the posttest. However, significant interaction effect was found between students' pretest score and their treatment condition (*i.e.*, the highlighting in their video), indicating differential effects of the highlighting techniques for students with different levels of prior knowledge (both $p_s < 0.05$).

To further investigate the interaction effect, we contrasted the learning gains of students with high (HPK) and low prior knowledge (LPK; Table 19.1). In case of HPK students (here defined as students with a pretest score of

Table 19.1 Pre- and post-test scores of high and low prior knowledge students across the three highlighting conditions implemented in the tutorial videos (control, static, dynamic).

	Control		Static		Dynamic	
Group	Pre	Post	Pre	Post	Pre	Post
High prior knowledge	0.87 (0.03)	0.86 (0.07)	0.87 (0.03)	0.82 (0.07)	0.88 (0.03)	0.85 (0.06)
Low prior knowledge	0.48 (0.07)	0.59 (0.06)	0.48 (0.07)	0.65 (0.11)	0.46 (0.06)	0.68 (0.09)

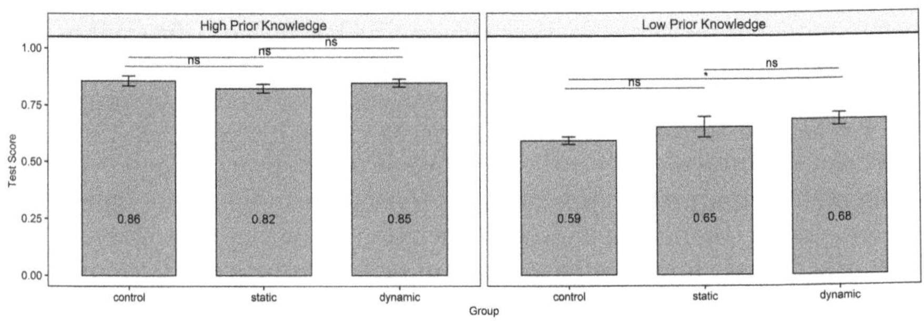

Figure 19.3 Post-test scores of high (left) and low (right) prior knowledge students, separated by highlighting group. Bars indicate mean values, error bars indicate standard errors. (Pairwise group comparisons: ns—non-significant, *—$p < 0.05$.)

one standard deviation above the overall mean score), no differences in the posttest were found, $F(2,30) = 0.80$, $p = 0.46$, $\eta^2 = 0.05$, indicating that the different levels of highlighting had no effect on learning gains for these students (Figure 19.3, left). For LPK students (with a pretest score of one standard deviation below the overall mean score), however, a significant difference was found, $F(2,26) = 3.59$, $p = 0.042$, $\eta^2 = 0.22$ (Figure 19.3, right). Pairwise comparisons with Bonferroni-adjusted p-values showed significantly higher posttest scores for the dynamic group compared to the control group ($t(26) = 2.63$, $p = 0.042$, $r = 0.46$), but no differences between control and static ($t(26) = 1.45$, $p = 0.478$, $r = 0.27$) or static and dynamic ($t(26) = 0.76$, $p = 1$, $r = 0.15$). Therefore, the results indicate that LPK is compensated by the corresponding highlighting while these techniques do not hinder HPK students.

Regarding students' evaluation of the videos in terms of comprehensibility and demand, a comparable pattern of results was found. HPK students rated the comprehensibility ($M = 3.52$, S.D. = 0.47) and demand ($M = 1.76$, S.D. = 0.56) on a similar level, regardless of the specific highlighting implemented in the video (comprehensibility: $F(2,30) = 1.34$, $p = 0.28$, $\eta^2 = 0.08$; demand: $F(2,30) = 0.07$, $p = 0.93$, $\eta^2 = 0.01$; Figure 19.4). While LPK students also rated the demand of the videos similarly across groups (demand: $M = 2.39$,

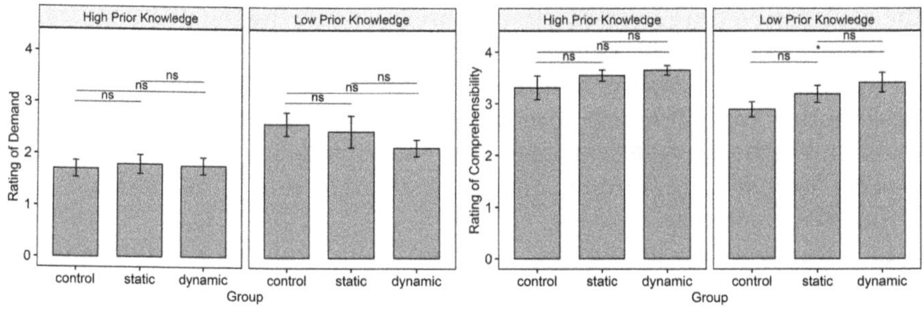

Figure 19.4 Students' ratings of demand (left) and comprehensibility (right) on a four-point Likert scale. Bars indicate mean values, error bars indicate standard errors. (Pairwise group comparisons: ns—non-significant, *—$p < 0.05$.)

S.D. = 0.69, F(2,25, Figure 19.3) = 1.19, $p = 0.31$, $\eta^2 = 0.09$), their ratings regarding the comprehensibility differed significantly ($F(2,26) = 2.72$, $p < 0.05$, $\eta^2 = 0.17$). Here, students watching videos with a dynamic highlighting rated the comprehensibility significantly higher ($M = 3.41$, S.D. = 0.62) than students who watched videos with no highlighting ($M = 2.89$, S.D. = 0.52; $t(26) = 2.32$, $p = 0.029$, $r = 0.41$). Ratings from students watching videos with the static highlighting were in between the other two groups ($M = 3.19$, S.D. = 0.41), but did not differ significantly ($p_s > 0.05$).

19.6 Discussion

The topic of reaction mechanisms in organic chemistry is known to be complex. Therefore, instructional material including illustrations and visualizations for deep understanding is needed in order for students to overcome conceptual and visual difficulties. Based on a series of three purposefully designed videos on nucleophilic substitution reactions, significant learning gains from pre- to post test were observed in our sample of students from beginner organic chemistry courses. This finding is in line with prior studies that used a comparable setting in terms of presenting purposefully designed videos of comparable length to a larger cohort of students, albeit in different domains and age groups (*e.g.*, 8th grade high school physics,[5] and introductory statistics at university[25]).

Based on the randomized controlled trial, learning gains were comparable for all three different highlighting conditions implemented in the different videos. However, the dynamic and the static highlighting seem to differentially support LPK students. Assuming that the provision of highlighting in videos serves to focus the students' attention to the relevant parts of the displayed structural formulae, the highlighting seemed to be most appropriate for LPK students, while HPK students were not affected, neither positively nor negatively. HPK students might already be prepared to connect the explanation to the presented mechanism, while the highlighting has an additional value for LPK students.[19] Given the already high pretest scores of HPK students, with regard to the short duration of the intervention, learning gains in this group may not have been captured in a sufficiently differential manner. Further insight is needed to evaluate in detail how these students process the video, *e.g.*, through qualitative approaches such as think-aloud or retrospective interviewing.[24]

In a similar vein, the different highlighting techniques had little impact on students' evaluation of demand, while students with LPK found videos with highlighting more comprehensible. While not all pairwise comparisons between the different groups were significant, the pattern of scores and ratings indicates that both students' learning and their perception are influenced by the different levels of highlighting in the expected order. However, the effect of the highlighting seems to be substantially smaller compared to the effect of the verbal explanation on students' learning gains (as evidenced by the small to medium effect on students' learning gains from pre- to post-test in all groups). This result concurs with findings from Kulgemeyer.[5] In his study, he varied the quality of the explanation (but not the visualization, as in our study) between videos and found larger learning gains in students who watched explanation videos with a higher explanation quality compared to students watching videos with a lower explanation quality. Taken together, purposefully designing both the explanation and the visualization seems to positively impact students' learning and to make the video accessible to larger student audiences.

19.7 Implications

Tutorial videos are a valuable learning resource for students. While the COVID-19 pandemic has increased the frequency in which both students and instructors have made use of this resource, its usefulness both in traditional and online teaching formats will certainly persist in the future. The present chapter aims to discuss the optimal structure of instructional videos in the field of organic chemistry and the teaching of reaction mechanisms. By providing explicit insights into the making of videos on the specific topic, the approach of how to design such videos may help other chemistry educators who teach organic chemistry and researchers who wish to investigate the effect of such videos in this complex topic. To our perception, such studies are rare.[5]

In our study, videos with different highlighting techniques were assigned to students randomly (to answer the research questions within the project). When results of a pretest are available before watching the videos, the assignment of videos could be based on students' test results. Also, students could be encouraged to select the highlighting that best matches their perceived performance themselves. Both assignment procedures would allow a matching of students and highlighting technique that is more adaptive to students' (perceived) prior knowledge.

Besides different highlighting techniques, adaptations to the explanations provided in the video could address differences in students' prior knowledge. On the one hand, these adaptations could be made with regard to the content, for instance, by increasing or decreasing the level of details or abstractness provided in the explanation or by adding or removing additional elements in the representations that are displayed.[26] While HPK students can often compensate more sparse information by filling in missing details from their background knowledge, LPK students often benefit from more explicit and detailed descriptions. Adaptations to the language itself could address differences in students' language capabilities, for instance, by rephrasing important parts of the explanations, *e.g.*, to provide students with more opportunities to capture the meaning of the provided explanations. While this adaptation would conflict with the coherence principle, students who learn the material in a second language might still benefit. The same expectation can be found with regard to providing subtitles (although conflicting with the redundancy principle).

When implementing videos in teaching, students' learning is also supported when being provided with specific tasks that guide their processing of a video. This might also include, for instance, recall questions so that students are encouraged to re-watch the video or specific scenes.[25] Also, prompts to generate summaries or self-explanations are powerful means of increasing student engagement to systematically recapitulate the content of a video.

The combination of purposefully designing or selecting videos and embedding these videos in targeted learning scenarios increases the likelihood that students actually learn from this resource.[5,13,25] While students often value the use of videos for completing homework or to prepare for exams, videos are not effective for learning *per se*. Embedding videos in constructive learning activities is necessary in not only achieving higher levels of student engagement, but also in emphasizing that "watching" is not the primary or even the only student activity associated with the use of videos when, in fact, "learning" is the intended outcome.

Acknowledgements

We would like to thank the German Research Foundation DFG (Deutsche Forschungsgemeinschaft) for funding this research (project number: 329801962).

References

1. C. Kulgemeyer, *Stud. Sci. Educ.*, 2019, **54**(2), 109.
2. R. E. Mayer, *Multimedia Learning*, Cambridge University Press, Cambridge, United Kingdom, New York, NY, 2021.
3. N. L. Schroeder and A. L. Traxler, *J. Sci. Educ. Technol.*, 2017, **26**(3), 269.
4. C. J. Brame, *CBE Life Sci. Educ.*, 2016, **15**(es6), 1.

5. C. Kulgemeyer, *Res. Sci. Educ.*, 2020, **50**(6), 2441.
6. S. Rönnebeck, S. Bernholt and M. Ropohl, *Stud. Sci. Educ.*, 2016, **52**(2), 161.
7. *Multiple Representations in Chemical Education*, ed. J. K. Gilbert and D. F. Treagust, Springer Netherlands, Dordrecht, 2009.
8. See Chapter 21: M. Milner-Bolotin and V. Milner, in *Digital Learning and Teaching in Chemistry*, ed. Y. J. Dori, C. Ngai and G. Szteinberg, Royal Society of Chemistry, United Kingdom, 2023.
9. See Chapter 17: J. Y. Han and F. M. Fung, in *Digital Learning and Teaching in Chemistry*, ed. Y. J. Dori, C. Ngai and G. Szteinberg, Royal Society of Chemistry, United Kingdom, 2023.
10. M. Kardas and E. J. O'Brien, *Psychol. Sci.*, 2018, **29**(4), 521.
11. J. Sweller, *Cognit. Sci.*, 1988, **12**(2), 257.
12. J. Sweller, *Educ. Psychol. Rev.*, 2010, **22**(2), 123.
13. J. Wittwer and A. Renkl, *Educ. Psychol.*, 2008, **43**(1), 49.
14. H. Sevian and L. Gonsalves, *Int. J. Sci. Educ.*, 2008, **30**(11), 1441.
15. G. Bhattacharyya, *J. Chem. Educ.*, 2013, **90**(10), 1282.
16. N. P. Grove, M. M. Cooper and K. M. Rush, *J. Chem. Educ.*, 2012, **89**(7), 844.
17. O. M. Crandell, M. A. Lockhart and M. M. Cooper, *J. Chem. Educ.*, 2020, **97**(2), 313.
18. N. Graulich, *Chem. Educ. Res. Pract.*, 2015, **16**(1), 9.
19. M. Rodemer, J. Eckhard, N. Graulich and S. Bernholt, *Int. J. Sci. Educ.*, 2021, **43**(17), 2707.
20. M. Rodemer, M. A. Lindner, J. Eckhard, N. Graulich and S. Bernholt, *Appl. Cognit. Psychol.*, 2022, **36**(4), 852.
21. N. E. Bodé, J. M. Deng and A. B. Flynn, *J. Chem. Educ.*, 2019, **96**(6), 1068.
22. N. Graulich and M. Schween, *J. Chem. Educ.*, 2018, **95**(3), 376.
23. I. Caspari, D. Kranz and N. Graulich, *Chem. Educ. Res. Pract.*, 2018, **19**(4), 1117.
24. J. Eckhard, M. Rodemer, S. Bernholt and N. Graulich, *J. Chem. Educ.*, 2022, **99**(6), 2231.
25. K. K. Szpunar, H. G. Jing and D. L. Schacter, *J. Appl. Res. Mem. Cognit.*, 2014, **3**(3), 161.
26. A. J. Dood, J. C. Dood, D. Cruz-Ramírez de Arellano, K. B. Fields and J. R. Raker, *J. Chem. Educ.*, 2020, **97**(10), 3551.

CHAPTER 20

Digital Tools for Equitable In-person and Remote Chemistry Learning

R. PULLEN*[a], A. MOTION[a], S. SCHMID[a], S. GEORGE-WILLIAMS[a], S. WILKINSON[a] AND S. LEACH[a]

[a]School of Chemistry, University of Sydney, Sydney, NSW, Australia
*E-mail: reyne.pullen@sydney.edu.au

20.1 Introduction

In this chapter, we aim to provide a resource to guide practitioners and researchers in future implementation or evaluation of digital or remote laboratories. While extensive research has been disseminated regarding the use of digital technology in STEM education for teaching,[1] this is a constantly changing space within Higher Education (and more broadly) and requires regular re-examination. Most recently the rapid change to online teaching necessitated by the COVID-19 pandemic has led to a period of introspection and a global review of how teaching can occur in online and remote learning environments.

For chemistry educators, while teaching as a whole has been impacted, laboratory learning poses a unique challenge given the historical emphasis on students physically engaging with experiments and developing hands-on-skills. In common with colleagues from across the globe, the changes to our laboratory program have prompted us to consider the digital tools that can positively impact learning and create an equitable student experience in the teaching laboratory. Further, we consider the potential for their inclusion

Advances in Chemistry Education Series No. 11
Digital Learning and Teaching in Chemistry
Edited by Yehudit Judy Dori, Courtney Ngai and Gabriela Szteinberg

Published by the Royal Society of Chemistry, www.rsc.org

beyond the pandemic, as tools to improve the effectiveness and accessibility of laboratory learning.

20.2 Theoretical Framework

In order to consider the interactions between students and technology when learning we have applied Laurillard's Conversational Framework to each of our highlighted examples.[2] The Conversational Framework was developed to critically analyse the use of digital technologies in student learning. This is accomplished through a description of the various forms of interactions and learning that can occur within a classroom or activity, which have been synthesised into six distinct forms of learning: (1) acquisition (*e.g.*, engaging with static digital content that is one-way in nature such as reading articles or listening to recordings); (2) inquiry (*e.g.*, utilizing digital resources or tools to explore questions of interest); (3) practice (*e.g.*, using interactive tools to practice or develop skills); (4) production (*e.g.*, the creation of digital content to demonstrate an understanding of concepts); (5) discussion (*e.g.*, using tools that enable discussion and exchanging of ideas); and (6) collaboration (*e.g.*, using tools that facilitate collaborative work or artefacts).

While this framework is content-dense, we believe this is a useful tool to capture and describe a diverse range of digital tools and laboratories. In the following sections, we will use these six forms of learning to highlight key features and strengths of the digital tools and laboratories we have included in this chapter for discussion.

A full review of all innovations in digital learning and remote laboratories would require far more depth than feasible for a chapter of this length. We have instead selected published examples that are of interest to the authors and exemplify aspects of the Conversational Framework. We believe this selection offers a range of distinct and innovative approaches to digital learning that could inspire further innovation and or reflection on our practice as a community. We also point interested readers towards Brinson's detailed 2015 review of the impact of digital tools on learning outcomes in the teaching laboratory.[3]

20.3 The Purpose of the Chemistry Teaching Laboratory

The primary purpose of a chemistry teaching laboratory has not changed substantially from the inception of the modern format in the early 1800s by von Liebig,[4,5] despite considerable changes to the mindsets of practitioners and the organisational structure of laboratory programs. Through the teaching laboratory, it has primarily been intended that students will develop laboratory techniques, hands-on skills and scientific thinking.[6]

In this chapter, we have chosen to base our discussions on the more recent learning outcomes emerging from an extensive review by Reid and Shah[7] which explores the role of laboratory work within tertiary-level chemistry.

The review summarises four broad learning outcomes that are achievable through a laboratory experience: (1) skills relating to learning chemistry; (2) practical skills; (3) scientific skills; and (4) general skills.

Within the highlighted examples that follow, each incorporates learning outcomes (1) and (4), but within a digital context, we believe incorporation of (2) and (3) are the most challenging. We have grouped the selected articles into two sub-sections. The first features digital tools designed to prepare or support students in the laboratory but to run contiguously with in-person practical sessions (although some were recently adapted as online-only offerings); and the second features digital tools that have been designed to replace in-laboratory learning in chemistry.

Equity and inclusion consideration are central to effective and *affective* teaching and research. We have therefore included an accompanying note for readers that considers how equity and inclusion has or could be more explicitly addressed in each of the highlighted examples.

20.4 Preparative and Supportive Digital Tools

20.4.1 Pre-laboratory Instruction

First year chemistry undergraduate laboratories are a place where vast amounts of new information are presented to students. In addition to the procedures, underlying theoretical concepts and analyses required to complete any given experiment, students must also contend with the physical layout of the laboratory, new glassware and instruments being used and health and safety considerations among others. The sheer amount of information that novices need to absorb, and process can be overwhelming. The origin of this problem can be brought back to the limited capacity of working memory (7 ± 2 concepts),[8] *i.e.*, if too much information is presented, some will be judged as unimportant and filtered out, or will be discarded from working memory without integration. Therefore, instruction needs to take account of existing knowledge if integration is to be facilitated, especially in situations where a large amount of new information is presented.

It has been shown that conceptual understanding developed prior to the laboratory session influences students' ability to process information in the laboratory.[9] Pre-laboratory work, if well designed, serves to pre-construct a scaffold that the students can use to help integrate laboratory-presented information into their existing knowledge structures. So, the principal reason for using pre-laboratory work is that exposure to related theoretical concepts and experiments increases deep learning and performance in the laboratory. However, there are other benefits. Pre-laboratory work eases the transition into new laboratory experiments by allowing students to familiarise themselves with the experiment and gain a clearer understanding of what is expected of them in the laboratory. In addition, effective preparation reduces anxiety while increasing student confidence. This produces a more productive and a more positive learning experience for the student.[10]

Pre-laboratory work, incorporating both introduction to the physical layout of the laboratory as well as introduction to the theoretical and practical aspects of the scheduled laboratory exercise has been implemented effectively in a number of contexts. In one such example from Dalgarno and coworkers,[11] a 3D virtual chemistry laboratory, allowed students to familiarise themselves with the real laboratory space and to explore procedures and apparatus. Students reported that they felt better prepared due to being familiar with the environment prior to attending in-person laboratory sessions and less anxious. More recently a similar idea was employed within a more realistic environment, where the simulation is based on a combination of photographs from the physical laboratory that enable a 360° interactive lab tour.[12] The results from this larger study agree broadly with the earlier example, supporting the utility of such introductory modules.

Similarly, simulations that allow students to work through concepts and step through the procedures required to carry out an experiment, greatly enhance performance, even if prior instruction has occurred. For example, Schmid and Yeung[13] reported that students with limited high-school chemistry background performed as well as students who had completed pre-university level high school chemistry, in a titration assessment, after they worked through a simulation on standard solution preparation. While the complexity of this simulation did not match those described in the following section, a trade-off between sophistication and cognitive overload may occur (Table 20.1).

20.4.2 Interactive Simulations

Many pre-laboratory tasks contain quizzes and videos (as previously discussed) which are passive tasks. It could be argued that simply asking a student to read the laboratory manual and answer a range of questions would have limited impact on their ability to visualise (or better yet, 'practice') a given methodological step. A potential solution to this limitation is the use of interactive technique-based simulations that would, theoretically, allow a student to 'practice' a given technique without significant safety concerns.

One such set of simulations was considered by Blackburn, Williams and Villa-Marcos[14] as provided by the company Learning Science Ltd and implemented at the University of Leicester. These simulations were incorporated into the students' pre-laboratory materials but were not directly assessed nor

Table 20.1 Pre-laboratory instruction as discussed in light of Laurillard's conversational framework.

Acquisition	Yes, students acquire knowledge of physical space and laboratory techniques.	Discussing	No
Inquiry	No	Production	No
Practicing	Yes, the simulations were designed to let students 'practice' outside of the laboratory.	Collaboration	No

compulsory to complete. An example is shown in Figure 20.1 in which students are given access to a simulation focused on the use of a rotary evaporator.

The Learning Science Ltd simulations are generally free-form; students can click on any of the given options (*e.g.*, turning the water bath on or off) in any order they choose. If a student clicks 'check' with an incorrect setup, the simulation will often display issues that this could cause in the lab. For example, an unsupported condenser can fall over, or an untethered water hose can detach pouring water outside of the internal glassware. In response to an incorrect setup, a hint is displayed to the student who is provided with an opportunity to rectify their initial choices (see Table 20.2).

In Blackburn, Williams and Villa-Marcos' study, student engagement with the simulations over a semester was shown to be high with a class of 99

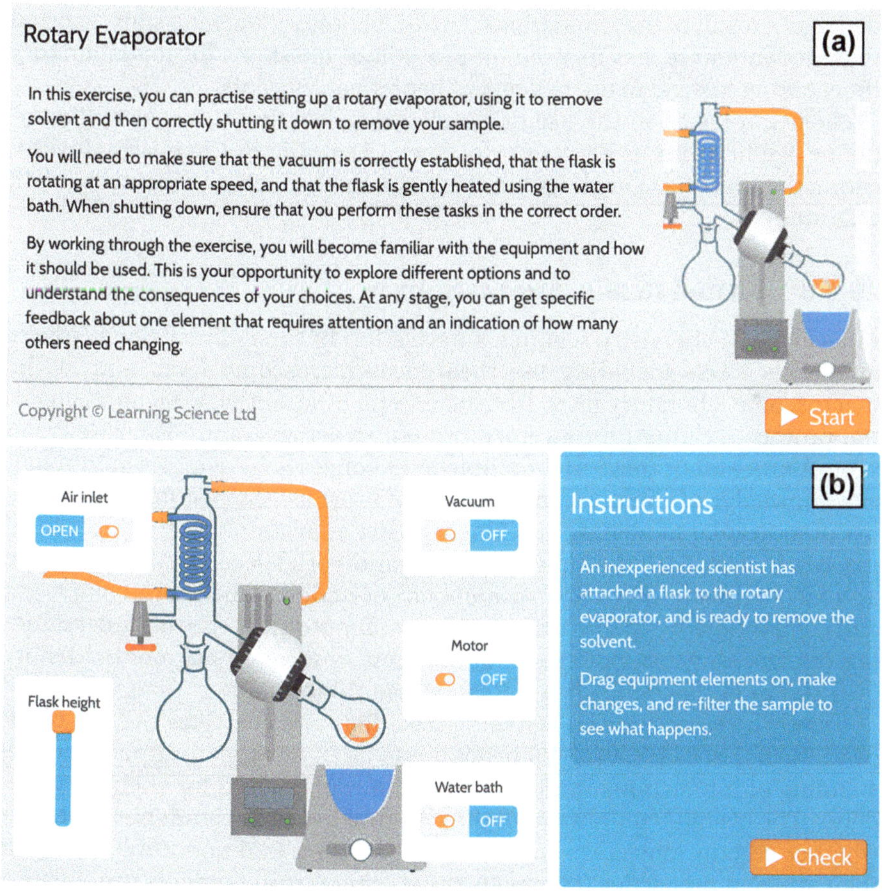

Figure 20.1 The Learning Science Ltd simulation of a rotary evaporator showing (a) the introductory screen and (b) the simulation itself. Reproduced from ref. 14 with permission from American Chemical Society, Copyright 2019.

Table 20.2 The learning sciences resources as considered through the lens of Laurillard's conversational framework.

Acquisition	Yes, the students acquired knowledge of laboratory techniques.	Discussing	No
Inquiry	No	Production	No
Practicing	Yes, the simulations were designed to let students 'practice' outside of the laboratory.	Collaboration	No

students accessing the simulations more than 4000 times over the semester.[14] Questionnaire data also indicated a positive uptick in both student confidence and their belief that the pre-laboratory materials provided were suitable for preparation. These results were also confirmed through a student focus group wherein students raised similar feelings of increased confidence as a result of the simulations. Lastly, laboratory teaching staff reported that students were less focused on procedural questions during laboratory time, opting instead to query deeper, theoretical concepts.

Those interested in the use of simulations, may find Chapter 21 (Smartphone Applications as a Catalyst for Active Learning in Chemistry: Investigating the Ideal Gas Law)[15] within this book to be of interest for an in-depth case-study.

20.4.3 Virtual Reality Simulations

Knowledge of chemistry requires a navigation of three-dimensional spaces at various levels including the theoretical/microscopic level, and at the macroscopic laboratory level. Consider steric hindrances, axial and equatorial protons, Felkin-Anh trajectory, protein structures, and Fisher/Newman projections—all of these microscopic-level concepts have a common requisite knowledge of chemical space and have provided the justification for the use of molecular modelling kits for chemistry education. Similarly, practical macroscopic skills in the lab require navigation of a space to gain experience at 'real' chemistry through manipulation of various laboratory equipment and completion of experiments. One way to support this type of understanding of three-dimensional space for students is through augmented reality (AR) and/or virtual reality (VR) methods (see Table 20.3).

A study by Ferrell and co-workers[16] describes and evaluates the use of VR (with the use of a HTC VIVE VR headset and iMD software) to support understanding of interactions in the dynamic molecular world for chemistry students. The overall requirement of the experiment was for students to predict whether certain molecules could move through a C60 nanotube based on their size and properties (Figure 20.2). To achieve this, students were placed into a virtual room and provided with the opportunity to interact with the nanotube and methane (and other) molecules. The students were then able to view the molecules through various models such as space filling and electronic fields which enriched their predictions.

Table 20.3 The VR chemistry learning experience as considered through the lens of Laurillard's conversational framework.

Acquisition	Yes, the students acquired an understanding of 3D chemical properties.	Discussing	No
Inquiry	Somewhat, the students are provided the opportunity to investigate properties of molecules to enhance their understanding, but the overall lesson is highly structured.	Production	No
Practicing	Yes, students were provided with tools to practice with prior to the prediction.	Collaboration	No

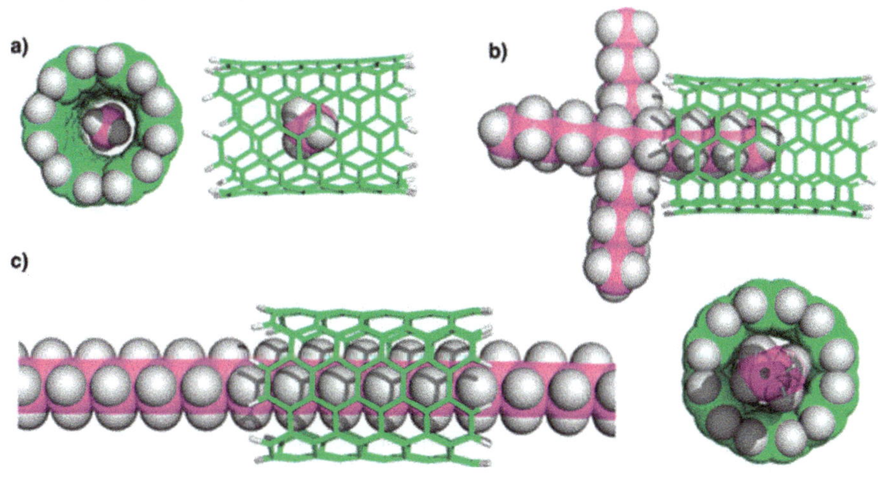

Figure 20.2 Example of a problem where students were required to predict whether molecule x (pink; examples (a), (b) and (c) showing three molecule options) would fit through a C60 nanotube (green). Reproduced from ref. 16 with permission from American Chemical Society, Copyright 2019.

The authors conducted an analysis of the student responses to the program to determine how well the experience improved their ability to predict chemical interactions, and their perceptions of the value of the experience. In general, results indicated a positive increase in predictive ability when compared to the control group, demonstrating the efficacy of the supporting tool. Similarly, the qualitative student responses were positive; students perceived the visualisation aspect to be a highly valuable contributor to their understanding.

> "There are only benefits to using VR as a visual tool because textbooks can only show flat images. Even using dashes and wedges can only be so helpful, whereas VR allows for better spatial representations." One student response.

While valuable, there are limitations to this learning method. First, it requires an expensive technology setup and knowledge on integration of virtual reality software. The hardware required to run this lesson, as detailed in the supporting information, includes a VR headset ($500–$800 USD) and a computer with a competent processor, RAM, and graphics card. Specifically, the authors use a NVIDIA GTX 1080 Ti graphics card which is approximately $699 USD—notwithstanding the price of the remaining parts of the system. Second, some students indicated that the virtual reality instruction gave them motion sickness or caused them to feel nauseated and generally uncomfortable.

Those interested in the use of VR or AR, may find Chapter 17 (Applications of Digital Technology in Chemical Education)[17] within this book to be of interest for an in-depth case-study.

20.5 Replacement Laboratories/Experiences

20.5.1 'Kitchen Chemistry' at Home

The provision of a laboratory experience from student homes is one option when access to a physical laboratory is not possible. These approaches are fraught with logistical challenges but when successful, can offer rich learning opportunities to those studying remotely or unable to physically attend laboratory spaces due to accessibility requirements or unforeseen circumstances such as the COVID-19 pandemic. Often, these activities are pared back or simplified to compensate for a lack of experimental equipment or safety considerations but are designed for students to learn key experimental skills or techniques (Table 20.4).

One such example of a 'kitchen chemistry' approach was provided by Schultz, Callahan and Miltiadous[18] as a direct response to the in-person restrictions in place at the time of the COVID-19 pandemic. This approach described the development and implementation of several introductory level chemistry activities conducted at home rather than a traditional teaching laboratory. A large first-year cohort (~1000) of students who had experienced little to no chemistry background in high school were required to undertake several experiments in their homes, targeting several key hands-on skills

Table 20.4 'Kitchen chemistry' activities as considered through the lens of Laurillard's conversational framework.

Acquisition	Yes, the students acquired knowledge of laboratory techniques.	Discussing	No
Inquiry	At times, depending on the nature of the experiment being undertaken.	Production	Yes, students are producing short videos.
Practicing	Yes, students were practicing skills and learning through observation.	Collaboration	No

including observations and data collection/handling. The authors noted that the simplicity of the activities served the dual purpose of providing a logistically feasible approach (providing resources, safety, on-site or online support) for such a large cohort and, being equitable and accessible. Expanding on the second point, the authors noted that a significant challenge in bringing laboratory activities into student homes is the lack of information educators possess regarding the places that students live and their access to the necessary materials or equipment to undertake the activity. By assuming students have access to the 'bare bones' (*e.g.*, basic kitchen facilities such as containers, water, hot plates or kettle, *etc.*; basic pantry or cleaning goods such as salt, vinegar, *etc.*; access to electricity), the risk of estranging students is minimised.

To this end, students undertook two 'kitchen chemistry' experiments: (1) an exploration of colligative properties using household materials (sugar, salt, water, *etc.*) to connect macroscopic observations with microscopic and symbolic understanding; and, (2) a two-part experiment including an expansion of the classic red cabbage indicator experiment with the flexibility for students to investigate other coloured natural products (berries, turmeric, flowers, *etc.*) and an investigation into gas evolution stoichiometry.

20.5.2 Remote Activated Labs

Replacement distance laboratories often rely upon simulations which, while useful, do not completely describe the relevant experience in chemistry so do not provide an equitable experience to students who cannot participate in person. To solve this, researchers have looked towards remote activated laboratories. These types of laboratories utilise experiments set up for students who, while assisted by technology, can manipulate and complete an experiment in a real laboratory at a distance. Remote experimental setups typically involve a student logging into software *via* the internet and manipulating an experimental apparatus that is set-up to be controlled at a distance. The student simultaneously receives real-time audio-visual feedback of their experiment progress (Table 20.5).

Table 20.5 The use of remote laboratory electrochemistry experiments as considered through the lens of Laurillard's conversational framework.

Acquisition	Yes, students are provided with detailed theory and practice material.	Discussing	No
Inquiry	Both, students are provided structured experiments *and* are provided opportunities to inquire with comparative experiments.	Production	No
Practicing	Yes, students are provided with opportunities to run simulations and experiments multiple times outside of a real lab.	Collaboration	No

Saxena and Satsangee[19] describe a remotely activated laboratory setup where students can participate in an array of electrochemical experiments. These include a dopamine cyclic voltammetry experiment, a DPASV analysis of a Yamuna River sample for heavy metal contamination, an EIS of a failing steel surface coating, and a comparative study of electrochemical solutions (Figure 20.3). Practically, once students have booked a time, they can input instructions through the server to the machine which conducts the experiment and sends them audio-visual feedback and relevant data for them to analyse.

This method can have significant limitations depending on the level of experiment and assistance provided by instructors. For example, the requisite equipment can be expensive and difficult to set up and requires regular maintenance. Additionally, the set-up of such an experiment can be complicated and may require coordination with an engineering department. The overall experiment therefore requires various pieces of hardware and software, including:

- software (server and websites)
- sensors to monitor and track experiment for feedback to students

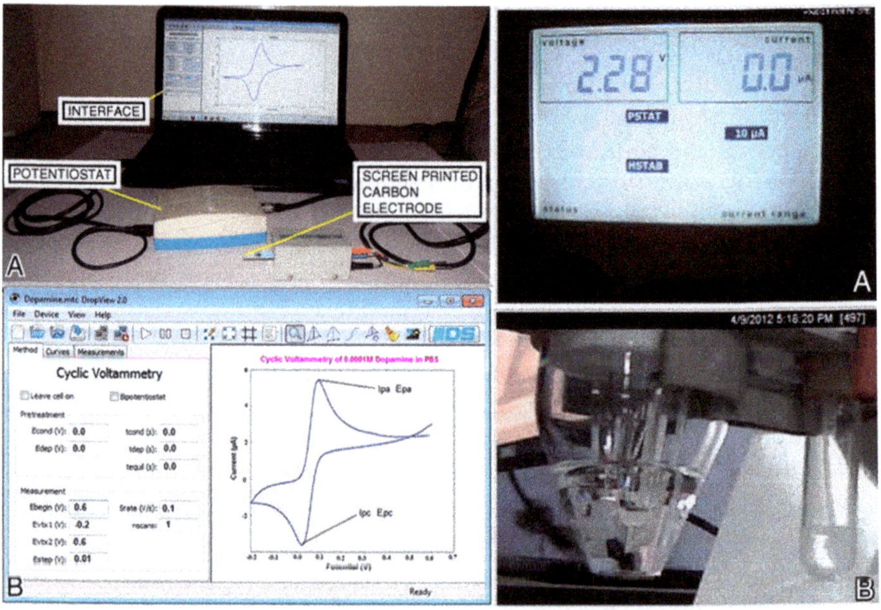

Figure 20.3 Photos of example electrochemistry experiments performed *via* remote access. Left (A) and (B) show the instrument set up shown remotely *via* webcam and accompanying output. Right (A) and (B) demonstrate simultaneous viewing of a potentiostat and electrochemical cell *via* webcam. Reproduced from ref. 19 with permission from American Chemical Society, Copyright 2014.

- input/output hardware to mediate the instrument and software
- coordinating research group with access to instruments.

Despite the limitations, the positive outcomes for this research are clear. First, the array of experiments available highlights the possibility of significant diversity for remote activated chemistry experiences. Their website details the broad range of experiments that are available which, notably, are not limited to electrochemistry but contain other areas including spectroscopy, and colloids and surfaces (https://www.vlab.co.in/broad-area-chemical-sciences). Second, these experiments provide opportunities for equitable educational experiences in distance chemistry globally, especially through the initiatives for access to experiments in remote communities described throughout the paper. Lastly, Saxena and Satsangee's intervention has been repeatedly successful across multiple institutions within their network which is indicative of the success programs generalisability across various environments and different student learning environments.[19]

20.5.3 Videos of Experiments

Laboratory teaching is driven by guided participant-centred enquiries leading to epistemological and procedural understandings. These connections are primarily forged through visual observations experienced by students during laboratory sessions. In our view, video footage of laboratory experiments offers the closest approximation to observing an experiment while present in the laboratory and was a common strategy, adopted by laboratory educators in their transition to online laboratory lessons during the COVID-19 pandemic.[20] A carefully scripted video allows the instructor to focus on communicating the important elements of an experiment through clearer images, close-ups, multiple angles, slow-motion, time-lapse and by skipping laborious steps. It can link to other resources to explain new concepts and allows students to learn difficult concepts at their own pace. However, video footage alone does not support the same level of individual inquiry supported in the laboratory, which can lead to instructor-centred lessons that lower engagement. This balance can be swayed by a number of supporting technologies (*e.g.*, resource repositories such as Canvas or other learning management systems, third-party asynchronous communication routes such as discussion boards of Padlet, simulations to familiarise or provide interactivity such as Learning Sciences as discussed earlier in this chapter, *etc.*) now available to educators.

Wang and Ren[21] describe a number of assessment strategies paired with laboratory videos to assess student understanding of laboratory content. Their first example embeds various question types (multiple choice, hotspots) within the videos that students must answer in order to progress. The questions are strategically placed to question a student's understanding of an observation or their rationale for a particular step of the procedure. The findings from Wang and Ren's study indicated that embedded assessments promoted engagement and guided students to connect practical applications to chemistry theory.[21] The authors note that this approach is best accompanied in a way that is analogous to traditional on-campus laboratory delivery, with a teaching assistant available in a synchronous 'virtual' laboratory lesson to guide students through the experiment and answer student queries.

The authors complete their lesson by assigning post-lab reports which question the rationale behind specific steps of the procedure and require the students to interpret data, mirroring traditional laboratory lessons. As teaching students how to manipulate equipment or perform laboratory techniques is not possible in a digital realm, the authors have instead focused efforts on developing a deeper understanding and intuition of laboratory procedures in students (Table 20.6).

In line with the trend of synchronous lecture delivery, Woelk and Whitefield[22] opted to forego asynchronous laboratory videos and instead live-stream laboratory experiments to their students. Their synchronous lab demonstrations allow students to ask questions or alter the experiment in real-time (live inquiry). Students also witness unedited footage of data collection (data authenticity) which may involve the observation and treatment of result abnormalities—an important aspect of practical chemistry, which is often edited from asynchronous videos. The authors highlight the importance of good camera control and placement. For example, moving the camera too fast can blur live-stream footage or an errant camera angle may miss a crucial observation of the experiment or receive interference from lab lighting and/or audio (*e.g.*, fume cabinet humming). The authors include waiting times (*e.g.*, stirring/refluxing or waste disposal), opting to present a full laboratory experience including the time management of certain procedures. While synchronous delivery has the advantage of real-time exchange between students and educators, it raises some considerations for equity and inclusion. For example, content delivery to students with poor internet connectivity or who reside in different time zones could pose challenges for participation.

Table 20.6 Use of experiment videos as considered through the lens of Laurillard's conversational framework.

Acquisition	Yes, students are provided with practical procedures, visual cues, and experiment data.	Discussing	Yes. Breakout rooms or assigned groups can be used to discuss and interpret observations and results.
Inquiry	No	Production	Yes, students produce post-lab assessments to confer their understanding.
Practicing	No	Collaboration	Yes. Groups can be assigned to work through the results and generate an output.

20.6 Equity, Diversity, and Inclusion Considerations

One of the significant barriers to traditional laboratory sessions is the requirement for students to be present within the classroom in-person and at a specific time. Digital technologies as outlined above, while not an equal replacement for in-person experiences that develop hands-on skills, enable flexibility in the location and timing of these activities. For example, the simulations used by Blackburn, Villa-Marcos and Williams,[14] can be accessed at all hours of the day and do not require significant internet bandwidth. The examples discussed within this chapter have been framed primarily as asynchronous, but it is important to note that if used synchronously, as with any activity, barriers for accessibility may result (*e.g.*, timing depending on student geolocation or employment or caring responsibilities).

One barrier posed by the use of any digital technology is access to the internet or required hardware to operate or engage with each activity. Ferrell and co-workers'[16] use of virtual reality simulations is a prime example of this. Examples such as Schultz, Callahan and Miltiadous'[18] 'kitchen chemistry' approach can be utilised to overcome this issue, by allowing students to complete experiments asynchronously and with inexpensive resources. There may of course be situations where access to an appropriate space might not be possible, and this can be partly mitigated by sending out equipment or reagents if required.

In common, with significant shortfalls in chemistry education more broadly, digital technologies depend heavily on visual processes and language. While language can be subsidised with translated artefacts (*e.g.*, using alternative text) or auxiliary means of support for students, the visual dependence can create a barrier for people who are blind or have low vision. The pandemic shone a light on the different ways that we can work to make learning more accessible for all students within both digital and in-person contexts. We have already instigated work to make our in-person laboratories accessible for students who are blind or have low vision inspired by the work of Supalo, Isaacson and Lombardi[23] and are working to improve accessibility for all student needs in our teaching.

20.7 A Reflection from the University of Sydney

To place this reflection within our context, we will briefly introduce the School of Chemistry at the University of Sydney. Central to our research and teaching is a vibrant and multidisciplinary team of academic and research staff, well supported by professional, technical and administrative staff. The school comprises more than 25 academic staff leading research groups, with >50 postdoctoral research fellows and >100 postgraduate students, including a large cohort of international students. Similarly, our undergraduate enrolments average >2700 students per semester from diverse academic, cultural, and socioeconomic backgrounds. First-year chemistry acts as a pipeline beyond chemistry majors to a large variety of tailored degree programs.

The school remains one of the largest chemistry departments in Australia and is well supported with modern laboratory facilities and instrumentation. When combined with excellent access to major international research facilities, this provides an outstanding opportunity for our students to engage with a wide variety of instrumentation and laboratory techniques as they progress through their degrees.

20.7.1 Teaching Laboratories Before the Pandemic

Prior to COVID-19, the teaching laboratories in the School of Chemistry pre-laboratory support tools mainly focused on short pre-laboratory quizzes. Some cohorts were supported with technique-focused videos with plans for wider roll-out, and additional tools such as technique simulations were also slated for inclusion. The structure of the laboratory activities was also being remodeled to include more context and inquiry-based activities to align with modern literature, and to address student feedback. Laboratory practical sessions were exclusively an on-campus activity, with assessment either undertaken in the laboratory itself (first-year) or in the form of scientific reports (higher-years).

20.7.2 Teaching Laboratories During the Pandemic

In response to the COVID-19 pandemic, as many educators across the world experienced, a sudden shift to emergency online teaching was required. As lecturers and laboratory directors within the School of Chemistry, we led many of the necessary changes. We recognised that the simplest and most expedient pivot was to provide data from our existing experiments to students acknowledging that this would not provide a rich learning experience comparable to the face-to-face equivalent.

In first-year courses, the original interventions included recorded videos and online quizzes supported by the Learning Sciences Ltd simulations. Within a cohort of 2000 students, simulations were accessed over 20 000 times during the semester coinciding with the start of the COVID-19 pandemic. Questionnaire data and interviews showed a similar trend to the University of Leicester results, with these data being analysed for publication.

As restrictions eased, we adopted a hybrid mode of labs that accommodated in-person labs for local students while delivering purely online labs for off-shore international students. To maximise equity, we adopted more authentic assessment tasks which focused on data analysis, technique understanding and science communication. On-campus students were able to collect data themselves in the labs while remote students were provided pre-recorded videos of identical experiments with data provided within videos or supplementary material. In both on-campus and remote labs, a laboratory teaching assistant was present to guide students through the experiment and data interpretation and technique simulations accompanied the labs as preparative activities.

20.7.3 Citizen Science: Digital Tools to Support Authentic Research in the Laboratory

The most significant limitation to many of the approaches highlighted in the chapter, and in common with many traditional laboratories, is the overuse of heavily controlled experimental parameters. When the problem, methodology, and outcome are all known and understood, the potential for students to engage with a laboratory experience that reflects authentic practices as graduates is limited. In our own laboratory teaching, authentic research experiences for students have been realised through citizen science.[24,25] Building on our pre-pandemic experiences of homegrown citizen science projects, we turned to an online citizen science project for inspiration.[26] The Foldit citizen science project enables participants to contribute to scientific research into protein structure by playing a crowd-sourced, online game. Citizen scientists playing Foldit have contributed to a number of important discoveries, for example the correct prediction of the Mason–Pfizer monkey virus, which shares structural similarities with HIV and the design of *de novo* proteins that could be successfully expressed in the research laboratory.[27] The data used to "score" proposed structures are based on modelling and experimental data. This allows participants to engage in a cycle of inquiry by proposing structures based on prior knowledge before iteratively improving these with strategic changes to later proposals. Notably, Foldit had released new challenges relating to the design of proteins that could bind to the coronavirus spike protein, thus creating an opportunity for authentic learning directly linked to the cancellation of usual laboratory teaching. We piloted activities and assessments scaffolded around the coronavirus puzzle for an advanced first year cohort and then adapted this for second year students. Students worked in small groups and were first prompted to download the Education version of Foldit which contains a series of tutorials designed to familiarise them with both the software and the basics of protein folding. Optional Zoom sessions were scheduled with teaching-assistants to support students through this training and students engaged with their chosen 'puzzle', trying a variety of different approaches to achieve the highest 'score' (as generated by algorithms that Foldit determine for each problem based on theoretical modelling). Through this process, students took careful notes of each attempt, recording any common approaches that resulted in high scores.

Assessment included a peer- and self-evaluation and completion of an infographic to communicate their findings to a general audience, and an individual reflection to explore student understanding of protein folding, the role of citizen science in addressing global issues, and their own experience and understanding of authentic research. This activity is freely available (along with many others) within the ACDS digital repository that can be found at https://www.acds.edu.au/teaching-learning/resource-repository/.

20.7.4 Post-pandemic and into the Future

In common with colleagues and peers across the globe, we are now considering how to move forward with laboratory learning. These decisions will be supported by the digital tools trialed in our teaching, and the experiences that others have shared so openly at conferences during the pandemic and that are emerging in the literature. While we are still undertaking this review, it is clear that digital tools will play a more prominent role in our teaching in the future. In their implementation we must keep our students and their learning needs at the forefront and be particularly mindful of the equity, diversity and inclusion considerations highlighted in this period.

In our own practice, we intend to continue using simulations as pre-laboratory preparation whereas recorded experiments will likely shift from a replacement of the on-campus activities to a pre-laboratory supporting role. Additionally, citizen science will continue to be investigated and incorporated as they provide unique research experiences that complement the on-campus activities.

Outside of our own practice, the use of the remote-activated laboratories or the 'kitchen chemistry' at home will continue to be vital for rural or remote institutions. Students with decreased on-campus access or those with other competing responsibilities, can benefit greatly from such asynchronous experiences. While undoubtedly urban institutions have students in similar situations, we feel that digital tools should supplement rather than supersede on-campus activities where possible as we believe they provide the most impactful and positive learning experience.

Acknowledgements

The authors would like to acknowledge the substantial contributions of both our colleagues and students. Without their collective experiences, progress in our practice would not occur.

References

1. C. A. Cañizares and Z. T. Faur, *IEEE Trans. Educ.*, 1997, **40**(3), 166.
2. D. Laurillard, *Teaching as a Design Science: Building Pedagogical Patterns for Learning and Technology*, Routledge, New York, NY, 3rd edn, 2012.
3. J. R. Brinson, *Comput. Educ.*, 2015, **87**, 218.
4. S. L. Bretz, *J. Chem. Educ.*, 2019, **96**, 193.
5. M. K. Seery, *J. Chem. Educ.*, 2020, **97**(6), 1511.
6. J. B. Morrell, *Ambix*, 1972, **19**, 1.
7. N. Reid and I. Shah, *Chem. Educ. Res. Pract.*, 2006, **8**(2), 172.
8. G. A. Miller, *Psychol. Rev.*, 1956, **63**, 81.
9. A. H. Johnstone, *J. Chem. Educ.*, 1997, **74**, 262.
10. B. P. Koehler and J. N. Orvis, *J. Chem. Educ.*, 2003, **80**, 606.

11. B. Dalgarno, A. G. Bishop, W. Adlong and D. R. Bedgood, *Comput. Educ.*, 2009, **53**(3), 853.
12. T. D. Clemons, L. Fouché, C. Rummey, R. E. Lopez and D. Spagnoli, *J. Chem. Educ.*, 2019, **96**, 1491.
13. S. Schmid and A. Yeung, *Res. Dev. Higher Educ.*, 2005, **28**, 471.
14. R. A. R. Blackburn, B. Villa-Marcos and D. P. Williams, *J. Chem. Educ.*, 2019, **96**(1), 153.
15. See Chapter 21: M. Milner-Bolotin and V. Milner, in *Digital Learning and Teaching in Chemistry*, ed. Y. J. Dori, C. Ngai and G. Szteinberg, Royal Society of Chemistry, United Kingdom, 2023.
16. J. B. Ferrell, J. P. Campbell, D. R. McCarthy, K. T. McKay, M. Hensinger, R. Srinivasan and S. T. Schneebeli, *J. Chem. Educ.*, 2019, **96**(9), 1961.
17. See Chapter 17: J. Y. Han and F. M. Fung, in *Digital Learning and Teaching in Chemistry*, ed. Y. J. Dori, C. Ngai and G. Szteinberg, Royal Society of Chemistry, United Kingdom, 2023.
18. M. Schultz, D. L. Callahan and A. Miltiadous, *J. Chem. Educ.*, 2020, **97**(9), 2678.
19. S. Saxena and S. P. Satsangee, *J. Chem. Educ.*, 2014, **91**(3), 368.
20. S. George-Williams, A. Motion, R. Pullen, P. J. Rutledge, S. Schmid and S. Wilkinson, *J. Chem. Educ.*, 2020, **97**(9), 2928.
21. L. Wang and J. Ren, *J. Chem. Educ.*, 2020, **97**, 3002.
22. K. Woelk and P. D. Whitefield, *J. Chem. Educ.*, 2020, **97**, 2996.
23. C. A. Supalo, M. D. Isaacson and M. V. Lombardi, *J. Chem. Educ.*, 2014, **91**(2), 195–199.
24. A. Motion, *Chemistry World*, 2019, accessed 22 March 2022 at: https://www.chemistryworld.com/opinion/how-teenagers-are-disrupting-drug-discovery/3010027.article.
25. Y. N. Golumbic and A. Motion, *Citiz. Sci.: Theory Pract.*, 2021, **6**(1), 31.
26. J. A. Miller, F. Khatib, H. Hammond, S. Cooper and S. Horowitz, *Nat. Struct. Mol. Biol.*, 2020, **27**, 769.
27. A. Motion, *Chemistry World*, 2019, accessed 22 March 2022 at: https://www.chemistryworld.com/opinion/bringing-video-games-into-the-protein-fold/3010769.article.

CHAPTER 21

Smartphone Applications as a Catalyst for Active Learning in Chemistry: Investigating the Ideal Gas Law

M. MILNER-BOLOTIN*[a] AND V. MILNER[b]

[a]Department of Curriculum and Pedagogy, University of British Columbia, Vancouver, V6T1Z4, Canada; [b]Department of Physics and Astronomy, University of British Columbia, Vancouver, V6T1Z4, Canada
*E-mail: marina.milner-bolotin@ubc.ca

21.1 Introduction and Chapter Goals

Since the launch of Sputnik in 1957, and consequently the birth of the science education research movement, scholars have accumulated ample evidence pointing to active engagement as one of the key factors in successful science, technology, engineering and mathematics (STEM) learning at both K–12 and post-secondary levels.[1,2] By active science engagement, we mean pedagogical approaches that involve students in doing science, such as building models, predicting experimental results, collecting and analyzing data, and discussing different explanations for scientific phenomena, as opposed to passively receiving the information provided by the instructor.[3–5] Various *active engagement* pedagogies[4,6,7] were proposed to implement these findings into practice. Many of them are being discussed and analyzed in this book. With the rapid development of novel educational technologies in

Advances in Chemistry Education Series No. 11
Digital Learning and Teaching in Chemistry
Edited by Yehudit Judy Dori, Courtney Ngai and Gabriela Szteinberg

Published by the Royal Society of Chemistry, www.rsc.org

recent decades, these STEM active learning strategies spurred the utilization of data collection and analysis tools, such as sensors[4] and video analysis software,[8] numerous instant feedback and classroom response systems;[9] as well as computer simulations and visualizations.[10–13] The technology-rich STEM learning environments have positively affected STEM learning for thousands of students,[14] yet their high cost has made access to them prohibitive for less affluent K–12 schools and post-secondary institutions.[15] Moreover, the availability of these technologies is too often restricted to schools, thus limiting student STEM learning to the classroom.

The recent challenges created by the COVID-19 pandemic have underscored the importance of designing remote hands-on student STEM experiences.[12,16–18] This is especially relevant to learning chemistry and physics, as these subjects are fraught with abstract concepts that many students and even future science teachers find challenging.[8,19] Understanding these concepts often requires extra time, more advanced mathematical knowledge, as well as the ability to connect algebraic, verbal, and visual representations.[20] These concepts include, but are not limited to, fluid pressure, heat, temperature, pH level, audio and optical spectra, sound level and intensity.[21,22] For example, pH level (the ***power (potential) of Hydrogen***[23]) is an important concept in the secondary chemistry curriculum.[24] It is defined as the negative logarithm of the hydrogen ion concentration in a water-based solution. Many science curricula across the world, such as in British Columbia, Canada, curriculum[24] suggest introducing pH scale to grade 10 students, who at that point are unfamiliar with the mathematics required to interpret it meaningfully, as logarithms are only introduced in grade 12. So how can students build a deep understanding of this concept as opposed to memorizing pH levels of various solutions? A number of fundamental science concepts, such as the ideal gas model, gas pressure and temperature, suffer the same fate, forcing students to memorize, as opposed to apply and make sense of new science ideas.[19] Understanding these abstract concepts can be significantly aided by applying them to collect, visualize, and interpret data using educational applications (apps) freely accessible on students' smartphones.[25–27]

In light of these findings, this chapter has three goals. First, to examine how smartphones, novel technologies that 21st century students already have in their pockets, can be used to facilitate active engagement inside and outside of the chemistry classrooms. Second, to consider an example of an ideal gas law and suggest relevant pedagogical approaches that use smartphone apps at secondary and post-secondary levels. We chose the ideal gas law exploration as it is found in many secondary science curricula both in chemistry and in physics. Third, to propose how freely available smartphone apps that already have ample teaching resources developed for their use in STEM classrooms can be incorporated in chemistry (and STEM in general) teacher education to (a) make future teachers aware of the possibilities offered by these powerful devices and (b) help them experience science learning with smartphones firsthand. Instead of considering smartphones to be a distraction or an obstacle to STEM learning, we call on teachers to uncover the unlimited possibilities offered by these omnipotent gadgets already found in students' pockets, thus engaging students in authentic chemistry learning be it in school or at home.

21.2 Research-based Science Smartphone Applications

To achieve this chapter's goals, we first introduce two research-based smartphone apps that offer wide-ranging opportunities to engage students in active STEM learning.[26] We decided to focus on two different apps for the following reasons. First, they can support two pedagogically distinct types of student active engagement with STEM. Second, both apps are freely available and do not require ongoing internet access. Third, these apps are based on STEM education research and have undergone significant testing with various student populations, and yet they offer different ways to engage students. Fourth, these apps have teaching communities associated with them, that have produced and tested numerous activities, lesson plans, virtual labs, and authentic experiments. The first app is a suite of PhET simulations that offers students an opportunity to model and visualize various STEM phenomena (see Figure 21.1(a)).[28] The second app allows students to collect and analyze data in real time, as well as transfer it to their computers for additional in-depth analysis (see Figure 21.1(b)).

21.2.1 PhET Simulation Suite as a Smartphone App

PhET (Physics Educational Technology) Interactive Simulations is a project founded by Carl Wieman in 2002. Since then, the PhET team has developed hundreds of interactive STEM simulations for K–12 and post-secondary learners. PhET is designed to promote student conceptual understanding of STEM fields by actively engaging them in scientific exploration and increasing their interest in science.[29] Thus, PhET simulations are designed to be engaging, interactive, connected to the real world, and yet representing the phenomena under study as accurately as possible.[3] Unlike many other STEM visualizations, PhET simulations are based on numerical solutions of underlying equations, governing the behavior of the physical system. For example, in the Gas Properties STEM simulation,[30] the size of the container is assumed to be 10 nm, to allow for the realistic number of particles in that volume. At the same time, pressure and temperature are calculated as macroscopic parameters of the system (see Figure 21.2).

While the simplest of the simulations—*Ideal* (see Figure 21.3) allows students to modify only three parameters: the number of particles, the types of particles, and the volume of the container, at the next level—*Explore*, the students can heat up the gas and even overheat it, such as the gas pressure will grow sufficiently to blow off the container's lid (see Figure 21.3). Moreover, in every type of the simulation, one can keep some parameters constant

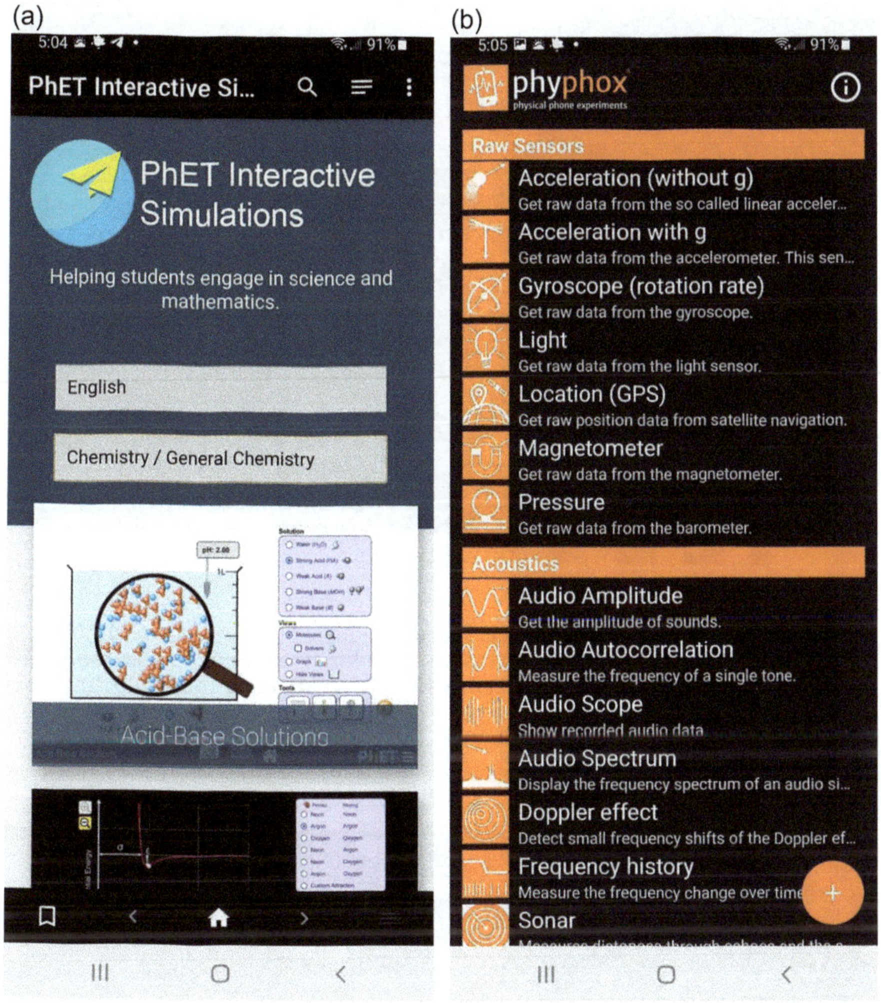

Figure 21.1 Screenshots of (a) PhET and (b) Phyphox smartphone apps. Reproduced from https://phet.colorado.edu/en/, and https://phyphox.org/, under the terms of the CC BY 4.0 license https://creativecommons.org/licenses/by/4.0/.

(*i.e.*, volume, temperature or pressure), thus isolating the parameters of interest and investigating how they are interrelated.

PhET simulations also use multiple representations, address known student conceptual difficulties and misconceptions, incorporate game elements, as well as allow teachers to increase the virtual experiment complexity within the same group of simulations (see Figures 21.2 and 21.3). As a result, this platform allows students to learn about the real-world physical systems through the virtual experience of cleverly-designed interactive simulations that allow students to visualize abstract phenomena, build experimental setups, make

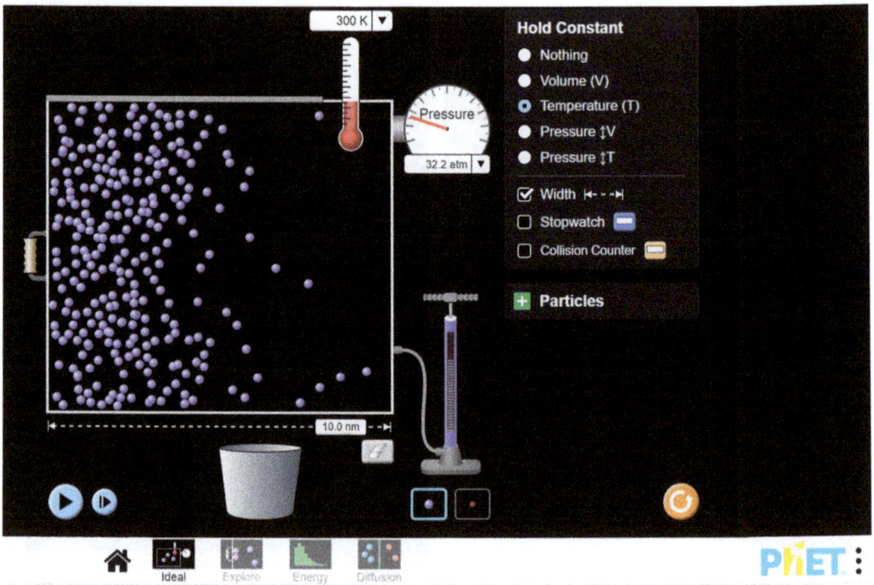

Figure 21.2 A screenshot of the Gas Properties: Ideal PhET interactive simulation. Reproduced from https://phet.colorado.edu/en/, under the terms of the CC BY 4.0 license https://creativecommons.org/licenses/by/4.0/.

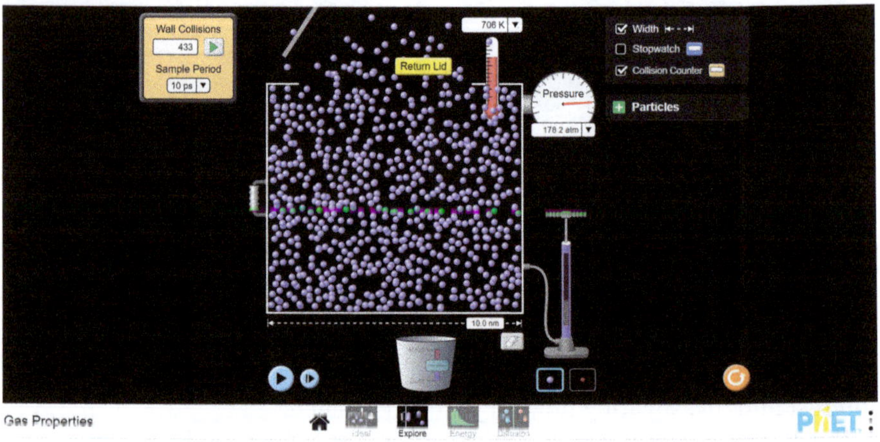

Figure 21.3 A screenshot of the Gas Properties: Explore PhET interactive simulation. The pressure of the gas as a result of heating blown off the lid of the gas container. Reproduced from https://phet.colorado.edu/en/, under the terms of the CC BY 4.0 license https://creativecommons.org/licenses/by/4.0/.

predictions and test them using virtual experiments.[31,32] While PhET simulations were originally built as computer programs, in the last years, PhET has been expanded to work on iPads, tablets, as well as smartphones.[33,34]

The PhET environment has numerous advantages, yet it is a relatively closed system and does not allow students to collect real time data. This can be done with the smartphone app, called *Phyphox*, discussed below.

21.2.2 Phyphox Smartphone App

Phyphox (**phy**sics **pho**ne experiments) app was designed at the University of Aachen, Germany and was spearheaded by Sebastian Staacks and his colleagues[35] (see Figure 21.1(b)). Like PhET, Phyphox was originally designed to support physics learning, but it can be expanded to other science fields, such as chemistry, biology, earth science, mathematics, *etc.*[36] Phyphox is also based on STEM education research[25,37] as it aims at promoting student active engaging through data collection and analysis, but it is much more open than PhET. While PhET has a structured experimental setup, where the students have limited degrees of freedom for its manipulation, Phyphox allows students to use sensors already available in their smartphones, such as motion sensors, accelerometer sensor, pressure sensors, light sensors, a gyroscope, magnetic field sensors, *etc.* for their own science experiments. The experiments are not limited by the Phyphox app but are widely open to students' imagination. The only limitations are the variables that can be measured and the range and sensitivity of the measurement. For example, students can measure acceleration, rate of rotation, ambient pressure, magnetic field, illuminance, frequencies and intensities of sound, *etc.* Then the students can export data into their favourite software (*e.g.*, MS Excel) and conduct the data analysis on their computers. In addition, Phyphox can connect wirelessly to the computer through its remote access option. Hence, while the smartphone works as the measuring device, the data can be simultaneously displayed on students' computers as will be demonstrated below.

21.3 Smartphone-enabled Investigations of Ideal Gas Laws

Let us examine how chemistry teachers can use smartphones to facilitate their learning goals, using the example of the ideal gas law, which is found in the secondary and post-secondary chemistry curriculum.

21.3.1 Examining Ideal Gas Laws with PhET Simulations

The ideal gas model (IGM) is one of the fundamental abstract models in both chemistry and physics encountered by students in their secondary and later post-secondary chemistry and physics courses.[38] It is used to construct the

Molecular Kinetic Theory of Gases, as well as to explain phenomena such as heat and energy transfer, thermal insulation, various thermodynamic processes, the relationships between temperature, volume and pressure of gasses, and finally, the laws of thermodynamics. IGM is one of the first models where students have to analyze an enormous number of miniscule invisible objects on both microscopic (or even sub-microscopic) and macroscopic levels. Understanding this model helps students differentiate between fundamentally different science concepts as heat and temperature, energy and work, thus building a foundation for studying thermodynamics.[39]

Consequently, IGM poses a number of conceptual challenges to the learners. In addition to being often invisible, even a small volume of gas, such as air, contains an extremely large number of particles, thus making the analysis of their behaviour appear more complex. For example, one breath of air at standard pressure and temperature will contain about 10^{22} molecules. Students, as well as many science teachers, do not have much experience with numbers of this large order of magnitude.[19] Furthermore, IGM makes important assumptions about the behavior of the gas particles located in a closed container. (1) They have to be point particles, *i.e.*, with no internal degrees of freedom, such as vibrations or rotations. (2) The collisions between the particles are elastic, so the particles do not lose energy as a result of these collisions. (3) The total volume of the particles is much smaller than the volume of the gas container. (4) There are no long-range intermolecular forces between the particles or between the particles and the container walls. Thus, IGM is a microscopic model that allows us to predict the behaviour of an extremely large number of gas particles that obey these assumptions. For its understanding, IGM also requires some prior knowledge (*i.e.*, momentum and energy conservation laws) as well as the ability to visualize a complex system.[39] Therefore, it is not surprising that many secondary students lack a deep understanding of this model and face difficulties applying it to describe properties of gases and the relationships between them.[40]

The microscopic IGM also helps explain the macroscopic properties of gases, such as their pressure, temperature and volume and the relationships between them, described by the ideal gas law:

$$PV = Nk_BT \tag{21.1}$$

where P is pressure, V is volume, N is the number of gas particles, k_B is the Boltzmann's constant, and T is temperature.

To gain a meaningful understanding, students not only have to be able to make sense of IGM, but also to bridge the microscopic and macroscopic models. This will allow to describe each one of these macroscopic gas properties using the microscopic IGM. For example, gas pressure can be defined as the average force gas particles exert on the unit area of a rectangular container's wall as a result of the collisions with it:

$$P = \frac{F_{ave}}{A} = N\frac{m\overline{v^2}}{3Al} \tag{21.2}$$

where F_{ave} is the average force exerted on the area A of the container's wall, l is the length of the container, m is the mass of an individual ideal gas particle and $\overline{v^2}$ is the mean square of the particle's speed (p. 595).[38] However, as in the previous case, the mathematical concepts involved here might exceed secondary students' knowledge. This is the place, where a teacher might introduce PhET suite of computer simulations that the students can freely install on their smartphones, or other electronic devices (see Figures 21.3 and 21.4).

Correia and her collaborators[33] have studied the effects of using PhET simulations on secondary students' learning of the ideal gas-related concepts. Specifically, they examined secondary students' comprehension of the gas behaviour at the sub-microscopic level as a result of engaging with PhET simulations. Using the pre- and post-conceptual surveys, they found that student comprehension of the material significantly increased as a result. One of the reasons identified by the researchers was not only the opportunity to observe the behaviour of particles of an ideal gas, but the ability to change the variables and observe the effect of this change on the system (see Figure 21.3). The students were able not only to see the "invisible" particles, but to ask and answer the "what-if questions" through conducting relevant measurements. For example, they could measure how the changes of the gas temperature affect its pressure. Even more importantly, the students were able to connect microscopic (the motion of individual particles) and the macroscopic (gas pressure, temperature) properties of the gas and their mathematical and physical representations. To address a common misconception—students'

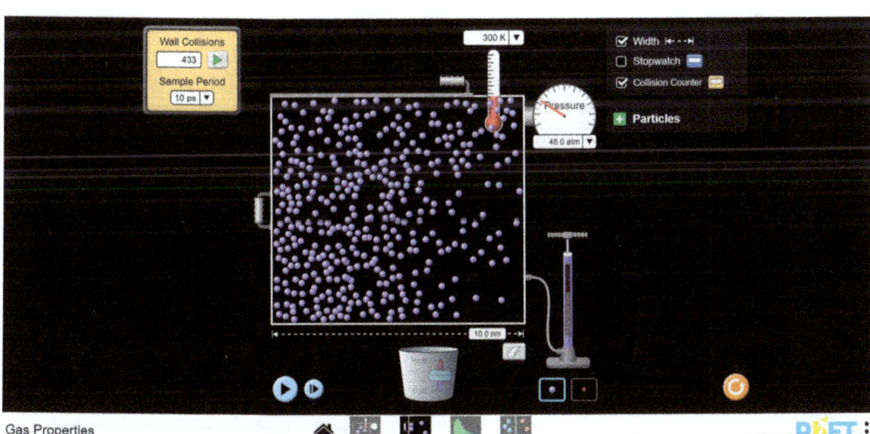

Figure 21.4 A screenshot of PhET interactive simulations Gas Properties: Explore. The simulation allows changing different parameters of the ideal gas system and observing and measuring the effects of these changes. For example, students can measure the number of wall collisions and connect them to the gas pressure. Reproduced from https://phet.colorado.edu/en/, under the terms of the CC BY 4.0 license https://creativecommons.org/licenses/by/4.0/.

confusion between heat and temperature,[40–42] they could use the energy part of this simulation (see Figure 21.5).

21.3.2 Examining the Ideal Gas Law with Phyphox

Complementary to examining IGM using PhET virtual experiments, students can also collect and analyze data using the Phyphox app (see Figure 21.6).[36] As one example, consider an investigation of the dependence of the gas pressure on temperature in the ideal gas law. A student can place their smartphone in a sealed plastic container and measure how the pressure of the gas changes if the container undergoes heating. The latter can be

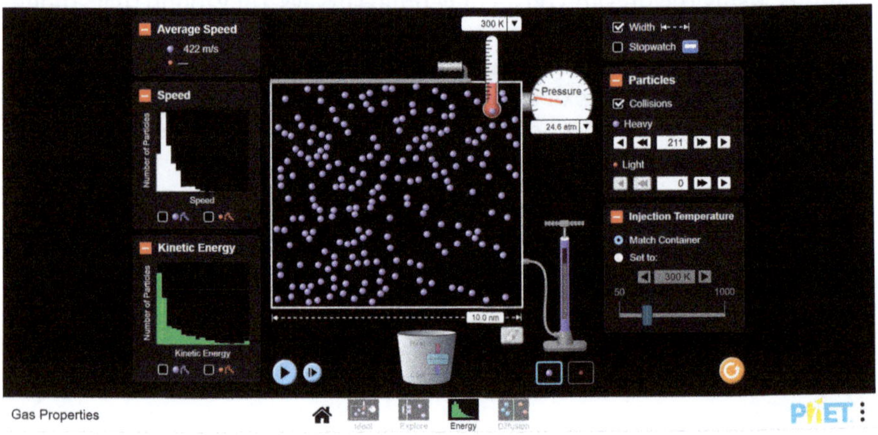

Figure 21.5 A screenshot of PhET interactive simulations Gas Properties: Energy. The simulation uses multiple representations (*e.g.*, energy and average speed bar charts) to help students examine different properties of IGM. Reproduced from https://phet.colorado.edu/en/, under the terms of the CC BY 4.0 license https://creativecommons.org/licenses/by/4.0/.

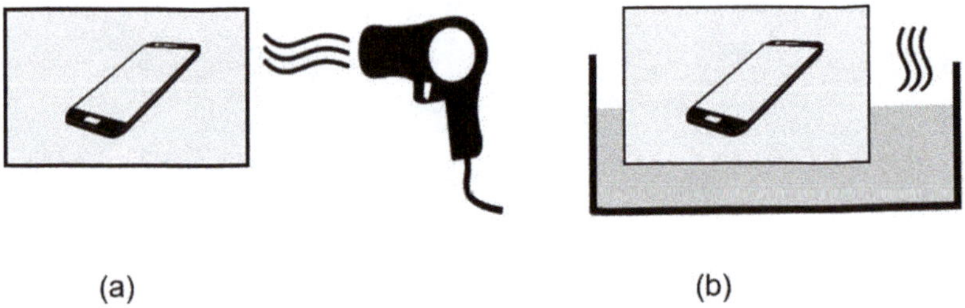

Figure 21.6 Examples of exploring the ideal gas law with a smartphone. A smartphone is measuring the air pressure in a sealed plastic container that undergoes heating (a) using a hairdryer and (b) by submerging the container in hot water.

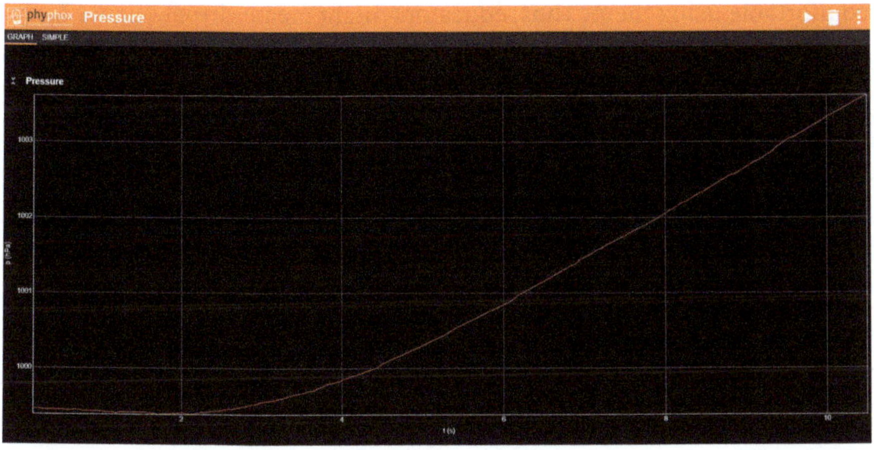

Figure 21.7 A screenshot of a computer remotely connected to the Phyphox app measuring ambient pressure as a function of time with a smartphone in a sealed container. At $t = 2$ s, an external hair dryer was turned on to warm up the container, hence raising the temperature of the air inside it. Reproduced from https://phyphox.org/, under the terms of the CC BY 4.0 license https://creativecommons.org/licenses/by/4.0/.

provided by exposing the sealed container to either hot air (*e.g.*, from a hair dryer) as shown in Figure 21.6(a) or hot water, as illustrated in Figure 21.6(b). Figure 21.7 demonstrates the result of such experiments. It shows the gas pressure gradually increasing with time as the air in the sealed container is warming up.

It is well known that smartphones cannot measure ambient temperature in a reliable way, because of the constant and irregular dissipation of heat from its own inner parts. Hence, a quantitative study would require a separate temperature sensor (*e.g.*, a simple household thermometer) to be placed in the sealed container, which would provide an independent temperature read-out. With that data available, the linear dependence of pressure P on temperature T could also be examined by the students using eqn (21.1).

Other examples of exploring the IGM include experiments on the dependence of the atmospheric pressure changes on altitude. Here, students can take their smartphones to various places located at different known altitudes, such as different floors of a high-rise building or a mountainous terrain.

21.3.3 Science Teachers' Motivations for Using Smartphones in the Classroom

There are many reasons why secondary science teachers might want to use smartphone-based technologies in their classrooms.[26,43] These technologies have already become parts of students' lives and the students are very comfortable using them. Thus, educators might create virtual science learning environments using the devices already familiar to the students. However, smartphones can also help teachers to level the educational inequality with regard to doing real science, where the science lab equipment in the school is rather limited. Engaging students in doing science is very different from asking students to read about science discoveries done by others. It also allows students to experience authentic process of scientific discovery from planning the experiment to implementing it and analyzing its results.

This claim was echoed by many British Columbia teachers whose students participate in annual hands-on Physics Olympics at the University of British Columbia.[44,45] Therefore, modern smartphone-based technologies have the potential to reduce inequality in technology access in our schools, as many students in remote and rural communities already own this technology, while their schools might have limited science equipment. With the use of sensors already embedded in students' smartphones, science educators can make another step towards meaningful student engagement: moving from virtual science learning environment towards doing science in an authentic real-world context.[25] However, to make this important step, science teacher educators have to support science teachers, who might have never experienced these learning environments as students.

21.4 Smartphone-enabled Investigations in Science Teacher Education

Teacher education is the first point of entry for many future STEM teachers to research-based and evidence-based education.[46] Few future teachers had an opportunity to experience active STEM learning as K–12 students and even fewer had an opportunity to use modern technology[8] or their own smartphones to carry out authentic and open-ended scientific investigations. This is important, as to help future STEM teachers adopt these innovative pedagogies and start thinking about how to incorporate technology deliberately into their lessons,[7] teachers have to gain their own experience in using these tools. This is the reason for incorporating modern technologies, such as the smartphone applications into STEM teacher education.

We have been doing it with the secondary science teachers for the last decade. Their science methods course assignments include science experiments that future teachers can conduct with their smartphones—the apparatus already available to them and potentially to their future students. For example, the growing collection of video experiments, created by the secondary science teacher-candidates, features many experiments that can be done on a shoe-string budget or with the devices already in students' disposal.[47] We are convinced that with the eminent expansion of the science-oriented smartphone applications, such as PhET and Phyphox, the range of this experiments will inevitably grow. We call on STEM teacher educators to incorporate these experiments in their courses.

21.5 Conclusions and Lessons Learned

This chapter examines how common smartphones, already available to many students and teachers, can be used to conduct authentic and open-ended chemistry (and science in general) investigations either at school or at home. We have shown two examples of powerful smartphone applications relevant to K–12 and post-secondary science learning. The PhET suite of computer simulations and a corresponding smartphone app allow students to conduct investigations in a virtual lab that reflects many aspects of a real science laboratory, found at a university. The Phyphox application allows students to collect data with their phone and then transfer it to their computers for further analysis. This application turns students' smartphones into data acquisition devices, thus allowing students in schools with limited scientific equipment perform authentic scientific investigations. We have tested the power of the Phyphox app during the virtual Physics Olympics at the University of British Columbia that took place in March of 2021.[45] The app allowed us to try and level the playing field by making sure that all students who wanted to participate in the event had equal access to science equipment to conduct their experiments—their own smartphones. This allowed us to engage students from all across British Columbia in creative science explorations, even when their schools might have had limited access to science equipment.

The example of the ideal gas law investigation illustrated how these apps can be used with the secondary or even post-secondary students to engage all students in doing science. This chapter discussed the strengths and the limitations of these applications in the context of chemistry learning. Finally, we proposed how smartphone applications can be incorporated into science methods courses for future teachers.

One of the prominent themes discussed during the 6th STEM in Education 2021 International Conference (https://stem2021.ubc.ca/) was technology-based professional development for practising science teachers. Despite the wide availability of these tools and ever-increasing student access, relatively few science teachers incorporate smartphone-enhanced inquiry in their science teaching. In order for STEM teachers to be open to using these powerful devices with their students, the teachers have to experience smartphone-enabled science experiments as learners. This can be done during teacher-education, as well as during teacher professional development. Thus, we call on science teacher educators to consider incorporating smartphone-enabled science activities in their professional development activities.

More than 60 years after the launch of Sputnik and the birth of the science education reform movement, too few students in our classrooms have an opportunity to experience "the pleasure of finding things out".[48] There is no better way to attract students into science than to provide them with this opportunity.

We hope that this chapter will encourage many chemistry (and science) educators to consider incorporating smartphone applications into their curriculum in order to engage all science students in meaningful scientific exploration.

References

1. J. D. Bransford, A. L. Brown and R. R. Cocking, *How People Learn: Brain, Mind, Experience, and School*, The National Academies Press, Washington, DC, 2002.
2. R. R. Hake, *Am. J. Phys.*, 1998, **66**, 64–74.
3. E. B. Moore, J. M. Chamberlain, R. Parson and K. K. Perkins, *J. Chem. Educ.*, 2014, **91**, 1191–1197.
4. M. Milner-Bolotin, *J. Chem. Educ.*, 2012, **89**, 477–481.
5. D. Martinovic and M. Milner-Bolotin, *STEM Educ.*, 2021, **1**, 279–298.
6. J. M. Fraser, A. L. Timan, K. Miller, J. E. Dowd, L. Tucker and E. Mazur, *Rep. Prog Phys.*, 2014, **77**, 17.
7. M. Milner-Bolotin, in *STEM Teachers and Teaching in the Era of Change: Professional Expectations and Advancement in 21st Century Schools*, ed. Y. Ben-David Kolikant, D. Martinovic and M. Milner-Bolotin, Springer, Cham, Switzerland, 2020, pp. 201–219.
8. M. Milner-Bolotin, O. Aminov, W. Wasserman and V. Milner, *Can. J. Phys.*, 2020, **98**, 571–578.
9. A. P. Fagen, C. H. Crouch and E. Mazur, *Phys. Teach.*, 2002, **40**, 206–209.
10. M. Milner-Bolotin and S. Nashon, *Protoplasma*, 2012, **249**, 25–30.
11. M. Milner-Bolotin, in *Creativity and Technology in Math Education*, ed. V. Freiman and J. Tassell, Springer, Cham, Switzerland, 2018, pp. 251–278.
12. See Chapter 20: S. Wilkinson, R. Pullen, A. Motion, S. Schmid, S. George-Williams and S. Leach, in *Digital Learning and Teaching in Chemistry*, ed. Y. J. Dori, C. Ngai and G. Szteinberg, Royal Society of Chemistry, United Kingdom, 2023.
13. See Chapter 18: D. F. Donnelly-Hermosillo, L. Gerard and M. C. Linn, in *Digital Learning and Teaching in Chemistry*, ed. Y. J. Dori, C. Ngai and G. Szteinberg, Royal Society of Chemistry, United Kingdom, 2023.
14. Y. Ben-David Kolikant, D. Martinovic and M. Milner-Bolotin, in *STEM Teachers and Teaching in the Digital Era: Professional Expectations and Advancement in 21st Century Schools*, ed. Y. Ben-David Kolikant, D. Martinovic and M. Milner-Bolotin, Springer, Cham, Switzerland, 2020, pp. 1–18.
15. *STEM Teachers and Teaching in the Digital Era: Professional Expectations and Advancement in 21st Century Schools*, ed. Y. Ben-David Kolikant, D. Martinovic and M. Milner-Bolotin, Springer, Cham, Switzerland, 2020.
16. O. Doyle, *COVID-19: Exacerbating Educational Inequalities?*, UCD Geary Institute for Public Policy, Dublin, 2020.
17. See Chapter 19: S. Bernholt, J. Eckhard, M. Rodemer, A. Langnerb, G. Asmussen and N. Graulich, in *Digital Learning and Teaching in Chemistry*, ed. Y. J. Dori, C. Ngai and G. Szteinberg, Royal Society of Chemistry, United Kingdom, 2023.
18. J. Y. Han and F. M. Fung, in *Digital Tools for Equitable in Person and Remote Chemistry Learning*, ed. Y. J. Dori, C. Ngai and G. Szteinberg, Royal Society of Chemistry, London, UK, 2022, p. 20.

19. M. Milner-Bolotin and R. Zazkis, *LUMAT: Int. J. Math. Sci. Technol. Educ.*, 2021, **9**, 336–365.
20. J. Hansen and L. E. Richland, *CBE—Life Sci. Educ.*, 2020, **19**, ar61.1–ar61.15.
21. M. Üce and İ. Ceyhan, *J. Educ. Train. Studies*, 2019, **7**, 202–208.
22. A. K. Prodjosantoso and A. M. Hertina, *Int. J. Instr.*, 2019, **12**, 1477–1488.
23. M. Francl, *Nat. Chem.*, 2010, **2**, 600–601.
24. British Columbia Ministry of Education, in https://curriculum.gov.bc.ca/, Government of British Columbia, Victoria, British Columbia, Canada, 2020.
25. S. Staacks, S. Hütz, H. Heinke and C. Stampfer, *Phys. Educ.*, 2018, **53**, 045009.
26. T. Maciel, *APS News*, 2015, **24**, 5–6.
27. R. Vieyra, C. Vieyra, P. Jeanjacquot, A. Marti and M. Monteiro, *Sci. Teach.*, 2015, **82**, 32–40.
28. C. E. Wieman, W. K. Adams, P. Loeblein and K. K. Perkins, *Phys. Teach.*, 2010, **48**, 225–227.
29. W. K. Adams, S. Reid, R. LeMaster, S. B. McKagan, K. K. Perkins, M. Dubson and C. E. Wieman, *J. Interact. Learn. Res.*, 2008, **19**, 397–419.
30. PhET Research Team, ed. K. Perkins, University of Colorado at Boulder, 2021, https://phet.colorado.edu/.
31. N. D. Finkelstein, W. K. Adams, C. J. Keller, P. B. Kohl, K. K. Perkins, N. S. Podolefsky, S. Reid and R. LeMaster, *Phys. Rev. Spec. Top. – Phys. Educ. Res.*, 2005, **1**, 010103.
32. W. K. Adams, S. Reid, R. LeMaster, S. B. McKagan, K. K. Perkins, M. Dubson and C. E. Wieman, *J. Interact. Learn. Res.*, 2008, **19**, 551–577.
33. A.-P. Correia, N. Koehler, A. Thompson and G. Phye, *Research in Science & Technological Education*, 2019, pp. 1470–1138.
34. K. Perkins, W. Adams, M. Dubson, N. Finkelstein, S. Reid, C. Wieman and R. LeMaster, *Phys. Teach.*, 2006, **44**, 18–23.
35. S. Staacks, S. Hütz, H. Heinke and C. Stampfer, *Phys. Teach.*, 2019, **57**, 112–113.
36. C. Stampfer, H. Heinke and S. Staacks, *Nat. Rev. Mater.*, 2020, **5**, 169–170.
37. M. Milner-Bolotin, V. Milner, A. M. Tasnadi, H. T. Weck, I. Gromas and P. D. Ispanovity, *GIREP – Physics Education Conference 2019 Proceedings*, 2021.
38. R. Hawkes, J. Iqbal, F. Mansour, M. Milner-Bolotin and P. Williams, *Physics for Scientists and Engineers: An Interactive Approach*, Nelson Education, Toronto, 2nd edn, 2018.
39. N. Erceg, I. Aviani, V. Mešić, M. Glunčić and G. Žauhar, *Phys. Rev. Phys. Educ. Res.*, 2016, **12**, 020139.
40. R. Leinonen, E. Räsänen, M. Asikainen and P. E. Hirvonen, *Eur. J. Phys.*, 2009, **30**, 593–604.
41. A. G. Harrison, D. J. Grayson and D. F. Treagust, *J. Res. Sci. Teach.*, 1999, **36**, 55–87.
42. E. L. Lewis and M. C. Linn, *J. Res. Sci. Teach.*, 1994, **31**, 657–677.
43. K. Jochen and V. Patrik, *Eur. J. Phys. Educ.*, 2013, **4**, 16–27.

44. M. Milner-Bolotin, T. Liao and J. McKenna, *Int. Newsl. Phys. Educ.*, 2019, **70**, 5–6.
45. T. Liao, J. McKenna and M. Milner-Bolotin, *Physics in Canada*, 2017, vol. 73.
46. M. Milner-Bolotin, *Front. Educ.: STEM Educ.*, 2018, **02**, 14.
47. G. Tembrevilla and M. Milner-Bolotin, *Phys. Educ.*, 2019, **54**, 025008–025018.
48. R. P. Feynman, *The Pleasure of Finding Things Out: The Best Short Works of Richard P. Feynman*, Helix Books–Perseus Books, New York–London, 1999.

CHAPTER 22

Theme Introduction: Digital Assessment

C. NGAI*

Office of Undergraduate Research and Artistry, Colorado State University, USA
*E-mail: Courtney.Ngai@colostate.edu

22.1 Assessing Student Learning in the Classroom

Historically, measuring what students know and can do *via* assessment has served an important role in education. At its core, assessment is a means of providing feedback for teachers and students. When implemented throughout a course, assessment can be used *for* learning and provide formative feedback to students as they chart their progress through the course content (*e.g.*, a check-in question at the beginning of each class period).[1,2] A final exam often serves as summative feedback for students about their overall performance in a course and is an example of an assessment *of* learning.[1,2] The variety of assessment types enables teachers to assess students' knowledge and thinking, close the gap between where students are and where they should be, and guide teachers in their lesson plans while making adaptations in response to students' difficulties and diversity.[3,4] As teachers implement assessment through science assignments in an ongoing process, students achieve higher gains on both formative and summative assessments while actively engaging and creatively demonstrating knowledge of concepts, processes, and scientific skills.[5]

Advances in Chemistry Education Series No. 11
Digital Learning and Teaching in Chemistry
Edited by Yehudit Judy Dori, Courtney Ngai and Gabriela Szteinberg

Published by the Royal Society of Chemistry, www.rsc.org

Assessment, then, can be a useful means for supporting students' learning. What is assessed communicates to students the content and skills instructors and teachers value in their course and discipline, and how students perform during these assessments determines whether they are "successful" in the course.[6] How the content and skills are assessed has the potential to increase students' engagement with the material and reinforce their learning. Correspondingly, a shift towards learner-oriented assessment, which emphasizes adapting assessment to the learners' needs and context, has been observed in both secondary and higher education.[7,8]

Assessing student learning using digital measures was growing in application prior to the COVID-19 pandemic. Technology has enabled new methods for implementing existing assessments (*e.g.*, quizzes and tests with closed-ended questions) and new types of assessment (*e.g.*, interactive assignments, virtual games, and online peer assessments).[9] Digital platforms, such as Blackboard, Canvas, and Moodle, provide a way to organize and distribute content and assessments that can vary in their structure and delivery. These are becoming more integrated in courses and at institutions following the COVID-19 pandemic, which has increasingly prompted instructors to turn to digital means of assessment. Instructors and teachers were challenged to translate their original forms of assessment to digital formats or to develop entirely new assessment modes that could be implemented virtually while still providing students with meaningful feedback.

22.2 Digital Assessment Theme Chapters

The chapters in this theme consider the role of digital assessment in learning chemistry. The authors explore features that are unique to digital assessment, such as varying the method and timing of providing feedback to students.[10–13] Other authors explore whether teachers feel prepared to create and implement digital assessment and how to support teachers in developing these essential skills.[14] The chapters in this theme ground their work in frameworks such as Community of Inquiry and on teachers' knowledge types based on the work of Shulman.[15] These chapters provide examples of applications of digital assessment that can serve as a model for educators and teachers wishing to implement the various techniques and methods.

Examinations are one of the most prevalent forms of assessment, but their structure and implementation can vary. Two-stage exams, where students complete an individual component and then work in small groups on the group component, have been proposed as a structure that (1) decreases student anxiety, (2) exposes students to other approaches for solving problems, and (3) helps students reflect on their own processes while solving problems. In Chapter 23, Stewart shares best practices for effectively utilizing two-stage exams in chemistry courses to increase student collaboration and support learning.[10] She draws on the Community of Inquiry (CoI) framework, which describes three "presences": teaching, cognitive, and social. These presences are important to attend to so that learning occurs when engaging in collaborative inquiry. Stewart examines the implementation of two-stage exams through the lens of each presence. Her recommendations are grounded in her own experiences teaching a virtual, high-enrolment, lecture-style organic chemistry course and research on implementing the two-stage exams in this course. This chapter demystifies the use of this type of exam structure and its guidance lowers the activation barrier for other instructors.

In Chapter 24, Wilkinson *et al.* recognize the potential of using digital tools to provide formative feedback to students.[11] However, they note that as the number of digital tools increases, it becomes ever more necessary to evaluate the effectiveness of these tools for providing feedback. The authors define important characteristics of formative feedback and review a number of digital tools that can be utilized to provide this type of feedback. The tools they review include Electronic Laboratory Notebooks, methods for delivering transmissive feedback, and adaptive learning systems.

Wilkinson *et al.* provide an extensive review of a learning analytic software called the Student Engagement Relationship System (SRES).[11] SRES can be used to collect data about student learning, enabling instructors to customize their feedback based on the data. The authors argue that the power of SRES comes from its ability to support giving personalized feedback to students and automating some of the processes associated with feedback for instructors. The authors then provide a series of case studies that demonstrate how SRES was used for grading laboratories using rubrics and exemplify the power of the SRES platform.

With the COVID-19 pandemic forcing teachers to develop and implement digital assessment as classes are moved online, they had little to rely on in terms of resources and training. In Chapter 25, Herscovitz *et al.* investigate what it takes to prepare future teachers to teach effectively using digital means.[14] They draw from the well-established Shulman's framework of content knowledge (CK), pedagogical knowledge (PK), and pedagogical content knowledge (PCK) and the latter construct—technological, pedagogical, and content knowledge (TPACK) of Mishra and Koheler that integrates technology into the former framework.[16] Herscovitz *et al.* proposed that these types of knowledge be augmented with assessment knowledge (AK). They suggest that a lack of training around developing TPACK is a major barrier to using technology for in-person and virtual education in general and for assessment in particular. Thus, they investigated a training workshop during which participants evaluated existing online chemistry assignments and then developed their own. Participants were asked to complete reflections and answer a survey about their TPACK, and the researchers analyzed the online chemistry assignments the participants developed using a TPACK rubric. The researchers found that participants ranked their technology knowledge lowest and identified several challenges related to using technology in the classroom. However, participants also pointed out advantages related to student engagement and learning that resulted from using technology in education. Finally, when examining the online chemistry assignments that the participants had developed, the researchers found that participants demonstrated higher levels of PCK and TK when compared to AK. This suggests that more support and training may be needed to increase teachers' ability to develop online assignments that assess students' chemistry knowledge and skills.

Low-stakes formative assessments can support student reflection on their learning. Their impact is increased when the feedback is immediate. Muteti and Mutambuki share research on the use of the Muddiest Point activity,[12] a type of low-stakes, formative, classroom assessment technique (CAT) designed by Angelo and Cross.[17] While CATs have been widely used to provide formative assessment opportunities for students, Muteti and Mutambuki explored the virtual implementation of the Muddiest Point CAT. They compared two general chemistry courses that had the same curriculum, learning outcomes, and assessments, where in the intervention course, the instructor implemented the Muddiest Point CAT at the end of every chapter. Through the Muddiest Point CAT, students anonymously identified the concepts that were most confusing to them and were able to see each other's responses and upvote them. The instructor and learning assistants then used these responses to inform the problems they assigned and went over with students, as well as the content of future classes. Muteti and Mutambuki found statistically significant differences between the three midterm exams given in the courses for all student groups. Furthermore, when asked about the impact of the Muddiest Point CATs on their learning experience, students reported a variety of ways in which the activities improved their learning and increased their confidence and sense of belonging. These findings suggest that adding activities like the Muddiest Point CAT using technology that enables anonymous feedback can be beneficial for all students learning chemistry.

In Chapter 27, Lawrie looks beyond a single assessment opportunity and conducts a deep investigation into providing feedback using digital means.[13] She highlights the benefits of feedback for student learning, including increased engagement with the content and better self-regulation of learning. Lawrie points out, however, that to be effective, feedback must satisfy a whole host of requirements,[10] including that it must be adaptive, discursive, interactive, and reflective. Teachers often give feedback in-the-moment and use visual cues, such as students' postures and expressions, to assist in gauging their level of understanding. As this type of dynamic interaction is less feasible with an online platform, integrating digital feedback must be more intentional. To investigate the most effective means for providing feedback to students using a digital platform, Lawrie presents a study of a large, first-year blended (*i.e.*, combining virtual and in-person learning) chemistry course that increases the opportunities and types of feedback provided over several years. Different types of feedback were provided digitally for at least three different modules of the course, with the goals of helping students connect their prior knowledge to new content and increasing peer–peer and peer–instructor interactions. Lawrie found that students reported preferring feedback that verified whether their thinking was correct or incorrect and valued feedback from the instructor and learning assistants more than feedback from their peers. However, students also reported gaining important insights and increasing their understanding through the activities where they were asked to give feedback to their peers. The chapter contains detailed examples of methods that are useful for novice and experienced chemistry instructors and teachers alike for increasing both the amount of feedback provided in a hybrid course and students' understanding of what constructive feedback looks like.

22.3 Final Thoughts

These chapters provide many useful examples of utilizing digital assessment to promote learning in chemistry. However, there are many areas that warrant further investigation. As noted by Herscovitz *et al.* in Chapter 25,[14] there is a lack of resources and training to equip chemistry teachers to develop and implement their own digital assessments. Utilizing a framework such as TPACK and AK in teacher education and professional development programs as well as designing training on how to use these resources may make it more likely that teachers will utilize digital assessment effectively in their own contexts.

While face-to-face assessments can be implemented virtually, this does not always mean they should be. In this theme, translating two-stage exams (Stewart, Chapter 23) and muddiest point activities (Muteti & Mutambuki, Chapter 26) into digital formats were presented.[10,12] The authors present evidence that these activities work effectively in a digital format to support student learning, but instructors and researchers should be careful as they translate other assessments to a digital format to make sure they are well suited to this environment. Whether other forms of assessment are suited to a digital format could be a source of future investigation.

Finally, while many of the chapters in the Digital Assessment theme identify ways in which implementing digital assessment is inclusive of a broader range of students, this is an area that could be investigated in its own right. The authors note that digital assessment can be more flexible in its format and delivery, such that it can be modified to meet students' needs. However, further exploration of how assessment provided *via* digital methods can address a greater diversity of student needs would be beneficial for instructors so that they can tailor their methods to the needs of the students in their context.

References

1. M. Birenbaum, K. Breuer, E. Cascallar, F. Dochy, Y. Dori, J. Ridgway, R. Wiesemes and G. Nickmans, *Educ. Res. Rev.*, 2006, **1**, 61–67.
2. J. W. Pellegrino, *Commission on the Skills of the American Workforce*, 2006, 1–15. https://ncee.org/wp-content/uploads/2017/03/Rethinking-and-Redesigning.pdf.

3. E. M. Furtak, M. A. Ruiz-Primo, J. T. Shemwell, C. C. Ayala, P. R. Brandon, R. J. Shavelson and Y. Yin, *Appl. Meas. Educ.*, 2008, **21**, 360–389.
4. Y. J. Dori and S. Avargil, in *Encyclopedia of Science Education*, ed. R. Gunstone, Springer-Verlag Berlin Heidelberg, 2015.
5. Y. J. Dori, *J. Res. Sci. Teach.*, 2003, **40**, 34–52.
6. R. L. Stowe, L. J. Scharlott, V. R. Ralph, N. M. Becker and M. M. Cooper, *J. Chem. Educ.*, 2021, **98**, 2490–2495.
7. E. F. De Waard, G. T. Prins and W. R. Van Joolingen, *J. Chem. Educ.*, 2022, 2991–2998.
8. D. Pereira, M. A. Flores and L. Niklasson, *Assess. Eval. High. Educ.*, 2016, **41**, 1008–1032.
9. N. Wengrowicz, Y. J. Dori and D. Dori, *Assess. Eval. High. Educ.*, 2017, **42**, 607–624.
10. See Chapter 23: J. Stewart, in *Digital Learning and Teaching in Chemistry*, ed. Y. J. Dori, C. Ngai and G. Szteinberg, Royal Society of Chemistry, United Kingdom, 2023.
11. See Chapter 24: S. Wilkinson, R. Pullen, P. Rutledge and S. Schmid, in *Digital Learning and Teaching in Chemistry*, ed. Y. J. Dori, C. Ngai and G. Szteinberg, Royal Society of Chemistry, United Kingdom, 2023.
12. See Chapter 26: C. Z. Muteti and J. Mutambuki, in *Digital Learning and Teaching in Chemistry*, ed. Y. J. Dori, C. Ngai and G. Szteinberg, Royal Society of Chemistry, United Kingdom, 2023.
13. See Chapter 27: G. A. Lawrie, in *Digital Teaching and Learning in Chemistry*, ed. Y. J. Dori, C. Ngai and G. Szteinberg, Royal Society of Chemistry, United Kingdom, 2023.
14. See Chapter 25: O. Herscovitz, M. Varsano, O. Faris and Y. J. Dori, in *Digital Learning and Teaching in Chemistry*, ed. Y. J. Dori, C. Ngai and G. Szteinberg, Royal Society of Chemistry, United Kingdom, 2023.
15. L. Shulman, *Harv. Educ. Rev.*, 1987, **57**, 1–23.
16. P. Mishra and M. J. Koehler, *Teach. Coll. Rec.*, 2006, **108**, 1017–1054.
17. T. Angelo and P. K. Cross, *Classroom Assessment Techniques: A Handbook for College Teachers*, Jossey-Bass, San Francisco, 2nd edn, 1993.
18. D. Nicol, *Assess. Eval. High. Educ.*, 2010, **35**, 501–517.

CHAPTER 23

The Community of Inquiry Framework as a Guide to Implement Inclusive Collaborative Two-stage Exams in Chemistry

J. J. STEWART*

2036 Main Mall, Vancouver, British Columbia, V6T 1Z1, Canada
*E-mail: jstewart@chem.ubc.ca

23.1 Online Collaborative Two-stage Assessments

Imagine a situation in which students are together in groups of three or four, huddled around a document, focused on a task. As you look around the room, students are gesturing with their hands and exclaiming when they have completed a task. The volume of chatter in the room is quite loud—louder than it is has ever been during an engaging collaborative activity during a class session.[1] As students leave the room at the end of the session, they say, "thank you!" and many have smiles on their faces. What happened here? Would you believe me if I said it was an exam? This dynamic and engaging collaborative assessment approach is commonly referred to as the "two-stage exam". Two-stage exams harness students' keen attention and high motivation to enable collaborative learning.

Advances in Chemistry Education Series No. 11
Digital Learning and Teaching in Chemistry
Edited by Yehudit Judy Dori, Courtney Ngai and Gabriela Szteinberg

Published by the Royal Society of Chemistry, www.rsc.org

Chemistry learning opportunities underpinned by digital technology are increasingly likely to rely upon corresponding digital approaches for formative and summative assessments. Many chemistry educators are familiar with digital platforms for formative assessments such as homework, and many are newly aware of the challenges and opportunities of online summative assessments due to the wide-scale remote teaching that occurred during the COVID-19 pandemic.[2,3] While substantial work on digital assessments in chemistry had been underway, the pandemic required many more educators to adopt digital assessments. Online assessment in chemistry is uniquely challenging due to the multi-modal representation of knowledge required in the chemistry domain, including text, drawings, mathematical formulae, and equations.[4] The development of digital chemistry assessments that support meaningful collaboration should consider inclusion and accessibility to promote equity. We have taken up these challenges in our implementation of collaborative exams in online chemistry courses.[5]

In this chapter, an adapted community of inquiry (CoI) framework is used to reflect on our implementations of two-stage exams in online chemistry courses. How the theoretical framework can be used to guide design decisions about two-stage assessment to promote meaningful inclusive collaborative learning is demonstrated. The framework is also used to identify opportunities for future research to make online two-stage exams a more equitable assessment practice for all.

23.1.1 An Introduction to Two-stage Exams

In a typical two-stage exam, students attempt a test individually (first stage), and then work in a group to respond to the same or similar sets of questions (second stage). Usually, students' overall assessment grade is a weighted average of the individual and group scores with the individual weighted much higher (*e.g.*, 85%). Instructors design several aspects of two-stage exams, including the proportion of total time in the individual and group stages, whether the group stage occurs immediately or following a delay, and whether the group exam content is the same, a portion of, or more advanced than, the individual exam. Instructors' goals for using collaborative exams and practical issues such as available class time guide these decisions in practice.

Two-stage assessments can be used for reviewing course prerequisite content,[6] for "readiness assurance tests" in team-based learning pedagogy,[7] for formative assessments such as quizzes,[8] or for higher-stakes midterm or final examinations.[9,10] It is possible to provide instantaneous feedback during the group stage, using tools such as the immediate feedback assessment technique (IFAT) scratch cards.[11] IFAT cards contain scratchable areas for students to indicate their answer to a multiple-choice question. A small star is revealed when a student scratches the correct choice. The immediate correct or incorrect feedback allows students to attempt a question until they arrive at the correct answer. Students report that the immediate correct or incorrect feedback they receive from the IFAT card is helpful to their learning.[12]

Two-stage exams have been implemented in science courses at our institution for many years in traditional lecture-based courses and highly structured active-learning courses.[10,12] The collaborative process of a two-stage exam shares features with the process of peer instruction[13] where students think about a problem and respond individually using clickers and then discuss in pairs or small groups before responding again. Thus, two-stage exams are a natural extension of the learning process used in many active learning chemistry classes. In this way, the peer instruction activities become learning opportunities for the group process required for a productive two-stage exam experience. Active learning instructional methods such as informal group worksheets, peer-led team learning (PLTL), and process-oriented guided inquiry learning methods will likely prepare students for collaborative assessments such as two-stage exams.

23.1.2 History of and Research on Two-stage Exams

Mathematics educators described a multi-stage "pyramid exam" with individual, group, and whole-class stages in a 1995 publication.[14] This multi-stage design was described as "elaborate" by authors of a study of individual and group-staged exams in sociology.[9] Certainly, a whole-class stage is only likely to engage a significant portion of the class in small classes, which explains why the literature is dominated by the two-stage format. At our institution, the Carl Wieman Science Education Initiative[15] seeded the growth of two-stage collaborative exams in science courses, including chemistry. Variations on two-stage exams, including doing the group stage first, are being explored and investigated.[16]

Well-designed experiments have shown improved performance on delayed post-tests for content that was part of the second stage of the exam in geology[10] and physiology.[17] However, an experiment conducted in a mathematics course did not detect a difference in delayed post-test scores for students who did and did not experience a collaborative second stage.[18] Importantly, collaborative exams have been shown to reduce chemistry and nursing students' test anxiety.[8,19] Other research has found students enjoy the experience, value the ability to learn about their peers' approaches to test questions, and appreciate the immediate feedback.[12] A qualitative study that examined students' reflections after the individual and group stages found that students were able to generate "internal feedback" for themselves during the process.[20] While evidence of the learning effects of two-stage exams is mixed, the use of collaborative two-stage exams reinforces collaborative learning pedagogies[18] and the literature suggests they may have non-cognitive benefits for science students.

23.1.3 Group Formation and Collaboration

Allowing students to self-select their groups for two-stage exams is simpler than assigning groups. Instructors can ask students to self-assemble immediately after the first stage or after a short break. While this is happening, members of the teaching team can provide support to students who seem to be having difficulties forming a group. Anecdotal evidence suggests that when students are used to doing work in small groups during class sessions, the transition to the second stage of the exam goes quite smoothly.

Instructors can form groups randomly or purposefully, or students can select their groups out of convenience (who they are sitting near), or more strategically (by seeking out specific classmates). Liljedahl synthesized the literature to distinguish instructor's goals for strategic group formation, *educational goals* relating to pedagogy, productivity, and peacefulness and *social goals* related to diversification, integration, and socialization.[21] Liljedahl makes the case for using visibly random group formation in high school mathematics courses, and discusses the positive outcomes including, but not limited to, mobilizing knowledge, decreased reliance on the teacher for answers, improved task engagement, and increased enthusiasm.

One source of guidance about how to form groups for two-stage exams could be the vast group composition literature[22] that explores the impact of group makeup based on variables including demographics, personality, motivation, and ability. However, it is possible that in the short-term collaborative experience of a group exam, the impact of group selection process is different from the longer-term teamwork more commonly studied. For chemistry instructors who are assigning groups, from an inclusive teaching perspective it is advised to attempt to avoid single members from a historically excluded demographic.[23] Instructors implementing two-stage exams should use a method to form groups that aligns with their pedagogical approach and instructional goals.

Training on how to work effectively as a team, such as workshop-style activities about effective teamwork, has a positive impact on team performance for new and established teams.[24] Repetition of skills without a goal or feedback is less likely to produce performance gains than deliberate practice,[25] so it follows that to improve group performance instructors should provide guidance for how teams can improve how they work together rather than assume teams will naturally develop their skills for effective collaboration.

23.1.4 Implementation of Online Synchronous Collaborative Exams

Implementing online two-stage exams in a similar manner to in-person exams requires students to be able to communicate synchronously. In general chemistry and organic chemistry, we used videoconference breakout rooms to bring students together for the collaborative stage of their exams. In the organic chemistry course, we assigned students to groups, since that is more practical in an online learning setting. In the general chemistry course, the instructor (the author) planned the groups ahead of time and students worked together on in-class activities and problem solving for several weeks before the two-stage exam. In a subsequent offering of the organic chemistry course, groups were randomly assigned by the videoconference software during the transition from the individual stage to the group stage. Allowing students to form their own groups is practical for in-person two-stage exams, since students can easily form groups with peers sitting nearby. However, when videoconferencing with collaborative document editing, instructors need to assign students to a specific virtual space.

For online collaborative assessments, students can be asked to document their own responses on paper and upload a single document to the learning management system. This works well when a lower-technology approach is desired in drawing-heavy courses, such as organic chemistry. Using collaborative editing software (*e.g.*, Google Drive, Microsoft OneDrive) allows for group members to contribute to a single document simultaneously. It stands to reason that when students can all access and edit the test document, they may be less likely to disengage from the collaboration. Online questionnaire tools such as Qualtrics can be programmed to provide immediate feedback that allows for multiple attempts on multiple-choice questions, which creates a very similar experience to the IFAT cards. Offering the ability to attempt a question multiple times is an advanced feature in some learning management systems and may be a good option for multiple-choice assessments.

23.2 Adapted Community of Inquiry Framework

The CoI framework was originally applied in distance and online learning contexts, and has been the subject of validation studies.[26,27] The framework describes the important components for learning *via* the process of collaborative inquiry, during which learners develop personal meaning while engaging with each other to question ideas and revise their understanding of the course content.[28] Garrison uses a broad definition of inquiry, encompassing a cyclical process of problem identification, exploration, integration, and resolution.

The CoI framework is based on three "presences" that interact when students engage in collaborative inquiry. Briefly, the three presences are defined by Garrison[28] as follows: *teaching presence* directs the functioning of the learning community and involves the "design, facilitation, and direction of cognitive and social presences" (p. 61); *social presence*, according to Garrison (2011), is the "ability of participants to identify with the group or course of study, communicate purposely in a trusting environment, and develop personal and affective relationships progressively by way of projecting their individual personalities"[29] (p. 34); and *cognitive presence* is the process of "learning and thinking collaboratively"[28] (p. 34), and emerges based on the instructor's selection of the course material, activities, and assessment tasks and is influenced by the explicit goals of the learning environment.

The CoI framework is a comprehensive tool to "help educators cope with the theoretical and practical challenges of constructive collaborative thinking and learning transactions"[28] (p. 85). It is often invoked to inform the design of online courses, particularly those where collaborative learning occurs *via* asynchronous discussions. Here, the framework has been adapted in Figure 23.1 to

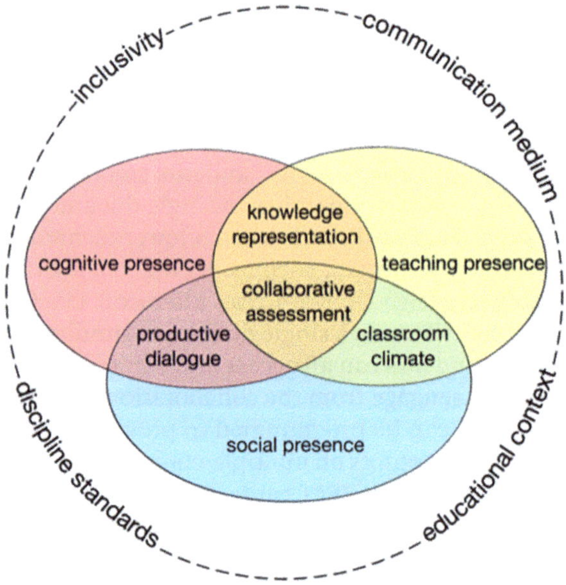

Figure 23.1 CoI framework emphasizing aspects of collaborative assessment. Adapted from ref. 28 with permission from Taylor & Francis, Copyright 2015.

uncover the complex design decisions and interactions necessarily for online two-stage exams to be collaborative and inclusive. The adapted CoI framework provides guidance and justification for collaborative assessment design decisions and reveals areas where chemistry educators would benefit from additional research to inform these decisions.

The presences dictate the importance of a supportive classroom climate, productive dialogue where students exchange ideas, and the involvement of all members in the group. Garrison's framework does not explicitly address diversity, equity, and inclusion, yet these important considerations are influenced by, and influence, all three presences. The framework's external ring represents situational and contextual factors that influence how the three presences emerge and interact. Inclusivity has been added to the external ring to highlight its over-arching influence on the learning process as a goal, mindset, instructional design consideration, contextual factor, and outcome of the learning process. The extent to which a group, course, or institution is inclusive of all students is an important contextual component to how the centre parts of the framework interact to foster collaborative learning during two-stage assessments. The CoI framework is centred on the collaborative inquiry taking place in a social context, thus the addition of inclusivity is a natural extension. In the following sections, the components of the CoI framework will be discussed in the context of the author's experiences implementing two-stage exams in chemistry courses, providing specific suggestions of how to enhance collaboration and inclusion.

23.2.1 Cognitive Presence in Online Two-stage Exams

The cognitive presence depends on the nature of the assessment tasks and how students discuss their thinking, question each other, and refine their understanding during the group stage of the exam. Higher-order cognitive tasks that require a decision with corresponding justification are ideal for collaborative assessments, such as multi-step syntheses in organic chemistry or conceptual questions that have common alternate conceptions as answer choices in general chemistry. In a study examining group dynamics in two-stage exams, researchers found that the group stage of the exam is positively influenced by an appropriate level of task challenge.[30]

Groups need to have sufficient time to discuss their reasoning and come to a consensus. A high-pressure environment, either through high-stakes grading associated with the assessment or unreasonable time constraints, will encourage behaviours that shut down collaboration, such as defaulting to the group member who seems to know the most or going with the majority without discussion. For this reason, many instructors who use two-stage exams reduce the number of test items for the group stage. There is greater learning potential when students discuss challenging course concepts rather than simple facts. Open-book two-stage exams encourage productive discourse since students can find information that may be stalling their progress. Open-book exams also reorient our courses away from memorizing facts toward the skilled use of existing knowledge.[31] We have moved from forbidding to encouraging the use of course notes and other materials in two-stage exams (both stages or just the group stage). Since our assessments contain a range of types of tasks, including higher-order tasks that require students to apply their understanding, having access to course notes has not hindered the lively collaboration we observe in the group stage of our assessments.

In our first implementation of online two-stage assessments we conducted an action research study to collect survey data to inform the design and implementation throughout the course. This organic chemistry course for students who were not chemistry or biochemistry majors was primarily lecture-based *via* Zoom software and had 424 students enrolled (zoom.us). The course used two-stage collaborative exams for both midterm exams, which also took place *via* Zoom. This work was approved by the University of British Columbia Behavioural Research Ethics Board (ID H20-01663) and will be reported in detail elsewhere.[5]

We included questionnaire items designed by physics colleagues who are investigating group interactions in two-stage exams.[32] The items probed students' perceptions of how their group arrived at their answers. Respondents selected the proportion of the exam content that was decided on with each of the four options in Figure 23.2. Of the survey respondents in the online course, 47% reported their groups discussed and reached consensus on most questions, whereas 24% percent reported they voted and went with the majority agreement with some discussion. Almost 17% reported they went with majority agreement without discussion. The results from the online course were very similar to the results collected in a face-to-face course before

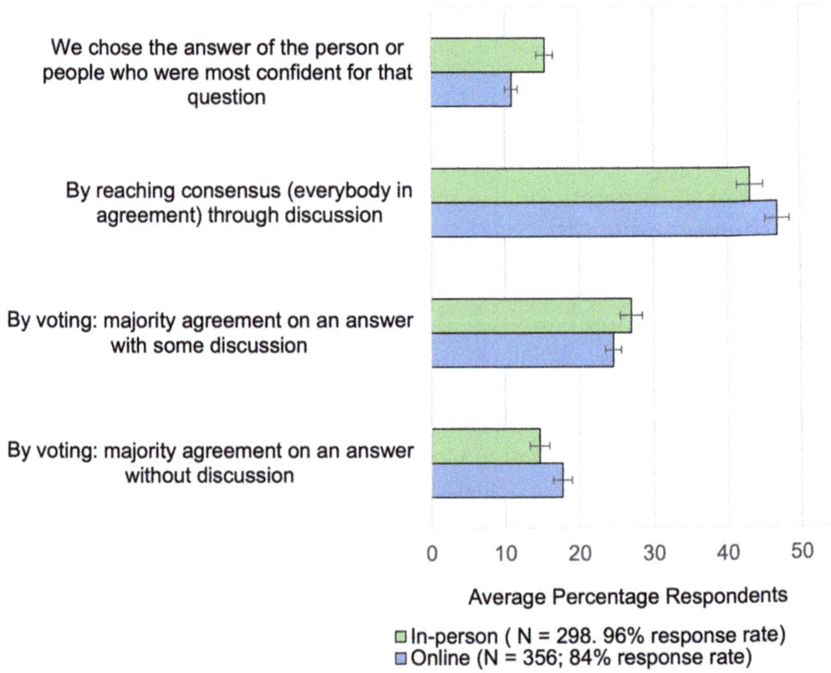

Figure 23.2 Comparison of face-to-face and online methods of responding to questions during the group exam, as self-reported on a course questionnaire. Error bars are the standard error of the mean.

the pandemic, which took place in 2018 (Figure 23.2). Interestingly, both the online and in-person organic chemistry course data showed slightly poorer collaboration than in physics courses reported in the literature.[12]

These preliminary results reassured us that the online format did not greatly negatively affect the nature of students' collaborative inquiry process. This is supported by qualitative research that compared discourse styles in online and in-person collaborative peer led team learning chemistry courses.[33] However, more research is needed to directly compare the process and outcomes of online two-stage exams to in-person exams.

23.2.2 Teaching Presence in Online Two-stage Exams

The teaching presence includes instructional design, instructor's facilitation of the learning experience, and peer-to-peer regulation of the inquiry process. The teaching presence can be enhanced by including collaboration and metacognition as explicit course goals, both of which can be linked to workplace skills. Making connections to future learning or work can help students "buy-in" to the two-stage exam process. For example, online collaborative assessment shares similarities with the day-to-day collaboration required in virtual teams in many workplaces.[34]

A course which minimizes competition by avoiding comparative grading ("scaling") is likely to be much better at fostering effective collaboration than an environment that appears to be mostly focused on grades. Leading up to the exam, we communicate to students that the goal of the second stage of the exam is to learn chemistry concepts and problem-solving strategies from each other. Explaining pedagogical choices is a category of "Instructor Talk" that communicates support for student learning and is a form of teaching presence.[35]

Clearly communicated instructions about the test logistics are an important part of teaching presence for online collaborative assessments. Instructors need to determine and communicate how students will move into breakout rooms and how their work needs to be documented and submitted. The questionnaires used in our action research study asked several questions about the logistics of the exam, for example about the clarity of instructions and whether students experiences issues with any of the technology. Three students reported a technical challenge so severe it prevented them from participating in the group stage of the exam. The vast majority (269 out of 375 questionnaire respondents) did not experience technical challenges. We used the results to inform the next offering of the assessment, including clearly stating what students should do in the event of some common challenges, such as being dropped from the Zoom call due to unstable internet.

For in-person collaborative assessments, those with accessibility accommodations, such as individual spaces to work and/or extra time for the exam are typically offered three choices for the second stage of the exam. One option is that they can decline to write the second stage, just like any other student (since we have a grading option for 100% individual exam score). To acknowledge the reality that students' grades are increased with the incorporation of the group scores, we calculate the students' exam grade using the current class average of the second stage scores for that exam. A second option is to write the second stage of the exam with other students who also have accommodations. A third option, and the most popular choice by far, is that we time the start of the second stage of the exam so that students can participate in the collaborative assessment with the rest of the class. This means that for students with additional time as their accommodation, they start the individual portion of the exam early enough to complete it and travel to the location of the group exam if in person. For online two-stage exams, it was seamless to provide accommodations to individuals and have these students participate in the two-stage assessment with the rest of the class since travel time was not needed.

23.2.3 Social Presence in Online Two-stage Exams

Building identities, trust, and relationships is challenging in a short-term group experience. Two-stage exams can be done with spontaneous short-term groups or in longer-standing teams, the latter are likely to launch into their discussions more quickly as they have developed relationships with each other. Inclusive collaborative assessments facilitate equitable group interactions. Approaches that facilitate equitable group interactions and welcome students to bring their whole selves to the classroom learning environment will lead to a positive classroom climate.[36]

In our courses, we have found that helping students see the value in working together helps to create a collaborative classroom climate. When students understand that the purpose of the two-stage exams is for everyone to improve their understanding, they develop a shared purpose which helps foster social presence.[37] If their discussions reveal that other students have similar chemistry challenges, the experience can normalize struggling and increase a sense of belonging.[38] When students have a strong sense of belonging, they are more likely to experience wellbeing and academic persistence.[39]

A growing body of research has identified that not all students experience in-class group work in the same way.[39] This likely also applies to collaborative assessments. Instructional strategies that aim to construct an inclusive classroom climate will lay the groundwork for more equitable participation and learning from two-stage exams. To do this, we lead by example and introduce ourselves with our social identities (*e.g.*, gender pronouns) on the course syllabus, learning management system, and in class sessions. Many instructors discuss why they decided to study chemistry to build instructor–student relationships.[35] To enhance students' perceptions of the value of teamwork, we show evidence regarding teams outperforming individuals, and explain that diverse groups with a broad range of life experiences and perspectives tend to perform better than non-diverse groups. Typically, this is done by sharing charts from research studies and discussing them during synchronous class time. Valuing teamwork and understanding diversity as a driver of excellence in science are important learning goals instructors can consider making explicit in their course goals.

23.3 Improving Collaborative Assessments With the CoI Framework

The CoI framework points to improvements to common implementations of two-stage exams, both in person and online. Striving toward a collaborative inquiry process will lead to more meaningful individual reflections and support groups to construct a deeper understanding of the material together. However, the work of a community of inquiry does not end once the assessment is over. Inquiry is a cyclical process where the results are used to prompt further learning discussions. The ideal two-stage assessment would involve students in repeated cycles of inquiry, which Garrison[28] defines as "a collaborative approach to problem resolution that transpires in the context of reflective discourse and interactive questioning" (p. 56). Garrison also states that Inquiry leads to the "questioning, interpretation, and transformation" (p. 56) of knowledge and experience which support deep and meaningful learning of chemistry concepts.

23.3.1 Supporting the Process of Collaborative Inquiry

To make two-stage exams more consistent with a cyclical collaborative inquiry process, an exam debrief could be added. During the debrief, students could respond to a structured worksheet containing questions such as "What was the hardest part about working together?" "Which questions did you need the contributions from multiple people on which to succeed?" A debrief activity will help create a climate that, in Garrison's words,[28] "encourages and supports open communication through a sense of belonging and trust" (p. 89). Groups could also reflect on the process they used during the assessment and set goals for improving on future collaborative assessments. Providing structure for these reflections is especially important online since the informal discussions about how the exam went are less likely to happen when students are not together in the same space. Asynchronous exam debriefs could be carried out with the mode of communication students tend to gravitate toward at a given time and location (*e.g.*, messaging software, social media, learning management system discussion boards). Synchronous exam debriefs could be carried out with voice or videoconferencing or in person. A worthwhile goal of active learning courses is to help students start to see each other as a source of knowledge and as learning supports and a debrief could help positively orient students towards future learning opportunities. As Lawrie discusses in Chapter 27 of this volume, it is common for chemistry students not to recognize peer feedback as useful, and exam debriefs may help with this.[40]

23.3.2 Enhancing Social, Teaching, and Cognitive Presences

Educators may want to consider enhancing the social presence of their two-stage exam by assigning roles to group members.[41,42] Instructors can make two-stage exams more inclusive by involving students as partners in test design to redistribute power and give students a voice. Offering choice to students in terms of tasks to demonstrate their learning can make assessment more equitable. Reducing the emphasis on scores and increasing the emphasis on growth and feedback, using "ungrading" approaches,[43] may increase intrinsic motivation for learning, reduce competition, and improve metacognition.

Leveraging technology can make assessments more authentic and realistic, capitalizing on the high levels of attention students typically have during two-stage exams. For online assessments, students are already at a computer, which allows for tasks that require them to use software or conduct an exploration. For example, students can collaborate to conduct an experiment with familiar simulation software if appropriate for the course. Two-stage exams with authentic tasks are likely to foster a high degree of cognitive presence.

23.4 Future Research Directions in Two-stage Exams

The initial and ongoing two-stage exam research in science education has revealed the positive impacts of this approach on learning in many contexts. Exciting new research is examining the cognitive processes that students engage in during collaborative exams.[20] To further refine this assessment strategy, and possibly explain some of the inconsistent effects on learning, future studies could use the CoI framework to investigate the distinct and overlapping contributions of teaching presence, cognitive presence, and social presence to foster collaborative learning. Additional discourse analysis studies could provide guidance in supporting the inquiry cycle of learning through problem identification, exploration, integration, and resolution and identify chemistry problem types that are well suited to two-stage assessments. We do not know how two-stage assessments impact students' metacognition nor shared metacognition.[28] To what extent can online two-stage exams play a role in developing learners' online collaboration skills related to workplace competencies? Additionally, chemistry educators would benefit from studies examining methods of scaffolding collaborative learning in two-stage exams. Specifically, what tasks and prompts help groups construct meaning and confirm their understanding, and what scaffolding helps groups move through the phases of inquiry? Clearly, there is a lot to like about two-stage exams, and a lot more to learn in this exciting area of pedagogical research.

Acknowledgements

I would like to acknowledge Joss Ives, for explaining how to set up Qualtrics to behave similarly to an IFAT card and for sharing some questionnaire items developed with Jared Stang; Jared Stang, for sharing these questionnaire items as well; Jason G. Wickenden, for collaborating on implementing two-stage exams in organic chemistry; and Kenny Lin, for creating Figure 23.2.

References

1. Two-Stage Midterm Exam for Large Class, http://blogs.ubc.ca/wpvc/two-stage-exams, accessed July 2021.
2. S. A. C. McDowell, *J. Chem. Educ.*, 2020, **97**, 3256–3259.
3. S. E. Lewis, *J. Chem. Educ.*, 2020, **97**, 3418–3422.
4. T. M. Clark, C. S. Callam, N. M. Paul, M. W. Stoltzfus and D. Turner, *J. Chem. Educ.*, 2020, **97**, 3413–3417.
5. J. J. Stewart and J. G. Wickenden, *Manuscript in Preparation*, 2022.
6. E. J. Maxwell, L. McDonnell and C. Wieman, *J. Coll. Sci. Teach.*, 2015, **44**, 48–52.
7. L. K. Michaelsen, A. B. Knight and L. D. Fink, *Team-based Learning: A Transformative Use of Small Groups in College Teaching*, Stylus Publishing, Sterling USA, 2004.
8. B. P. Rempel, M. B. Dirks and E. G. McGinitie, *J. Chem. Educ.*, 2021, **98**, 2527–2535.
9. J. F. Zipp, *Teach. Sociol.*, 2007, **35**, 62–76.
10. B. H. Gilley and B. Clarkston, *J. Coll. Sci. Teach.*, 2014, **43**, 83–91.

11. M. L. Epstein, A. D. Lazarus, T. B. Calvano, K. A. Matthews, R. A. Hendel, B. B. Epstein and G. M. Brosvic, *Psychol. Rec.*, 2002, **52**, 187–201.
12. G. W. Rieger and C. E. Heiner, *J. Coll. Sci. Teach.*, 2014, **43**, 41–47.
13. C. H. Crouch and E. Mazur, *Am. J. Phys.*, 2001, **69**, 970–977.
14. D. Cohen and J. Henle, *UME Trends*, 1995, **2**, 15.
15. C. Wieman, *Improving How Universities Teach Science*, Harvard University Press, Cambridge USA, 2017.
16. P. Belleville, S. A. Wolfman, S. Bradley and C. Heeren, *The 51st ACM Technical Symposium on Computer Science Education*, 2020, pp. 720–726.
17. S. P. Rao, H. L. Collins and S. E. DiCarlo, *Adv. Physiol. Educ.*, 2002, **26**, 37–41.
18. G. Kinnear, *Int. J. Res. Undergrad. Math. Educ.*, 2021, **7**, 33–60.
19. M. Lusk and L. Conklin, *J. Nurs. Educ.*, 2003, **42**, 121–124.
20. D. Nicol and G. Selvaretnam, *Assess. Eval. High. Educ.*, 2021, 1–16.
21. P. Liljedahl, in *Transforming Mathematics Instruction: Multiple Approaches and Practices: Advances in Mathematics Education*, ed. Y. Li, E. A. Silver and S. Li, Springer International Publishing, Switzerland, 2014, pp. 127–144.
22. I. A. G. Wilkinson and I. Y. Y. Fung, *Int. J. Educ. Res.*, 2002, **37**, 425–447.
23. N. Dasgupta, M. M. Scircle and M. Hunsinger, *Proc. Natl. Acad. Sci. U. S. A.*, 2015, **112**, 4988–4993.
24. D. McEwan, G. R. Ruissen, M. A. Eys, B. D. Zumbo and M. R. Beauchamp, *PLoS One*, 2017, **12**, 1–23.
25. K. A. Ericsson, in *The Cambridge Handbook of Expertise and Expert Performance*, ed. K. A. Ericsson, N. Charness, P. J. Feltovich and R. R. Hoffman, Cambridge University Press, Cambridge, 2006, ch. 38, pp. 685–705.
26. Y. J. Joo, K. Y. Lim and E. K. Kim, *Comp. Educ.*, 2011, **57**, 1654–1664.
27. G. Heilporn and S. Lakhal, *Comp. Educ.*, 2020, **145**, 103712.
28. D. R. Garrison, *Thinking Collaboratively: Learning in a Community of Inquiry*, Routledge, New York, 2015.
29. D. R. Garrison, *E-Learning for the 21st Century: A Community of Inquiry Framework for Research and Practice*, Routledge, London, 2nd edn, 2011.
30. J. Ives, M. de Jong Van Lier, N. K. Sumah and J. B. Stang, *Phys. Educ. Res. Conf. Proc.*, 2016, 1–4.
31. J. M. Holmes and K. E. Weaver, *New Dir. Teach. Learn.*, 2020, **164**, 57–64.
32. J. Ives, J. B. Stang, J. Arias-Bustamente, P. J. Dubois and A. Hofmann, *Phys. Educ. Res. Conf. Proc.*, 2020, 228–233.
33. J. Smith, S. B. Wilson, J. Banks, L. Zhu and P. Varma-Nelson, *J. Res. Sci. Teach.*, 2014, **51**, 714–740.
34. J. H. Dulebohn and J. E. Hoch, *Hum. Resour. Manag. Rev.*, 2017, **27**, 569–574.
35. C. D. Harrison, T. A. Nguyen, S. B. Seidel, A. M. Escobedo, C. Hartman, K. Lam, K. S. Liang, M. Martens, G. N. Acker and S. F. Akana, *CBE Life Sci. Educ.*, 2019, **18**, ar47.
36. C. Neill, S. Cotner, M. Driessen and C. J. Ballen, *Chem. Educ. Res. Pract.*, 2019, **20**, 197–203.
37. P. Rogers and M. Lea, *Behav. Inf. Technol.*, 2004, **24**, 151–158.

38. K. L. Lewis, J. G. Stout, S. J. Pollock, N. D. Finkelstein and T. A. Ito, *Phys. Rev. Phys. Educ. Res.*, 2016, **12**, 020110.
39. S. L. Eddy, S. E. Brownell, P. Thummaphan, M.-C. Lan and M. P. Wenderoth, *CBE Life Sci. Educ.*, 2015, **14**, 1–17.
40. See Chapter 27: G. A. Lawrie, in *Digital Learning and Teaching in Chemistry*, ed. Y. J. Dori, C. Ngai and G. Szteinberg, Royal Society of Chemistry, 2023.
41. S. L. Eddy, S. E. Brownell, P. Thummaphan, M.-C. Lan and M. P. Wenderoth, *CBE Life Sci. Educ.*, 2015, **14**, 1–17.
42. C. Flener-Lovitt, K. Bailey and R. Han, *J. Chem. Educ.*, 2020, **97**, 2519–2525.
43. C. Sorensen-Unruh, in *Ungrading: Why Rating Students Undermines Learning and What to Do Instead*, ed. S. D. Blum, UBC Press, Vancouver, 2020, ch. 9, pp. 140–161.

CHAPTER 24

Digital Formative Assessments for Learning

S. WILKINSON*[a], R. PULLEN[a], P. RUTLEDGE[a] AND S. SCHMID[a]

[a]School of Chemistry, University of Sydney, Sydney, NSW, Australia
*E-mail: shane.wilkinson@sydney.edu.au

24.1 Introduction

Through this chapter, we hope to provide a resource that both practitioners and researchers could utilise when considering their respective practice for future research directions. While numerous reviews and conversations have been disseminated regarding the use of digital technology for assessment,[1,2] innovations within this space are frequent and we must be vigilant in interrogating promising technologies while keeping the needs of our students foremost in mind. With this in mind, we have selected a number of digital technologies, both new and familiar, that we believe will be prominent in the global shift to online teaching and learning. We conclude this chapter by disclosing a bespoke digital tool (the student relationship engagement system, SRES) along with some case studies. The authors have utilised SRES at their institution to aid in the delivery of customisable student feedback in large classroom environments across a range of assessments.

Advances in Chemistry Education Series No. 11
Digital Learning and Teaching in Chemistry
Edited by Yehudit Judy Dori, Courtney Ngai and Gabriela Szteinberg

Published by the Royal Society of Chemistry, www.rsc.org

24.1.1 Formative Assessments

Before delving into the core of this chapter, we deemed it important to establish a clear definition of formative assessment. The first distinction between formative and summative assessment with regards to student learning was made by Bloom who described formative assessment as providing feedback at regular intervals through the teaching-learning process.[3] This is distinct from summative assessment that focuses on a single instance evaluation of progress and/or knowledge without any intention of contributing to further learning. Since this time, extensive research has been conducted to better understand and utilise formative feedback for learning but a significant barrier to this is the lack of a clear, consistent definition.[4] A definitive shift has been seen from a teacher-centric model of providing formative feedback to better understanding and facilitating student use of formative feedback for learning.[5]

In preparing this chapter and considering the implications of formative feedback and assessment for learning, we have used Carless and Boud's feedback literacy theoretical framework to guide our thoughts and conclusions.[6] Underpinning this framework, Carless and Boud state that "For feedback processes to be enhanced, students need both appreciation of how feedback can operate effectively and opportunities to use feedback within the curriculum" (p. 1315).[6] While the former lies in the communication of how feedback can be utilised, and therefore not the aim of this chapter, the latter can be facilitated through the use of digital technologies to provide and guide this learning.

24.1.2 The Rise of Digital Tools in Education

Technological advancements (such as high-speed internet and smart phone innovations) have resulted in the transition of the learning environment into digital modes. A review by Bernholt and co-workers,[7] while not extensive, details four ongoing examples of utilising digital technology to support chemistry education with associated positive outcomes relating to motivation, digital literacy, development of deeper understanding, and accommodating different modes of engagement for students. A review into the use of digital technology within mathematics education does offer some caution to ensure that the reason why digital technology has been implemented is clear and aimed towards specific outcomes.[8] However, these advancements may not be accessible by all students globally and consideration of infrastructure (power and internet) as well as socioeconomics (that may impact a student's access to these digital tools) need to be considered by the practitioner. From a teacher's perspective, one of the major appeals to this digital transition is the ability to observe the learning process through learning analytics (LA). Student scores, assessment submissions, communications, and engagement can be captured, stored and retrieved. Analysis of these data can be used to evaluate formative assessments and adapt later teaching opportunities to address learning needs. This is typically facilitated within a learning management system (LMS) such as Canvas, Moodle and Blackboard (to name a few) that house learning content, student interaction, assessment tools and reports of learning progress and student activities, across multiple digital platforms, all in one environment. One important consideration not mentioned is the importance of educators being digitally competent and the provision of regular professional development to support adaptation to the rapidly changing digital technology landscape.[9]

It is common to integrate or use digital tools that complement an LMS, allowing teachers to engage with students through multi-modal resources and approaches including videos, games and social media. If internet access is available, these resources can be accessed by students from anywhere and at any time, which has expanded accessibility to the content. This has been a critical factor for an increased uptake of digital tools during the COVID-19 pandemic where traditional face-to-face interactions were prohibited. When transitioning the delivery of a lesson or assessment into a digital form, it is important that pedagogical design such as the objective, assessment, and feedback are not compromised or lost. In this chapter, we have highlighted some of the digital tools commonly utilised by educators today, as well as some emerging platforms, and examine how they provide feedback to the student to close the loop. For a recent collection of how practitioners adapted to digital assessment through the COVID-19 pandemic, readers might like to peruse the 2020 *Journal of Chemical Education* special issue.[10]

24.2 Digital Assessment Options

There is an ever-growing list of digital tools available and attempting to review them all could be a textbook on their own! For the purpose of this chapter, the authors have focused on digital tools familiar to them either through anecdotal experience, peer-reviewed applications or wide-spread adoption in their field of teaching. In the following sections, the digital tools have been grouped into categories based on the type of assessment. Their pedagogical merits and contribution to formative assessments will be examined.

24.2.1 Digital Submission Flexibility and Assessment

The previously mentioned technological advances afford an opportunity for assessment of learning in areas not easily accessible historically. For example, current mobile phones all connect to the internet and usually house cameras that offer anywhere from standard definition to ultra-high-definition video quality. The cost and prevalence of mobile phones (even in remote areas of the world) means video assessments have become more viable than a decade before this publication. Alongside this shift, the move to digital submissions affords a wider range of artefacts that can be assessed beyond the traditional paper submission, including but not limited to reports, video submissions, websites, and electronic laboratory notebooks. This inherently affords greater accessibility through the provision of multiple methods for demonstrating proficiency. For example, students who may typically struggle with a written report based on their familiarity with writing in the expected language may find greater success in demonstrating their understanding when submitting a recorded video with primarily audio and visual explanations.

Electronic feedback tools incorporated into an LMS, such as Speedgrader in Canvas, allow practitioners greater flexibility when providing feedback or comments on electronic submissions. Assessment guides such as rubrics can easily be integrated within these systems, providing students a clear understanding of the expectations for the task and feedback mechanisms in one convenient location. More specific feedback can be annotated within the student submission to provide efficiency for assessors communicating feedback and clarity to students receiving the feedback. A unique addition available in some systems is the possibility for feedback to be uploaded as an audio or video file. This is especially impactful in situations where remote teaching is required, since it allows for a more personal connection of the teacher to the student.[11] With a little practice it can also become less time consuming for practitioners than written feedback of the same quality. The digital nature of these processes additionally provides long-term value through recycling of commonly given feedback and informing future teaching practices.

In Australia, the chemistry community formulated a set of threshold learning outcomes (TLOs) that all students majoring in chemistry will have attained on graduation. It has since been found that apart from content knowledge, many TLOs are not purposely developed throughout degrees, nor assessed in a meaningful way.[12] With digital literacy and (science) communication becoming common learning outcomes within the higher education sector, digital submissions have provided the means to grade and supply feedback to students about these outcomes across a range of digital media. Student-generated videos depicting science/chemistry concepts have become increasingly popular over the last decade.[13] Student video submissions may simultaneously meet a number of content-related learning outcomes, as well as develop transferable skills like digital literacy, communication, teamwork or problem solving.[14] Video tasks can also act as a facilitator for demonstrating student activities when engaging in remote laboratories. For example, students may complete simple laboratory exercises from their homes, posting videos of the experimental set-up and execution of the experiment for the instructor to comment on.[15]

Keeping a record of laboratory work is quintessential in any science discipline and instilling this practice through our undergraduate courses is pertinent in career readiness. Electronic laboratory notebooks (ELNs) are gaining popularity in industry and academia due to their audit trail, searchability, accessibility, shareability and, possibly most importantly, their multi-modal input allowing all aspects of an experiment to be amalgamated in one place (*e.g.*, videos, photos, spectra, spreadsheets, *etc.*). As such, many instructors are beginning to incorporate ELNs into their courses and consequently a number of ELN providers have tweaked their research products for educational integration which allow student assessment, feedback and LMS integration. Studies[16] have reviewed the various ELNs available, examples of curriculum integration with said ELNs and their associated outcomes. Interestingly, unexpected new lines of communication were observed with students posting questions within their ELNs instead of through conventional pathways (emails and face-to-face). This had the benefit of directing instructors specifically to the content the student needed assistance with and delivering specific feedback to guide that student. It was commented that the ability to "access a student's notebook remotely allowed instructors to more easily track a student's progress and offer guidance or (more immediate) feedback" (p. 187).[16] It is evident that the opportunities afforded by digital submission tools have the potential to broaden the range of traditional assessment items to include those with potential to develop TLOs other than content knowledge, including transferable skills that are highly sought after by 21st century employers.

24.2.2 Question-based Assignments and Quizzes Using Transmissive Feedback

There is a diversity of tools available for the delivery of online quizzes and other question-based assessments, with increasingly sophisticated feedback provision. As in all aspects of teaching, there are advantages and disadvantages to synchronous and asynchronous delivery of quizzes[17,18] and the choice will hinge both on the learning context and the preferences of the individual teacher. Predominantly, static quizzes or question-based assignments would fall into the transmissive feedback category by nature of offering static feedback not necessarily explicitly connected to subsequent assessment tasks.

There are many options for delivering asynchronous online quizzes with feedback. For most, the choice will be dictated largely by the higher-level choice of LMS, since each of these has its own integrated quizzing tools. With Canvas for example (the LMS with which we are best acquainted), the teacher has a dozen different question formats from which to choose, ranging from multiple-choice and true or false, through matching, numerical, formula, and essay questions. Detailed feedback can be written into the quiz for each response (correct or otherwise), tailored to address common misunderstandings or issues specific to the cohort, and linked to additional resources (*e.g.*, videos, lecture notes, or textbook content) to develop and deepen students' understanding. Each student gains feedback on their specific response set, and the teacher can review each student's answers, and thus gain insight at an individual level. However, it can be time-consuming to make effective use of this information, particularly in a large cohort, and integration with a personalisation/engagement tool such as the SRES (see later in a sub-section of this chapter) may be required to unlock the full potential of this feedback.

There are a number of live-polling tools such as Socrative,[19,20] Nearpod,[18] and Kahoot,[21] as well as a variety of web-based 'clicker' tools,[22] all of which can enable synchronous, real-time quizzing and feedback, enhancing active, discussion-driven pedagogy. Vallely and Gibson identify three "key multi-disciplinary strategies that can be supported by Mentimeter to engage students" by utilising different question formats within this tool: 'gauging opinion', 'engaging discussion' and 'voicing concerns'.[20] Mentimeter's word cloud tools also enable quick and facile co-creation of visually rich and engaging graphics, allowing students and teacher to compile their shared priorities, misunderstandings, or areas of concern across a disparate class.

Christianson highlights the power of real-time class polls delivered through Socrative as "a versatile way to incorporate real-time student participation even in a virtual environment" and a means for "keeping an active [learning] component in the online classroom" even while teaching remotely.[19] Socrative offers three quiz formats—'Instant Feedback', 'Open Navigation', and 'Teacher Paced'—each of which enables a different mode of feedback delivery and learning. There is also capacity to introduce a gamified aspect to quizzes using the 'Space Race' tool which sees students working in teams and racing the clock, and an 'Exit Ticket' feature to crosscheck student understanding as they leave a class.

The key advantages of these synchronous live polling tools are their ease-of-use (it is quick for the teacher to build the slides and other resources needed to integrate these tools, and students find them intuitive and easy to access), the relative anonymity (responses are pooled and presented collectively, creating a safe environment in which students can contribute without fear of judgement or direct criticism), and the instant, real-time feedback that is enabled (both feedback to students on their own understanding, and feedback to teachers on how well a concept has been understood). On the flipside, although these tools are easy to use, careful design planning is required to ensure effective integration and avoid the temptation to over-quiz students which could lead to disengagement. Also, the anonymity can enable non-participation—not all students will contribute, and those who do not deny themselves the opportunity for feedback on their understanding—and the pooling of responses precludes specific, personalised feedback to individuals. In addition, students can only contribute if they have an appropriately connected device—decreasingly an issue in our modern world, but something for practitioners to be mindful of, nonetheless.

24.2.3 Adaptive Learning Systems Using Corrective Feedback

One limitation of utilising online quizzes as formative feedback for students is that the content and difficulty remain constant regardless of the quality of feedback provided for each question—each student will encounter the same questions (or equivalent questions if banked or randomised). Adaptive learning systems (ALSs) address this using artificial intelligence algorithms to adjust the difficulty and type of content provided to students as they progress—each student receiving a different question and corrective feedback, based on their response to the previous question. ALSs address this using artificial intelligence algorithms to adjust the difficulty and type of content provided to students as they progress—each student receiving a different question based on their response to the previous question. There is a growing number of ALS software now available with many linked to textbook publishers that align with textbook content (ALEKS, Alta, LearnSmart, MasteringChemistry, MyLab and Mastering, and SmartSparrow were familiar to us at the time of writing).[23] ALSs identify each student's strengths and weaknesses on a topic, based on a student's response to a question before providing teaching resources and further formative assessments to target the identified student's weakness(es). This process continues until the system judges a student has mastered a topic (by predetermined criteria) and builds up the complexity of the topics and questions again.

While this may lengthen the time spent by students struggling with the content, the additional revision of fundamental topics is essential to fill gaps in students' knowledge prior to engaging with more complex concepts. Despite "more work" being assigned to these students, the shift to less complex content rebuilds a student's confidence and motivation before reintroducing more complex topics.[24] In contrast, academically high performing students proceed smoothly through fundamental questions and are instead offered more challenging content to keep them engaged and achieving higher order learning outcomes. Many ALSs also offer multi-disciplinary content allowing for engagement in a wide coverage of topics. For example, an inability to answer stoichiometry questions may not stem from a student's misunderstanding of the chemistry but instead result from a weakness in algebra. An ALS system could offer resources to strengthen this student's understanding of algebra before progressing back to stoichiometry.

A study by Richards-Babb and co-workers[25] compared "traditional" quiz feedback *vs.* adaptive learning systems as homework platforms for their General Chemistry unit and found that students who utilised the adaptive learning system on average, performed better in their final exam. However, students reported their frustrations when a question was incorrectly answered resulting in additional questions for a student to complete to master a topic. Richards-Babb and co-workers[25] reported that the benefits outweighed the cons by concluding that "active engagement and good study habits are important attributes that promote learning despite students' apparent displeasure at devoting time to learning." (p. 698).

RiPPLE is another ALS but differs from the aforementioned list by having its content created *via* crowdsourcing.[26] Users (students and educators) can create content to be added to the RiPPLE question repository and rate the quality of the content provided to them as they proceed through the ALS modules. As superior content (as determined by user feedback) is added to the repository, it will eventually replace the original recommended content for a particular topic. The ranking of students and content in RiPPLE is determined through an Elo rating system—as a student correctly answers a question, their rating goes up whilst the question's rating goes down. The dynamic rating of the content in RiPPLE means not only is the content adaptive to the student's needs, but the supplied content presented to the students is also adaptive to ensure the best rated content is always presented.

RiPPLE has a dedicated teaching mode to facilitate students' creation, moderation, and evaluation of content as part of the crowdsourcing aspect of the ALS before being integrated into the public repository. The content creation provides a deeper understanding of their topics by forcing students to consider the learning outcomes of the course, what are the common misconceptions of certain topics and being able to explain their understanding of the topic to their peers through the content feedback that will be revealed once a student has completed the question. Before the content is added to the repository, students are prompted to rate and provide feedback on their peers' content based on the content's ambiguity, correctness, and difficulty. Content creators can view and reflect upon this feedback, refining their content before adding it to the repository.

One last example can be seen with a pilot study by Herrington and Sweeder offering an example of an asynchronous approach with the use of text messaging.[27] This messaging aimed to prompt self-regulated study and provide formative assessment based on student responses. Their approach begins with students receiving a text message linking to a series of spot-check content-based questions that are relayed back to the instructor. The instructor can then utilise these responses to prepare collective formative feedback for the next formal learning session. While Herrington and Sweeder elected to take the collective formative feedback approach, an opportunity is present to extend this approach with the use of automated feedback depending on the question type or, for smaller classes, customised feedback from instructors or teaching assistants. Results from this study indicated that student reception was positive while uptake of this system was low, with the caveat that engagement would be increased if marks were attached to these activities.

24.2.4 Opportunities for Dialogic Feedback

Outside of crafted feedback opportunities such as quizzes, question-based assignments and ALSs as discussed earlier, one avenue for delivering informal feedback is using dialogic feedback through two-way communication channels between peers and, between peers and instructors. The importance of creating a space for social discourse in an online learning environment is an argument that has been long made and accepted.[28] The forms in which these can take are many with examples in these authors' experiences including both asynchronous modes such as discussion boards, instant messaging and third-party services such as Padlet, and synchronous modes such as video conferencing, Slack or Jamboard. Of these, discussion boards are perhaps the most prevalently used, often being integrated into the LMS. One recent example investigating the types of discussion seen within massive open online courses (MOOCs) has been completed by Tawfik and co-workers.[29] Within this study, they noted three types of discussion boards had been designed aiming to compartmentalise discussion into: (1) general discussion; (2) study groups; and (3) content relating to the weekly module. Both represented in this study and in our own experiences, valuable dialogue and feedback can be provided through these media though with the caveat that user engagement tends to drop off significantly through a semester leaving a smaller pool of students benefiting from this space.

While only one example has been provided here, the range and flexibility of these tools is expansive allowing a high degree of customisation for practitioners. In our own experience, one of the key roles these communication channels, both synchronous and asynchronous, have held is the opportunity for students to engage in a manner such that they feel safe and supported, and in an anonymous mode when advantageous to do so. This allows students who would perhaps not feel comfortable taking a risk on a question they do not know the answer to or are concerned with their knowledge of the language being used to engage actively.

24.2.5 Learning Analytical Tools for Digital Feedback

Feedback can be time-intensive and with ever growing class sizes and diversity, it can be impossible to provide timely and high-quality feedback through traditional measures. Increasing the number of teachers involved in the feedback process can result in differing, and sometimes conflicting, feedback being delivered across the cohort, which can confuse students and negate the primary intention of providing feedback in the first place. LMSs are equipped to capture and store large datasets of student assessments and even metadata of student engagement to present educators with a snapshot of their student's performance with dashboards. However, the real challenge becomes reaching out to all students to provide timely, useful and personalised feedback. There are now a number of emerging LA software that fills this void and empowers teaching with adequate tools to provide quality feedback on scale whilst also reducing practitioner workloads.

One such LA software that offers this means of support is the SRES.[30] Data can be input into the system using a variety of resources, from data synchronisations with LMSs or databases, spreadsheet uploads, or manually inputting the data through simple web app interfaces which can be designed for both teacher and student input. Once data is stored in an SRES list, spreadsheet-like data analysis can be performed. Analysis can range from using simple arithmetic manipulations to complex what-if logical expressions without the need of knowing coding language. Finally, the processed information can be outputted through a number of avenues. Customised information can be sent to all or selected students through mail-merger-type emails or *via* web page content which can be embedded into a student's LMS. The information could also be delivered to the teaching staff in the form of dashboards, reports or emails to the teaching staff who may convey feedback directly to the student(s). What makes this system such a powerful tool is that the logical decisions made by educators during the feedback process (during data analysis and constructing feedback) are captured as conditional statements in SRES and then applied *en masse* to the cohort. The educator has full control of the data analysis and how the feedback is delivered making the overall act of feedback a genuinely humanistic approach despite the automation. To better understand how this software is applied, we have composed a couple of case studies of how SRES is utilised in our own institution to facilitate the feedback process.

24.3 SRES: A Case Study for the Provision of Digital Feedback and Assessment

The SRES can be best described as an online database to capture and manage student data.[31] Its value comes from its specialised tools that help deliver personalised feedback to students and automate repetitive tasks for the teacher (*e.g.*, emailing students). We have outlined a few case studies below of where SRES has been utilised to deliver personalised feedback to students based on the outcome of an assessment in a large (~2000 students) student cohort.

24.3.1 Case Study 1: Rubric-to-Feedback

SRES is used extensively in our laboratory environment primarily to capture the grading of non-digital submissions or practical assessments that are difficult or impossible to mark *via* LMS options. These include assessments such as grading laboratory samples, marking competency or technique-based assessments (practical exams) or even oral presentations. This is due to the flexibility of SRES to accommodate basic computing logic including complex calculations to set correct parameters for each submission, personalised feedback generated based on each rubric criterion grade, and conditional (what-if) statements. Teaching assistants (TAs) are provided with a marking rubric which is captured on a mobile-friendly web interface generated in SRES. Whilst constructing the rubric, SRES allows you to assign specific feedback and grade to each criterion (see Figure 24.1).

TAs do not see the grades or feedback associated with each criterion and have only descriptor tiles to press when assessing work. Once a TA has assessed a student's work, the grades are immediately tallied by the system and presented to the student in an embedded webpage in their LMS. In addition, the feedback statements that were tied to each criterion selected by the TA, are concatenated to generate a detailed feedback account. Generally, this feedback highlights where and why the student lost their marks and presents rectifying actions or directions to further resources to improve in future opportunities. Where students scored full marks, congratulatory remarks were also included as positive reinforcement. The total score and feedback are generated immediately after the assessor submits their grading. In the past, this grade and feedback

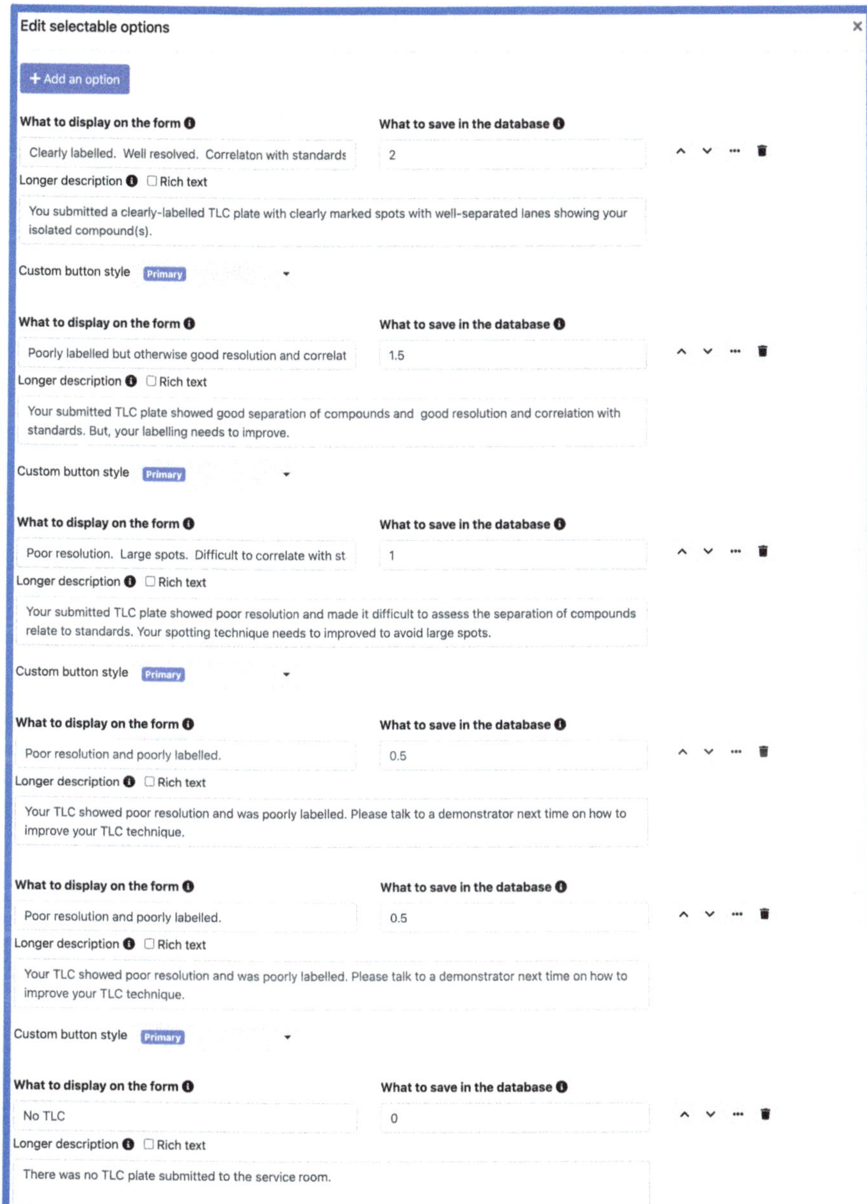

Figure 24.1 A marking criterion for the assessment of thin-layer chromatography (TLC) performed by students in the chemistry laboratory. "Text to display" shows the criterion descriptor presented to the assessor. "Value to save" is the respective grade (out of 2) associated with the criterion. "Longer description" captures the feedback statement presented to the student if that criterion is selected by the assessor.

was released to the students *via* an email from the lab coordinator, however it generated an unexpected amount of work when many students replied to the email wishing to express their gratitude for the time spent providing such a detailed account of their samples. In future lab classes, the results were issued *via* a web portal within the students' LMS (Figure 24.2). Although less personal than a "tailored" email, anecdotal and unpublished research data from students emphasised their gratitude for the detailed feedback. Through SRES, personalised detailed feedback that would have taken hours to write, for a cohort of over 250 students, took less than an hour by only writing criterion-specific feedback statements as part of the marking rubric creation.

24.3.2 Case Study 2: Peer Assessment

SRES was used to facilitate the peer assessments of student oral presentations as a strategy to engage students and foster critical thinking by involving them in evaluating the contents of peer presentations. In addition, we allowed 30% of a student's oral mark to be determined by the average mark given through peer assessment. Students used their mobile devices to access the marking rubric presented to them in a web app interface from SRES (Figure 24.3). As students presented, their peers (and the academic) were grading the presentation live.

Upon submitting their gradings, SRES averaged the peers' scores, tallied it with the academic determined grade, and posted it immediately on the presenter's LMS. In addition, what-if statements were applied to the average scoring of each criterion to construct appropriate feedback statements that were stitched together to provide an immediate overall feedback account of the presentation. Presenters had a final score and feedback available to them before they could sit down from their presentation. This instant feedback allowed students to immediately reflect on the experience and start considering their approach for future opportunities. In addition to summative and

Experiment 3

Online Results Mark: **6** / 8 marks
(This contributes 25% of this experiment's marks)

Sample Mark: **6.2** / 8 marks
(This contributes 25% of this experiment's marks)

Sample Feedback: Well done! All your samples were correctly labelled. For your major component; your sample colour looked great suggesting good purity. However, your sample was still a bit wet with solvent. Have a look in the lab manual on how to dry your compounds or speak to one of the demonstrators when you come to labs next time. Your submitted sample had a higher than expected yield. This indicates that the sample retained solvent or other impurity. Well done, you were able to isolate the minor component relate to standards. Your spotting technique needs to be improve to avoid large spots

Figure 24.2 Example of a SRES lab results page projected into a student's LMS. The grade of 6.2 was tallied and the "Sample Feedback" was constructed automatically and immediately after the assessor had graded this student's sample.

Figure 24.3 Example of a SRES web interface depicting a marking rubric for an oral presentation.

student peer feedback, students also experience self-feedback by having the opportunity to reflect on and critically evaluate the reliability of their data, quality of their (laboratory) work and performance in verbal communication against their peers.

Unlike traditional assessments, where more of the educator's time is required post-assessment with grading and supplying feedback, SRES shifts all the labour towards the planning phase of an assessment. The feedback is generated alongside the marking criteria which has an additional benefit of making the educators reflect upon their assessment structure and how they would advise students on improving different elements in the assessment. However, it is in the long run that efficiency is truly appreciated by the educator as the marking and feedback platform generated in SRES can be cloned and re-used from year to year. This allows for an equally content-rich feedback experience for the students with little time commitment by the educator.

24.4 Concluding Thoughts

The digital tools outlined in this chapter are not intended to propose the most suitable platforms nor offer an extensive list of all digital tools available to educators. Instead, this selection was compiled based on tools familiar to the authors or were of interest for future practice and had peer-reviewed accounts of their application to demonstrate what feedback was achievable for a particular genre of digital tool. For example, while we have found SRES to be an excellent tool for managing personalised feedback, OnTask[32] is another web-based ALS that, like SRES, also focusses on personalised student feedback and teacher insights. An often repeated, but important, mantra is that of considering the purpose of the task and the assessment. Regularly asking yourself questions such as "What is your assessment for?", "What is it measuring?", and "When is your feedback required?" will lead to reflection on how well a digital tool integrates into your curriculum.

The accessibility of these tools can often depend on the learning curve for practitioner and student use, and the financial and logistical considerations for implementation within your institution and practice. Further to this, understanding the unique conditions in which each of our students is studying is essential when crafting a learning environment (see also Chapter 20 by Pullen and colleagues).[33] Many digital tools rely upon regular, uninterrupted access to the internet or computers, which for some students can be an unreasonable ask. Additionally, digital tools that have a heavy emphasis on a single mode of communication (*i.e.*, visual-based, text heavy, grounded in a narrow content culturally) can be alienating for students who have learning needs relating to disabilities or may be in unfamiliar or uncomfortable settings as international or marginalised groups. Alongside these concerns, there are also the practical considerations of software compatibility given an entire cohort may not be using the same devices or operating systems. In conclusion, the most successful digital tools will be those that allow flexible feedback opportunities and evolve to the needs of their users to address the diverse and novel applications arising from contemporary pedagogical practices. Speaking as practitioners, we will be regularly interrogating our use of digital tools while continuing to listen closely to the feedback that both practitioners and students provide when operating within these digital learning environments and the associated feedback protocols in place.

List of Abbreviations

ALS	Adaptive learning system
ELN	Electronic laboratory notebook
LA	Learning analytics
LMS	Learning management system
MOOCs	Massive open online courses
SRES	Student relationship engagement system
TAs	Teaching assistants
TLC	Thin-layer chromatography
TLOs	Threshold learning outcomes

Acknowledgements

The authors would like to acknowledge both our colleagues and students who constantly contribute to the advances in our practice. Additionally, the authors would like to acknowledge the key individuals, A/Prof. Danny Liu and the Education Innovations team at The University of Sydney, who developed and maintain SRES.

References

1. L. Eyal, Digital Assessment Literacy — the Core Role of the Teacher in a Digital Environment, *Educ. Technol. Soc.*, 2012, **15**, 37–49.
2. M. Akbar, Digital Technology Shaping Teaching Practices in Higher Education, *Front. ICT*, 2016, **3**, 1.
3. B. S. Bloom, Learning for Mastery. Instruction and Curriculum. Regional Education Laboratory for the Carolinas and Virginia, Topical Papers and Reprints, Number 1, 1968.
4. R. E. Bennett, Formative assessment: a critical review, *Assessment Educ.*, 2011, **18**(1), 5–25.
5. E. Molloy, D. Boud and M. Henderson, Developing a learning-centred framework for feedback literacy, *Assessment Eval. Higher Educ.*, 2020, **45**(4), 527–540.
6. D. Carless and D. Boud, The development of student feedback literacy: enabling uptake of feedback, *Assessment Eval. Higher Educ.*, 2018, **43**(8), 1315–1325.
7. S. Bernholt, K. Broman, S. Siebert and I. Parchmann, Digitising Teaching and Learning – Additional Perspectives for Chemistry Education, *Isr. J. Chem.*, 2019, **59**(6–7), 554–564.
8. P. Drijvers, Empirical Evidence for Benefit? Reviewing Quantitative Research on the Use of Digital Tools in Mathematics Education, in *Uses of Technology in Primary and Secondary Mathematics Education: Tools, Topics and Trends*, ed. Ball L., Drijvers P., Ladel S., Siller H.-S., Tabach M. and Vale C., Springer International Publishing, Cham, 2018, pp. 161–175.
9. L. Amhag, L. Hellström and M. Stigmar, Teacher Educators' Use of Digital Tools and Needs for Digital Competence in Higher Education, *J. Digital Learn. Teach. Educ.*, 2019, **35**(4), 203–220.
10. *Insights Gained while Teaching Chemistry in the Time of COVID-19*, ed. Holme T. A., American Chemical Society, 2020.
11. J. Borup, R. E. West and R. Thomas, The impact of text versus video communication on instructor feedback in blended courses, *Educ. Technol. Res. Dev.*, 2015, **63**(2), 161–184.
12. M. Schultz, S. Schmid and T. Holme, *Technology and Assessment Strategies for Improving Student Learning in Chemistry*, American Chemical Society, Washington, DC, 2016.
13. M. Gallardo-Williams, L. A. Morsch, C. Paye and M. K. Seery, Student-generated video in chemistry education, *Chem. Educ. Res. Pract.*, 2020, **21**(2), 488–495.
14. M. A. Erdmann and J. L. March, Video reports as a novel alternate assessment in the undergraduate chemistry laboratory, *Chem. Educ. Res. Pract.*, 2014, **15**(4), 650–657.
15. M. Schultz, D. L. Callahan and A. Miltiadous, Development and Use of Kitchen Chemistry Home Practical Activities during Unanticipated Campus Closures, *J. Chem. Educ.*, 2020, **97**(9), 2678–2684.

16. K. Colabroy and J. K. Bell, Lab eNotebooks, *Biochemistry Education: From Theory to Practice*, ACS Symposium Series, American Chemical Society, 2019, vol. 1337, pp. 173–195.
17. M. A. Christiansen, A. M. Lambert, L. S. Nadelson and K. M. Dupree, TA Kingsford. In-Class Versus At-Home Quizzes: Which is Better? A Flipped Learning Study in a Two-Site Synchronously Broadcast Organic Chemistry Course, *J. Chem. Educ.*, 2017, **94**(2), 157–163.
18. J. A. Milligan, What Is the Value of Synchronous Engagement in Small Remote Organic Chemistry Classes? Analysis of Multiple-Choice Polling Data from the COVID-Impacted Spring Semester of 2020, *J. Chem. Educ.*, 2020, **97**(9), 3206–3210.
19. A. M. Christianson, Using Socrative Online Polls for Active Learning in the Remote Classroom, *J. Chem. Educ.*, 2020, **97**(9), 2701–2705.
20. K. S. A. Vallely and P. Gibson, Engaging students on their devices with Mentimeter, *Compass*, 2018, **11**(2), DOI: 10.21100/compass.v11i2.843.
21. J. Murciano-Calles, Use of Kahoot for Assessment in Chemistry Education: A Comparative Study, *J. Chem. Educ.*, 2020, **97**(11), 4209–4213.
22. K. M. Shea, Beyond Clickers, Next Generation Classroom Response Systems for Organic Chemistry, *J. Chem. Educ.*, 2016, **93**(5), 971–974.
23. M. Bagheri, Intelligent and Adaptive Tutoring Systems: How to Integrate Learners, *Int. J. Educ.*, 2015, **7**, 2.
24. J. M. Fautch, *Enhancing Retention in Introductory Chemistry Courses: Teaching Practices and Assessments*, ed. Kradtap S. and Gupta T., 2019.
25. M. Richards-Babb, R. Curtis, Z. Georgieva and J. H. Penn, Student Perceptions of Online Homework Use for Formative Assessment of Learning in Organic Chemistry, *J. Chem. Educ.*, 2015, **92**(11), 1813–1819.
26. H. Khosravi, K. Kitto and J. J. Williams, RiPPLE: A Crowdsourced Adaptive Platform for Recommendation of Learning Activities, *J. Learn. Anal.*, 2019, **6**(3), 91–105.
27. D. G. Herrington and R. D. Sweeder, Using Text Messages To Encourage Meaningful Self-Assessment Outside of the Classroom, *J. Chem. Educ.*, 2018, **95**(12), 2148–2154.
28. C. Brook and R. Oliver, Online learning communities: Investigating a design framework, *Australas. J. Educ. Technol.*, 2003, **19**(2), 139–160.
29. A. Tawfik, T. Reeves, A. Stich, A. Gill, C. Hong and J. McDade, *et al.*, The nature and level of learner–learner interaction in a chemistry massive open online course (MOOC), *J. Comput. Higher Educ.*, 2017, **29**, 411–431.
30. N. Arthars, M. Dollinger, L. Vigentini, D. Y. T. Liu, E. Kondo and D. M. King, Empowering Teachers to Personalize Learning Support, *Empowering Teachers to Personalize Learning Support*, Springer, 2019, pp. 223–248.
31. D. Y. T. Liu, K. Bartimote-Aufflick, A. Pardo and A. J. Bridgeman, Data-Driven Personalization of Student Learning Support in Higher Education, in *Learning Analytics: Fundaments, Applications, and Trends: A View of the Current State of the Art to Enhance e-Learning*, ed. Peña-Ayala A., Springer International Publishing, Cham, 2017, pp. 143–169.

32. Chapter 20: A. Pardo, K. Bartimote, S. B. Shum, S. Dawson, J. Gao and D. Gašević, *et al.*, OnTask: Delivering Data-Informed, Personalized Learning Support Actions, *J. Learn. Anal.*, 2018, **5**(3), 235–249.
33. R. Pullen, A. Motion, S. Schmid, S. George-Williams, S. Wilkinson and S. Leach, in *Digital Learning and Teaching in Chemistry*, ed. Y. J. Dori, C. Ngai and G. Szteinberg, Royal Society of Chemistry, United Kingdom, 2023.

CHAPTER 25

Online Assignments: Pre- and In-service Chemistry Teachers' Knowledge, Perceptions and Reflections

ORIT HERSCOVITZ*[a], MERAV VARSANO†[a], OLA FARIS†[a] AND YEHUDIT JUDY DORI[a,b]

[a]Faculty of Education in Science and Technology, Technion, Israel Institute of Technology, Haifa, 3200003, Israel; [b]Samuel Neaman Institute for National Policy, Haifa, Israel
*E-mail: orither@ed.technion.ac.il

25.1 Introduction

The COVID-19 pandemic and the resulting changes to remote and in-school learning programs forced teachers to transition instantly to synchronous, asynchronous, or hybrid online teaching that combines lectures with digital tools and assessment measures. Accordingly, teacher education programs that educate the teachers of the future should incorporate online tasks and teach their own students in a technological environment that facilitates multi-sensorial and relevant learning. Researchers claim that using technology can help in implementing a variety of methods to foster assessment for learning.[1,2]

†This study was carried out as part of the second and third authors Masters degrees as supervised by the two other authors.

Advances in Chemistry Education Series No. 11
Digital Learning and Teaching in Chemistry
Edited by Yehudit Judy Dori, Courtney Ngai and Gabriela Szteinberg

Published by the Royal Society of Chemistry, www.rsc.org

Ashburn and Floden describe meaningful learning in which the technology serves as an intellectual partner that helps students develop their thinking skills and gain a better understanding of the world.[3] Learning with technology can facilitate meaningful learning if the involvement of learners in the construction of knowledge is established through conversation, acquired knowledge, and reflective thinking. These learning processes can be achieved and improved through investigation, design, communication, writing, modeling, and visualization (see Chapter 16 in this book by Krajcik).[4] The combination of textual and visual means follows the second multimedia learning approach, according to which learning takes place *via* two channels in parallel: the visual and the textual.[5] Teaching abstract topics in an approach that integrates visual and textual representations caters to the needs of students with diverse learning needs. For example, teaching nanotechnology using computerized simulations and animation can improve learning and develop higher-order thinking skills.[6] A problem-based learning (PBL) web environment with inquiry activities has been shown to promote students' ability to mentally traverse the four levels of chemistry understanding: symbolic, macroscopic, microscopic, and process.[7] The visual–textual approach contributes to students' motivation and satisfaction. However, new pedagogies are not readily implemented, and STEM teachers mainly employ traditional teaching and assessment methods.[8,9]

Teachers must have several types of knowledge to utilize technology effectively in their classrooms. The goal of our research was to identify teachers' perceptions of the implementation of online assignments, assess what types of knowledge they need to implement them, and explore their reflections on the process of developing online assignments. In this chapter, we describe assessment of the chemistry teachers' pedagogical content knowledge (PCK), technological knowledge (TK), and assessment knowledge (AK). The knowledge type that focuses on application of educational technology is technological, pedagogical and content knowledge (TPACK). It requires deep understanding of several knowledge types—TK and PCK.[10] Our assumption is that if a teacher has high PCK, TK, and AK then their TPACK will also be high. TPACK is reflected in the current study through the process of the teachers' development and implementation of online assignments.

25.2 Teachers' Knowledge

The foundations of a framework for developing teachers' knowledge was introduced by Shulman in 1987.[11] According to Shulman, teachers must master the interaction between pedagogy and content to instill deep content understanding in their students. This is referred to as PCK.[11] PCK involves understanding the interplay between content and pedagogy and using pedagogical methods in addition to being a content expert.

Pedagogical uses of technology require development of a complex type of knowledge called TPACK, which focuses on the knowledge required to teach effectively using technology.[10,12] TPACK refers to the teachers' knowledge to integrate educational technology into teaching any content area, using appropriate pedagogical methods and technologies. TPACK consists of three main components of learning and teaching: content, pedagogy, and technology, as well as the overlaps between them.

We added to this framework the AK component, as it is an integral part of the teachers' everyday practice in their classrooms. AK is an important component in teachers' knowledge related to PCK and refers to teachers' ability to identify the dimensions of scientific learning that should be assessed, and to their ability to use and develop methods for assessment of learning.[8] However, some teachers do not feel adequately capable of assessing their students due to lack of effective pre-service exposure to different assessment.[13,14]

In recent years, the use of technology in education has been embedded in the teaching and learning process. However, the use of technology in the assessment process (*i.e.*, e-assessment) has not been extensively researched, even though assessment typically dictates how a subject should be taught and how technology should be used.[15] Online learning environments offer the potential for enhancing the quality of academic assessment, but is constrained by infrastructure, support, educator and student skills, and limited time.[16] Researchers suggested directions for ways teachers can integrate technology into their classrooms and create new learning opportunities for their students. The more pre-service teachers gain experience with developing technology-enhanced material during their professional training, the more confidence they acquire in using technologies.[17] Many teachers view lack of resources, such as time, Wi-Fi, as well as too little personalization, training, and support, as barriers to using technology in the classroom.[18]

25.2.1 Reflection

In the context of teaching and learning is related to introspection and thinking processes that aspire to gain new insights and draw conclusions for the future. The reflection process offers an opportunity to think about what we know and what we could have done differently and organizes new ideas in our minds that are based on previous ones. For teachers, the reflection process is a significant stage in the logical reasoning cycle employed by teachers.[11] Schön (1983) defined two types of reflection, in-action and on-action, to help teachers modify their decisions during and after teaching.[19] In this study, we asked the teachers to reflect on the process of designing their own online assignments on-action, *i.e.*, after the process.

25.3 Research Goal and Setting

The goal of our research was to identify teachers' perceptions towards implementation of online assignments, assess the knowledge types they need in order to implement these assignments, and explore their reflections on the process of developing online assignments.

The research objectives were to: (1) identify pre- and in-service teachers' perceptions about using online assignments for teaching chemistry and as a tool for assessing their students; (2) characterize the teachers' technological pedagogical and content knowledge (TPACK), as reflected by the online assignments they developed, and (3) characterize teachers' reflections following their experience developing online assignments.

25.3.1 Research Participants

The study comprised 45 participants, who were chemistry teachers divided in two groups: pre-service and in-service teachers. Their demographic information is described in Table 25.1. In Israel, about 30% of the citizens are minorities, which include Christian Arabs, Muslim Arabs, Druze, and Bedouins. In high schools, chemistry is a popular major among the Arab and Druse societies, since it opens doors and enables the graduates to study medicine, pharmacology, or chemistry and related fields in higher education. Both Jewish chemistry teachers and those from minority groups participate in most of the in-service teacher programs and workshops, since they are motivated to foster their students' learning outcomes.

25.3.2 Research Setting

The pre-service chemistry teachers attended the "Teaching Chemistry Methods" course held at the research university. The in-service teachers participated in a professional training summer workshop. While the participants of the two research groups attended two different courses, both groups developed the same kind of online chemistry assignments and discussed how to implement them in their classes.

Table 25.1 Participants' demographic information.

	Pre-service teachers	In-service teachers
N	20	25
Education background	30% were toward graduating with a B.Sc. The rest had a B.Sc. in chemistry or a related field and were studying for a B.Sc. in chemistry education.	30% had a B.Sc. 50% had a M.Sc. The rest had a Ph.D.
Teaching experience	Out of the pre-service, 70% had no teaching experience.	Out of the in-service, 50% had up to 10 years of teaching experience and the rest had 11 and more years of teaching.
Gender	100% female	88% female
Age range	70% under 30-years-old	80% over 30-years-old
Culture	50% Jewish; 50% Arab (Muslim, Christian, or Druze)	

The pre-service course lasts 13 weeks, for 4 h a week. The course focuses on the main topics in the chemistry curriculum, as well as on teaching methods. The pre-service teachers were exposed to several online chemistry assignments that had already been developed using Google Forms, and their final project was to develop their own online assignment for high-school chemistry students.

The in-service workshop took place at the research university and consisted of four days (7–8 h each day). During the workshop, the in-service teachers discussed how to use and develop online assignments on Google Forms, and how to assess students using this platform. Discussions on how to integrate online assignments into chemistry lessons, and what challenges teachers face while integrating Google Forms into teaching and learning were the main focus of the workshop. The final project was to work in teams to develop an online chemistry assignment for chemistry students.

Google Forms is basically a survey tool in which teachers can ask a variety of questions, including open-ended questions, multiple-choice questions and incorporate visualizations. They can integrate videos, website links, and images that illustrate the subject matter. An important feature of the platform is that it enables teachers to integrate answers and provide written feedback to students on their answers. Another feature is that it enables teachers to track each student's answers and generate statistical data.

25.3.3 Research Tools and Data Analysis

The research used a mixed methods approach that included both qualitative and quantitative tools.[20] The participants' perceptions were collected using a questionnaire (quantitative tool), and their TPACK was examined based on the online assignments they had developed (qualitative tool). Content analysis was conducted on their written reflections.

The questionnaire aimed to characterize teachers' perceptions about their ability and willingness to teach and assess their chemistry students using online assignments, and their perceptions about the advantages and challenges of teaching and assessing students using online assignments. The questionnaire consisted of three parts. Part I contained 13 items related to demographic information. Part II contained 28 Likert scale items (1 = strongly disagree, 5 = strongly agree) to identify teachers' perceptions and willingness to implement assignments using Google Forms. The items were based on the Teachers' Sense of Efficacy Scale questionnaire—TSES.[21] We adapted items related to classroom management and student engagement, and added items related to TK and AK.

Part II was divided into four categories: (1) student engagement—items related to teachers' ability to motivate their students; (2) PCK—items related to teachers' ability to teach chemistry using Google Forms; (3) TK—items related to teachers' ability to incorporate technology in their teaching materials; and (4) AK—three levels of assessment using Google Forms—using/modifying existing online assignments, developing new online assignments, and implementing online assignments.

We used Cronbach's Alpha statistical test to assess the internal consistency of the items in these categories. The Cronbach's Alpha score for all items was 0.821.

Examples of items in each of the categories and their Cronbach's Alpha scores are shown in Table 25.2.

When analyzing teachers' answers regarding the advantages and challenges of integrating online assignments, we found three major themes: the teacher aspect; the student aspect; and the technology aspect. We asked five researchers to categorize the teachers' answers based on the different themes. We compared our analysis with the researchers' analysis and found over 90% agreement. Table 25.3 presents examples of teachers' answers and how they were categorized.

We asked the teachers to work in groups to develop an online assignment using Google Forms that is related to one of the topics on the chemistry curriculum, such as energy, atom structure, chemical bond, *etc.* The assignment should include diverse types of questions (multiple choice, open-ended, *etc.*) as well as animations and images, to facilitate visualization of the chemistry content.

To analyze the assignments that the teachers developed, we used a rubric comprised of three main categories: PCK, TK, and AK (see Figure 25.1). Figure 25.2 shows examples of parts of teachers' assignments and their scores. The PCK category (score range 0–10) comprised of three criteria: chemistry concepts, thinking level, and adjustments to the student's level. Each criterion was given a score of 1–3, and an extra point was given for including a question that involves values, opinion, or relevance. The TK category (score range 0–10) comprised of two criteria: Google Forms visual appeal and use of the various tools supported by the Google Forms platform. We gave an AK score of 0–10 for use of evaluation tools embedded in Google Forms, where 0 means no evaluation at all and 10 means exceptional use. Figure 25.1

Table 25.2 Examples of items from the Likert scale section of the questionnaire.

Category	Number of items	Example	Cronbach's Alpha
Student engagement	4	I predict that my students will be excited by the use of online assignments as a class activity.	0.534
PCK	8	I am in favor of using online tasks as a pedagogical tool and I will definitely use them in my classes.	0.741
TK	5	I can answer students' technical questions about Google Forms while they work on the online assignment.	0.612
TPACK and AK	11	I plan to develop online assignments because I have acquired the proper tools for doing so. (Level 2—Develop) I will evaluate my students based on their answers on Google Forms assignments. (Level 3—Assess)	0.742

PCK Category
Chemistry concepts (3 out of 3)
☑ Accurate use of chemical language in most assignment sections.
☑ Sufficient use of representations and analysis of information using graphs and molecular structures.
☑ Three levels of chemistry understanding (symbolic, macroscopic, and microscopic).
Thinking level (3 out of 3)
☑ Type of questions: knowledge and understanding level.
☑ Questions that require expressing an opinion on a relevant issue.
Adjustment for the students' grade level (score=3 out of 3)
The assignment included questions sorted in ascending order from easy to difficult.
Question related to values, opinions, or of relevance to the student
☑ The assignment contained a question of this kind (1 out of 1).
TK Category
Google Form visuality (3 out of 5)
☑ Sections are divided logically.
☑ Question titles are clear.
☒ Different font types and sizes are used.
Usage of various tools provided by Google Forms (max. 5)
☑ A variety of open- and closed-ended questions and file uploads are used.
☑ A variety of visualization tools such as images, videos, and simulations are used.
AK Category
Evaluation tools (8 out of 10)
☑ Proper use of the evaluation tools provided by Google Forms, providing effective feedback for students' correct or incorrect answers.
☒ Although many questions provided good feedback, several questions did not include feedback at all.

Figure 25.1 Analysis of an assignment on covalent bonds.

presents an example analysis of an assignment on covalent bonds (total score 26 out of 30). Each online assignment was analyzed as a complete unit and each category could receive a maximum score of 10. Our rubric is a modification of the PCK and AK rubric,[8] and the TK rubric.[22]

The participants were asked to write their personal reflections for the online tasks they developed, from four perspectives: their goals when choosing the topic of the assignment, advantages and challenges related to incorporating the assignment in the learning process, and lessons learned for peers. The content of the reflections was analyzed from these four perspectives, and several categories were exposed for each perspective. For example, the categories exposed for goals were social and academic.

Table 25.3 Examples of teachers' answers regarding advantages and challenges, and a qualitative analysis of these answers.

	Theme		
Teacher's answers	Teacher aspect	Student aspect	Technology aspect
Advantages			
Google Forms enables a break from the routine classroom setting that mainly involves a teacher's lecture, especially if the form includes illustrative images, videos, and simulations. In addition, Google Forms can be used for summary assignments and for assessing students' performance (PST94897).	✓	✓	✓
Google Forms are easily accessible, editable, and saved on the drive, so that you can always add new participants and limit the time allocated for answering (IST39579).			✓
Challenges			
It will be difficult for students to upload files, as there are not enough computers in class and the internet connection is unreliable (although they can connect using their mobile phones). Watching videos can cause noise in the classroom (IST92676).	✓	✓	✓
Internet availability, not enough computers, the teacher does not have extra time and must cover all the topics on the curriculum (IST75930).	✓		✓

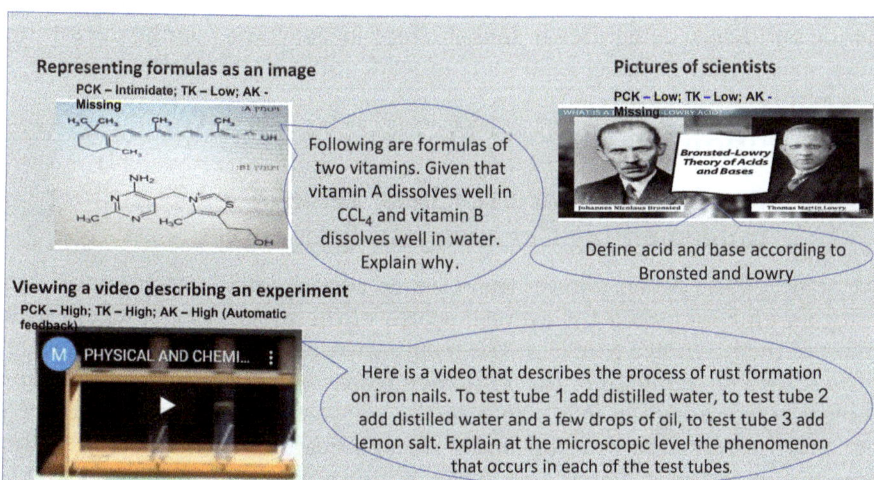

Figure 25.2 Examples of assignments and their analysis.

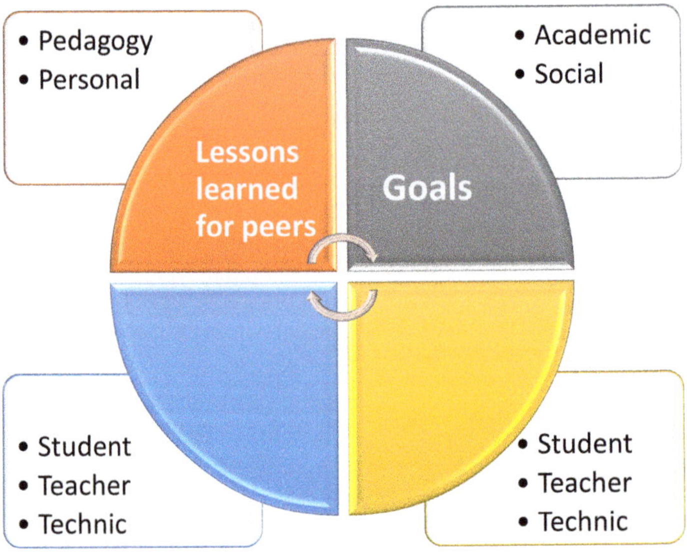

Figure 25.3 Perspectives and categories that emerged from the participants' reflections.

From the perspective of advantages and challenges, the categories were teacher, student, and technology, which coincides with the three themes we identified when analyzing the responses to the open-ended question on the perceptions questionnaire. From the perspective of lessons learned for peers, pedagogical and personal recommendations were mentioned. The categories and perspectives that emerged from the content analysis are shown in Figure 25.3, and examples of the participants' answers for the various categories are shown in Table 25.4. In the goals aspect, the difference between academic and social categories is that the former concerns cognitive learning outcomes, while the latter is about teamwork and peer interactions.

25.4 Findings

25.4.1 Pre- and In-service Teachers' Perceptions Regarding Implementing Chemistry Assignments

We analyzed the perceptions of the research participants in each knowledge category and compared the average scores of the pre-service teachers with those of the in-service teachers. The results are presented in Table 25.5.

As shown in Table 25.5, the in-service teachers' perceptions for all four categories were higher than those of the pre-service teachers. However, a t-test for independent samples found no significant difference between the pre- and in-service teachers ($p > 0.05$). Both groups ranked their TK ($t(43) = -0.482$) competence with lower scores than their competence in student engagement ($t(43) = -1.240$), PCK ($t(43) = -1.083$), and AK ($t(43) = -1.194$).

Table 25.4 Examples of participants' answers in the written reflection.

Aspect	Category	Example
Goals	Academic	"The teacher can incorporate videos and animation that can help the student understand the important topics covered in class." [PRPS12218]
	Social	"Working in pairs or in groups of three enables high-level acquisition of the new information." [IRIS3118]
Advantages	Student	"The students were generally very excited about it, and it interested them very much. It motivated them to learn, which is the most important thing." [IS1221119]
	Teacher	"The assignment can be given as classwork, and I can track each student's progress; they can work at their own, individual pace". [PRPS10218]
	Technical	"The assignment is available to the students from their mobile phones, whenever and wherever they want to access it." [IS3112119]
Challenges	Student	"The students use the technology, but they can also 'exploit' it." [PRPS921]
	Teacher	"It really depends on the teacher. Teachers who are more computer-oriented will spend less time on it. If they're less computer-oriented it will take them longer to understand." [IIS2112119]
	Technical	"There will probably be technical and communications problems, like power outages, lack of access to computers or mobile phones, or instable internet connections." [IRIS1118]
Lessons learned for peers	Pedagogical	"I recommend that the teachers fully understand all the possibilities and try them as students. That will help them find the potential problems." [IRIS12118]
	Personal	"It is very worthwhile for the teacher to diversify the levels and types of questions, using open-ended and multiple-choice questions." [PIPS5112219]

Table 25.5 Average scores for pre- and in-service teachers' perceptions for each knowledge category.[a]

Participant type	*N*	Student engagement *M* (S.D.)	PCK *M* (S.D.)	TK *M* (S.D.)	AK *M* (S.D.)
Pre-service	20	3.67 (0.73)	3.74 (0.66)	3.49 (0.80)	3.71 (0.66)
In-service	25	3.94 (0.70)	3.95 (0.61)	3.60 (0.77)	3.94 (0.61)

[a]*N* = number of participants; *M* = average score; S.D. = standard deviation.

To test whether there is a correlation between the scores given by teachers for their own knowledge and experience with computer applications on one hand and the TK category on the other hand, we carried out a Spearman test between those two variables. We found a significant correlation ($r = 0.461$, $p < 0.00$) indicating that teachers who declared having rich experience with computer applications also declared high capabilities in the TK category.

Although the average scores for pre-service teachers and in-service teachers were nearly identical in all four categories, as shown in Table 25.5, we compared how many pre-service teachers gave themselves high scores ("agree" and "strongly agree") *versus* in-service teachers. The results are presented in Figure 25.4.

The results presented in Figure 25.4 show that percentage of in-service teachers who declared positive perceptions toward implementing online assignments in all categories was higher than that of the pre-service teachers. The smallest difference between the two groups was in the TK category.

A comparison of teachers' scores for the items related to the AK category in each assessment level presented in Table 25.6 shows that the pre-service

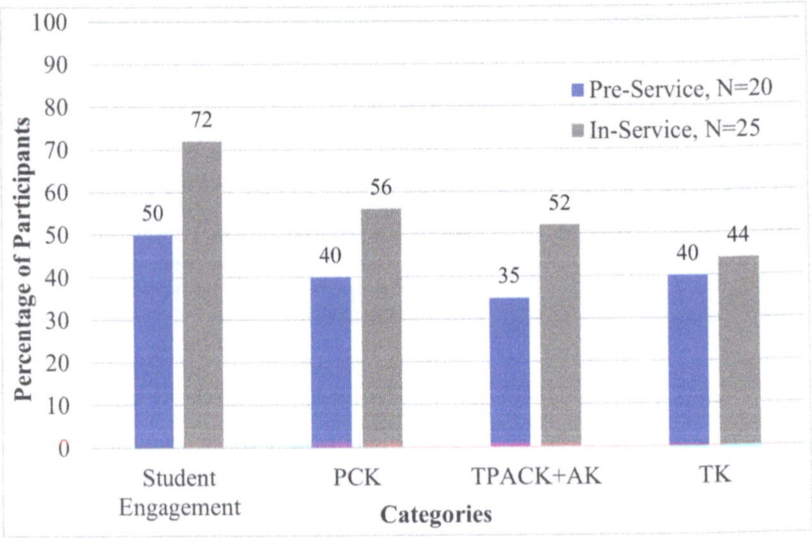

Figure 25.4 The percentage of participants who gave high scores in all four categories.

Table 25.6 Pre-service and in-service teachers' perceptions of their AK levels.[a]

Participant type	*N*	Level A—Use *M* (S.D.)	Level B—Develop *M* (S.D.)	Level C—Implement *M* (S.D.)
Pre-service	20	3.78 (0.68)	3.70 (0.86)	3.60 (0.84)
In-service	25	3.85 (0.64)	4.05 (0.75)	3.91 (0.67)

[a]*N* = number of participants; *M* = average score; S.D. = standard deviation.

teachers ranked their ability to implement online assessment as lower than their ability to develop and use these assignments, while the in-service teachers ranked their ability to use and implement online assignments as lower than their ability to develop them. The in-service teachers' perceptions of all three levels—use, development, and implementation—were more positive than those of the pre-service teachers.

The open-ended question in the questionnaire focused on the advantages and challenges of integrating online assignments into the teaching process. The participants discussed three themes: the teacher, the student, and technology. The advantages mentioned for the student theme included improving students' interest in learning, learning in a technological environment, improving thinking skills, and adaptation for student diversity. Learning in a technological environment is the norm for the current generation of learners. Moreover, it has been found to be critical during the COVID-19 pandemic, as it has enabled continued learning while students were confined to their homes. The advantages in the teacher theme were diverse teaching methods and ease of assessment. The advantages in the technology theme were the advantages of integrating simulations and animations, and the convenience and effectiveness of these assignments. Regarding challenges, the student theme included frustration and interference when using online learning assignments, and the dependency on technological knowledge. The teacher theme included challenges related to technological skills, assessment knowledge, and class management. The technology theme included difficulties with the technological infrastructure in schools and with scientific writing (such as chemical formulas), which is not supported by Google Forms.

Our analysis of the statements made by the pre-service and in-service teachers to describe the advantages and disadvantages for each theme shows that about 50% of the positive statements were made in reference to the student theme, while about 50% of the challenges were related to the technology theme. No significant differences were found between the two groups. The findings also show that the pre-service teaches were more concerned about the challenges this platform would present to their students (28%) than the in-service teachers (19%).

25.4.2 Pre-service and In-service Teachers' Knowledge Types as Reflected by the Online Assignments They Developed

Eight assignments were developed by pre-service teachers, and eight were developed by in-service teachers. We analyzed teachers' PCK, TK, and AK as reflected by the eight assignments developed by each group during the course and during the summer workshop, respectively. Figure 25.5, which presents the average score distribution for each of the assessed categories based on these assignments, shows that the average score in the AK category was lower than the average scores in the TK and PCK categories for both pre- and in-service teachers' assignments. In-service teachers received higher

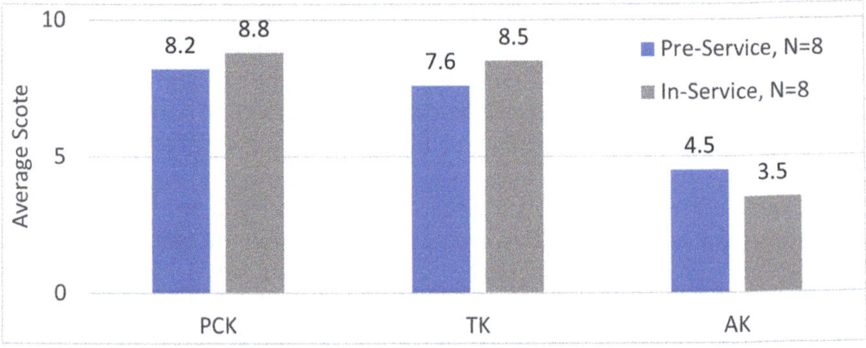

Figure 25.5 Distribution of average scores for each category for online assignments developed by teachers.

average scores than pre-service teachers in the PCK and TK categories, while the average scores in the AK category for the pre-service teachers' assignments were higher than those of the in-service teachers' assignments.

25.4.3 Pre-service and In-service Teachers' Reflections

The reflections written by the participants after they developed online assignments addressed four aspects, as shown in Figure 25.3.

25.4.3.1 Objectives of Integrating Online Assignments

Our analysis of the reflections revealed two categories: social and academic. Most of the participants addressed the academic category (95% of pre-service teachers, and all in-service teachers), and described educational goals as the reason for choosing the subject of the online assignment they developed. However, only 10% of pre-service teachers and 17% of in-service teachers mentioned the social aspect as a consideration.

25.4.3.2 Advantages of Integrating Online Assignments in the Classroom

Our analysis of the participants' reflections found three categories related to the advantages of online assignment, which matched the categories identified in the responses to the perceptions questionnaire:

- The student—referring directly to the students and their enjoyment and understanding.
- The teacher—referring to the teacher's work, teaching methods, pool of available assignments, and how the assignments are implemented.
- Technology—the tools needed to complete the assignments, school infrastructure, system quality, and more.

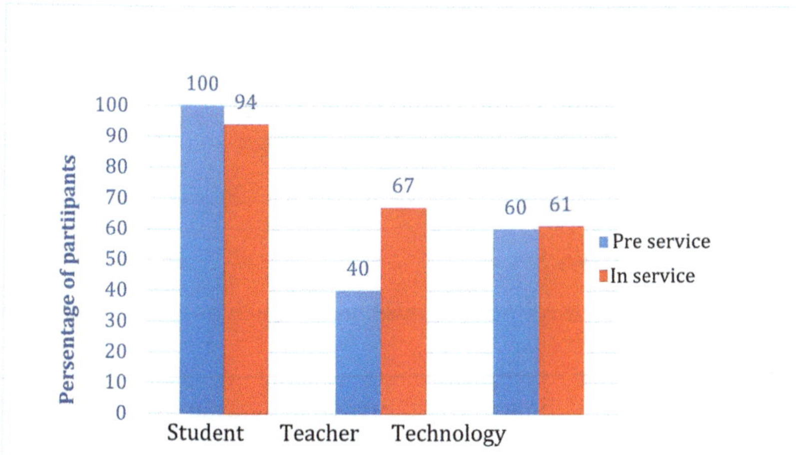

Figure 25.6 Distribution of the research groups per category regarding the advantages of integrating online assignments.

All the pre-service teachers mentioned the advantages of integrating online assignments for the students; 60% mentioned technological advantages, and 40% mentioned advantages for the teachers. A similar percentage of in-service teachers noted advantages from the technological and student perspectives, though 60% of in-service teachers found advantages for the teacher compared to 40% of pre-service teachers who mentioned advantages in the teachers' category, as shown in Figure 25.6.

25.4.3.3 Challenges Related to Integrating Online Assignments in the Classroom

Analyzing the challenges mentioned in the participants' reflections, we found that the three categories identified were the same as those found for the advantages: student, teacher and technology. Both groups' statements addressed the same categories to a similar extent. Pre-service and in-service teachers all identified the main challenge of integrating online assignments as related to the student and technology, while the teacher category was significantly less prominent. Despite the many challenges the teachers identified, they believed that they are capable of coping with them.

25.4.4 Recommendations for Teachers: Lessons Learned for Peers

Our analysis of the participants' reflections identified two categories related to the participants' recommendations: pedagogical and personal. The majority of participants in both groups mentioned pedagogical recommendations more frequently than personal recommendations. A significant difference was found between the groups regarding recommendations on the personal level. Pre-service teachers mentioned recommendations on the personal level almost twice as often as in-service teachers (20% *vs.* 11%). This may be related to the pre-service teachers' lack of experience and personal concerns about using these assignments.

25.5 Discussion

This study focused on three perspectives, from which its goals were derived:

- The participants' perceptions: identifying perceptions regarding the integration of online assignments in teaching and for student assessment.
- Knowledge type: describing the participants' knowledge types as reflected by the online assignments they developed for their students.
- Reflection and observations: the written reflections of the participants were documented after they developed the online assignments for their students.

Figure 25.7 demonstrates the structure and perspectives of this study.

Pre- and in-service teachers typically use MS Office applications and are exposed to social media. However, a common obstacle that limits integration of technology into the school curriculum is the teachers' lack of knowledge and skill.[15] These researchers found that the main obstacle for technology

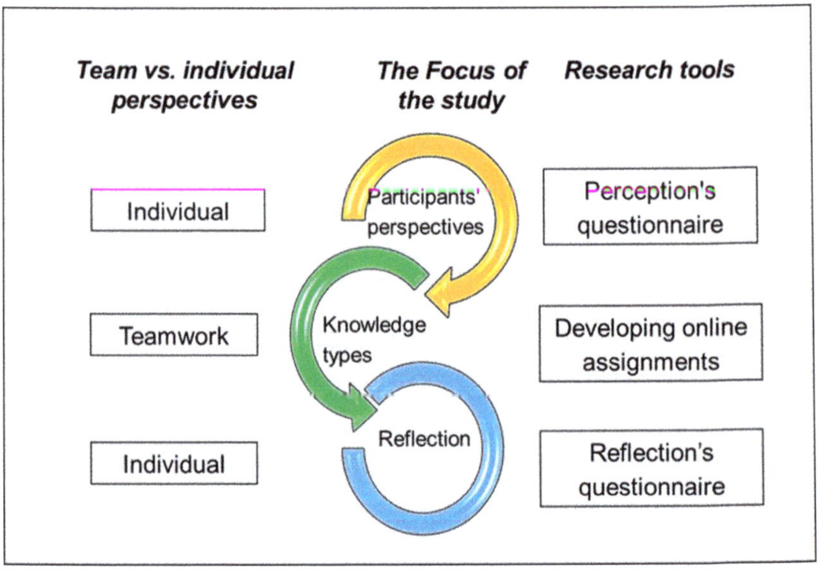

Figure 25.7 An overview of the research structure and perspectives.

in the classroom was the teachers' beliefs that technological tools are unstable or inconvenient. All the participants in this study ranked their technological knowledge as lower than that of their students, and lower than their pedagogical content and assessment knowledge. Chittleborough found that ongoing professional development can improve teachers' confidence in their ability to utilize technology.[17] Accordingly, we found that the perceptions of pre- and in-service teachers regarding the advantages of online assignments from the technological perspective, was less positive than their perspectives of its advantages for teachers and students. In both groups, more challenges were perceived in the technology category than in the other categories. The main advantages identified for integrating online assignments in both groups were related to the student perspective, particularly improving their learning skills by enhancing use of technology in online assignments with visualizations, as exemplified in Figure 25.2. The teachers thought that visualizations can help students to gain meaningful understanding of abstract concepts. Studies have found that the primary use of technology in chemistry classes is visualization.[23] Other advantages mentioned in the context of the student perspective were enjoyment, independent learning, and visualization aid. This coincides with Kopcha,[24] who claimed that teachers believe that technology is important for improving student learning.

Student assessment is the highest level of teachers' professional development.[25] Accordingly, we found that pre-service teachers rank their ability to assess students below their ability to develop online assignments. In-service teachers rank their technological ability to develop these assignments lower than their ability to teach or assess using online assignments. This is most likely related to the in-service teachers' teaching experience. Regarding pre-service teachers, the aspect of student assessment corresponds with other studies that found that assessment of students' knowledge requires suitable preparation so that teachers become familiar with diverse assessment methods.[13,14]

When analyzing the online assignments that the teachers developed, we designed a TPACK rubric divided into three categories: PCK, TK, and AK. Overall, the participants demonstrated higher levels of PCK and TK than AK. One possible explanation may be related to the additional technological knowledge required in order to integrate answers and feedback for students. This finding coincides with other studies that showed that technology-assisted assessment is limited by infrastructure, support, skill, and time.[16] We also found that although in-service teachers have a higher level of PCK and TK than pre-service teachers, the pre-service teachers had a higher level of AK than in-service teachers. This is apparently related to the fact that the pre-service teachers were more meticulous about integrating assessment tools in their Google Forms.

Our analysis of the participants' reflections showed that the pre- and in-service teachers incorporated diverse statements in their reflections that were related to pedagogical and personal aspects. We found that regarding the considerations and objectives for choosing the subject of the assignments developed, both groups made frequent mention of the academic category

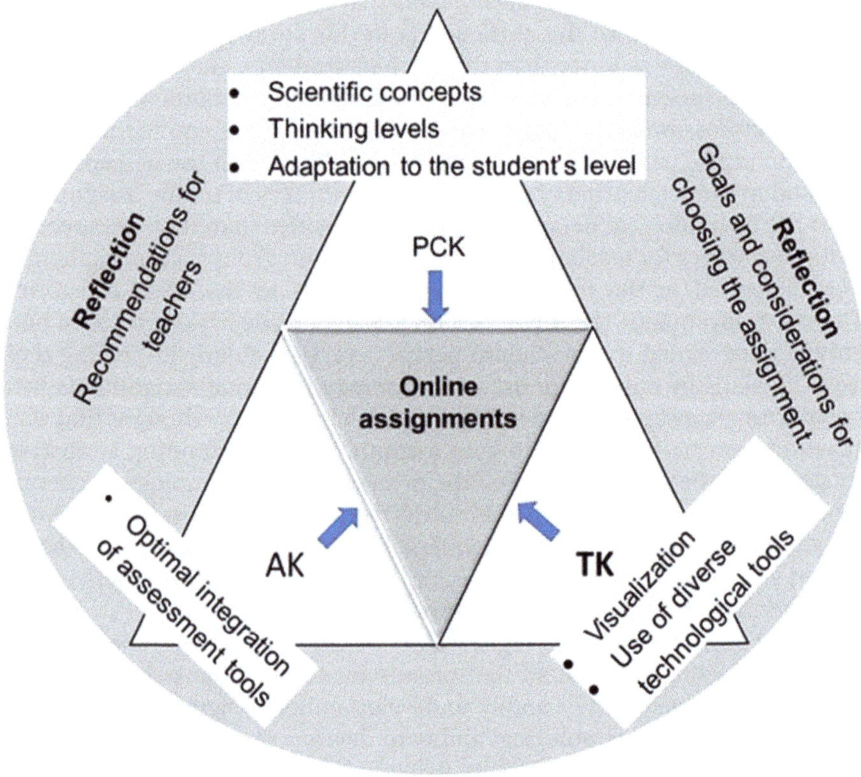

Figure 25.8 Three types of pre- and in-service teachers' knowledge as demonstrated by their online tasks and the three aspects of their reflections.

and significantly less frequent mention of the social category. The academic perspective was clearly a central goal for both pre- and in-service teachers, who found that integrating online assignments helped students understand the material and could also help improve their motivation and comprehension. From the social perspective, the participants reflected mostly upon the development of social student–student and teacher–student interactions, though this perspective is less prominent in the context of developing and applying online assignments.

Figure 25.8 summarizes the knowledge types of the pre- and in-service teachers, as shown in this study, alongside the three categories retrieved from the participants' written reflections.

25.6 Insights and Recommendations

Shulman claimed that CK and PCK are basic knowledge fields that teachers must acquire during their training.[11] AK is an important knowledge type that represents the highest level of professional growth.[8] This study has shown that the AK level demonstrated in the online assignments developed by the pre- and in-service teachers was at a lower level than TK and PCK, especially among in-service teachers (see also Figure 25.5). This finding is supported by previous studies,[25,26] in which AK was also found to be lower than PCK.

Therefore, we recommend that pre-service teachers' academic training and professional development programs for in-service teachers address the interaction between TK and AK and include practical exercises in technological platforms. In addition, establishing supportive communities for in-service teachers and for pre-service teachers during their training can improve their technological skills.

We asked our pre- and in-service teachers to develop online assignments in groups, based on the assumption that teamwork can improve the TPACK level. However, this assumption should be assessed in future research.

The study can help develop guidelines for pre-service teachers' training programs as well as for in-service teachers' professional development regarding online assignments and how they can be incorporated in the classroom. The criteria developed for assessing the participants' reflections can serve as a measure for thinking about thinking, *i.e.*, the meta-cognition of the teacher who wrote the reflection.

References

1. S. J. Deeley, *Assess. Eval. Higher Educ.*, 2018, **43**, 439–448.
2. See Chapter 26: C. Z. Muteti and J. Mutambuki, in *Digital Learning and Teaching in Chemistry*, ed. Y. J. Dori, C. Ngai and G. Szteinberg, Royal Society of Chemistry, United Kingdom, 2023.
3. E. A. Ashburn and R. E. Floden, *Meaningful Learning Using Technology: What Educators Need to Know and Do*, Teachers College Press, 2006.
4. See Chapter 16: J. Krajcik, in *Digital Learning and Teaching in Chemistry*, ed. Y. J. Dori, C. Ngai and G. Szteinberg, Royal Society of Chemistry, United Kingdom, 2023.
5. R. E. Mayer, in *The Cambridge Handbook of Multimedia Learning*, 2005, vol. 41, pp. 31–48.
6. Y. J. Dori, V. Dangur, S. Avargil and U. Peskin, *J. Chem. Educ.*, 2014, **91**, 1306–1317.
7. M. Barak and Y. J. Dori, *Sci. Educ.*, 2005, **89**, 117–139.
8. S. Avargil, O. Herscovitz and Y. J. Dori, *J. Sci. Educ. Technol.*, 2012, **21**, 207–225.
9. M. Barak, *J. Sci. Educ. Technol.*, 2014, **23**, 1–14.
10. M. Koehler and P. Mishra, *Contemp. Issues Technol. Teach. Educ.*, 2009, **9**, 60–70.
11. L. Shulman, *Harv. Educ. Rev.*, 1987, **57**, 1–23.
12. A. Cetin-Dindar, Y. Boz, D. Y. Sonmez and N. D. Celep, *Chem. Educ. Res. Pract.*, 2018, **19**, 167–183.
13. C. A. Mertler, *Improv. Sch.*, 2009, **12**, 101–113.
14. W. J. Popham, *Educ. Meas. Issues Pract.*, 2003, **22**, 45–48.

15. K. F. Hew and T. Brush, *Educ. Technol. Res. Dev.*, 2007, **55**, 223–252.
16. S. Bennett, P. Dawson, M. Bearman, E. Molloy and D. Boud, *Br. J. Educ. Technol.*, 2017, **48**, 672–682.
17. G. Chittleborough, *J. Sci. Teacher Educ.*, 2014, **25**, 373–393.
18. UNECSO, Scaling up digital learning and skills in the world's most populous countries to drive education recovery, https://en.unesco.org/news/scaling-digital-learning-and-skills-worlds-most-populous-countries-drive-education-recovery, accessed 26 January 2021.
19. D. A. Schön, *The Reflective Practitioner: How Professionals Think in Action*, Basic Books, New York, 1983.
20. J. W. Creswell, *Res. Des. Qual. Quant. Mix. Methods Approaches*, 2014, vol. 2014, pp. 3–24.
21. M. Tschannen-Moran and A. W. Hoy, *Teach. Teach. Educ.*, 2001, **17**, 783–805.
22. A. Oster-Levinz and A. Klieger, *Turk. Online J. Distance Educ.*, 2010, **11**, 47–71.
23. S. Gabby, S. Avargil, O. Herscovitz and Y. J. Dori, *Chem. Educ. Res. Pract.*, 2017, **18**, 214–232.
24. T. J. Kopcha, *Comput. Educ.*, 2012, **59**, 1109–1121.
25. S. Avargil, O. Herscovitz and Y. J. Dori, *J. Sci. Educ. Technol.*, 2012, **21**, 207–225.
26. M. Tal, O. Herscovitz and Y. J. Dori, *Chem. Educ. Res. Pract.*, 2021, **22**, 1003–1019.

CHAPTER 26

'I Felt Not So Alone': the Impact of Muddiest Point Activities on Student Learning Outcomes Through Top Hat Technology

CAROLINE Z. MUTETI[a] AND JACINTA M. MUTAMBUKI*[a]

[a]Department of Chemistry, Oklahoma State University, Stillwater, OK 74078, USA
*E-mail: jacinta.mutambuki@okstate.edu

26.1 Background

Classroom assessment techniques (CATs) are powerful formative assessment tools and have been used by many instructors in various disciplines across the United States (U.S.).[1] CATs are low-stakes, informal formative assessment tools aimed at providing immediate feedback to assess student learning throughout the semester by allowing students to reflect on their learning.[1] CATs are learner-centered but teacher-directed and thus, by engaging students in these reflections, the instructors can obtain useful feedback on how much and how well students are learning, improving the quality of instruction and meaningful learning. A variety of CATs and their potential benefits to learning and instruction have been described.[1–3] Examples of reported CATs enacted in the postsecondary chemistry courses are the one-minute paper,[4] constructive error climate,[5] and muddiest point.[6,7]

Advances in Chemistry Education Series No. 11
Digital Learning and Teaching in Chemistry
Edited by Yehudit Judy Dori, Courtney Ngai and Gabriela Szteinberg

Published by the Royal Society of Chemistry, www.rsc.org

One-minute paper involves asking students to summarize the most important points learned in each day's lesson or any questions that remain unanswered after the lesson. Students can either write their names on the paper if the instructor wishes to give them credit or the CAT can be completed anonymously.[1,3,4] Harwood's study (1996) study with 250 General Chemistry students showed that in addition to providing the instructor with insights into the concepts mastered and the difficulties students were experiencing on the lecture material, the one-minute paper CAT increased the instructor's rapport with students in the course and increased students' self-confidence as learners which opens up a door of communication since students can comfortably ask questions in class.[4]

Constructive error climate CAT uses errors as the central focus of mastering the concepts; hence, leading to more stable knowledge construction and improved performance.[5] He (2020) investigated the impact of constructive error climate in a General Chemistry course for two academic years. The CAT was implemented on three specific chemistry topics: naming ionic compounds, writing ionic compound formulas, and writing ground-state electron configuration.[5] The teaching of the "naming ionic compounds" was followed with students working in small groups of 2–4 to find the errors in given problem sets. According to He (2020), students discussed identified errors and misconceptions, the reasoning behind the errors, and how the mistakes could be solved rather than memorizing the correct answers. The groups shared their ideas with peers and compared their solutions to the instructor's feedback on correct answers and with the errors highlighted in different colors. Students developed confidence in problem-solving and actively participated during the class as a result of the CAT. Additionally, the CAT enhanced the students' development of higher-order thinking and increased the students' awareness of their level of mastery of the content which was beneficial to the instructor.[5]

Muddiest point is a low-stakes writing exercise in which students identify the most confusing concepts on given content material.[1] The "muddiest points" can be written on a notecard, a sheet of paper distributed and collected at the end of class, or administered *via* high technology student-response systems, such as *i*-clickers[6] and Top Hat.[7] The responses are anonymous, which allows students to report honest responses and promote equitable participation; thus improving performance, especially for marginalized student populations, such as first-generation students, women, and minority ethnic groups.[7] Muddiest point activities induce students to reflect on the learning experiences and identify difficult or confusing concepts.[1,3] The traditional approach to administering the muddiest point activities is providing index cards to students during the class to write responses to given prompts assessing specific content.[8] Then the instructor collects the written index cards, reviews and analyzes the responses for common themes, and provides feedback, including closing the identified gaps in learning.[1] The paper-based approach is, however, tedious and time-consuming.

King (2011) described an alternative approach to administering the muddiest point activities using clickers.[6] The author provided pre-chosen topics covered during the lesson for students to select the difficult concepts at the end of each lecture through clicker polling in a general chemistry course. The results of the clicker responses were then used as a source of a mini-review of the difficult concepts at the beginning of the subsequent lecture.[6] The caveat of *i*-clickers, however, is the inability to support open-ended student-generated responses beyond multiple-choice letter options.

This chapter describes the use of Top Hat technology, a student engagement platform for promoting active participation in and outside the classroom in which students use their own digital devices, in supporting a more manageable implementation of muddiest point activities in large General Chemistry courses,[7] and findings on the impact of the CAT on students' performance and learning gains in a General Chemistry lecture course.

26.2 Equity and Inclusion

STEM classrooms in the U.S. continue to be more diverse.[9] However, the literature indicates that a large proportion of underrepresented students (women and minority ethnic groups—black/African-American, Native Americans/Native Alaskans, Hispanic, Hawaiian/Pacific Islanders) leave STEM programs.[9] Factors attributed to switching from STEM majors include, but are not limited to "chilly" classrooms,[10] internalizations of stereotype threats,[11] and a lack of sense of belonging.[12,13] Classroom participation in STEM classrooms can be linked to affective measures that can be influenced by axes of identity that are negatively stereotyped, marginalized, or underrepresented, such as gender, race/ethnicity, and first-generation status.[12,14] For instance, Eddy *et al.* (2014) investigated the level of student engagement and gender gaps in 26 large introductory biology classrooms at an R1 university. Their findings revealed that although the class composition was 60% female, female voices were heard answering or asking questions only 40% of the class time.[12] Soria and Stebleton's (2012) study involving first-year students at a large research-based university found that first-generation students were less likely to ask questions in class or contribute to class discussion compared to non-first-generation students.[14]

Some reported benefits of student participation include improved sense of belonging,[15] decreased anxiety,[15] and a better understanding of the course material.[13] Because the muddiest point activities are usually anonymous and ungraded, they tend to minimize the feeling of intimidation, especially for shy and struggling students who may not be comfortable asking questions in front of their peers,[6,16] therefore, providing equitable access to learning by allowing all students. In particular, women, first-generation college students, and minority ethnic groups can comfortably voice their concerns on difficult concepts or areas of struggle.[7] The current study capitalized on the advantages of digital assessments to incorporate the muddiest point activities in General Chemistry through Top Hat technology. Like other digital assessments (*e.g.*, clicker questions), Top Hat allows students to take charge of their learning by providing a voice for everyone. Additionally, it is an effective way of presenting questions that demand open-ended responses, obtaining feedback on learning and instruction in large enrollment classes,[17] and a mechanism for anonymously displaying to the students the responses collected from assessments.[18]

Obtaining immediate feedback from the students on gaps in their knowledge or understanding of current material allows the teacher to take "*just-in-time*" corrective action to address learning challenges and, therefore improves performance[7] and students' learning experiences.[1] With the equity mindset, digital assessment through the Top Hat was accomplished through different digital devices such as laptops, smart and non-smartphones, and tablets to provide access to the same resources and opportunities, including highlighting the areas of struggle, to all students.

26.3 The Formative Assessment Cycle Framework

The formative assessment cycle framework[19] guided the implementation of the muddiest point activities. Specifically, the following seven components were considered: (1) identifying and developing the desired learning outcomes; (2) selecting the suitable CAT and developing learning activities; (3) performing the instruction; (4) gathering evidence of learning/collecting data—from the CAT activities using Top Hat; (5) analyzing students' responses to the CAT activity on each chapter; (6) providing feedback and closing the identified learning gaps; and (7) reflecting on, refining, and modifying the instruction to address students' needs in the subsequent lessons.[19] The instructor consciously sets the learning goals for a lesson *(Develop Learning Objectives)* and determines criteria for achieving these goals *(Select the CAT and Develop Learning Activities)* during the first three stages. The instructor then discusses the goals with the students as success criteria for learning and assigns learning assignments to them (*Perform Instruction*).

In step 4, the instructor uses appropriate ways to collect evidence of learning during instruction in order to achieve the intended learning objectives (*Collect Evidence of Learning*). Questioning techniques, monitoring instructional tasks (*e.g.*, explanations, problem-solving), mid-lesson checks (*e.g.*, low-technology response systems—thumbs up/down, ABCD/flashcards, chorus answers, and high-technology such as clickers), and writing the most confusing points (muddiest point) are just a few examples.[1,3] In the current study, although questioning style, clicker questions, and muddiest point activities were used to assess the students' learning progress, the latter received the most attention.

The Top Hat technological platform enabled the collection of learning evidence. The muddiest point activities were given at the end of each chapter and before introduction of a new chapter to allow students more moments of self-reflection on the topic and to resolve any gaps in learning that were not addressed during the lecture session.[7]

26.4 Methods

26.4.1 Research Design

A mixed-method approach was employed.[20] The quantitative research methodology was employed to compare performance between the treatment and the control groups as a result of the muddiest point activity. Specifically, a quasi-experimental design was employed to investigate the effect of the intervention on performance in the course. On the other hand, qualitative research methodology was used to supplement the quantitative findings by uncovering participants' experiences from the muddiest point activities, particularly in the treatment group.

26.4.2 Context of the Study

The study was approved by the institutional review board at Oklahoma State University. The study was conducted in the General Chemistry I lecture course. The course was taught in conjunction with a laboratory component. Class meetings met thrice a week. The same learning outcomes were developed and implemented by two instructors, one taught the treatment group and the other taught the comparison group. Similar traditional formative assessments (TFAs), such as homework, quizzes, and clicker questions were implemented in both groups, but the treatment group was also exposed to the muddiest point activities at the end of each chapter for ten chapters covered in the course.

26.4.3 Participants

Participants were STEM majors enrolled in the General Chemistry I course. 341 out of 370 students consented to participate in this study—224 students were from the comparison group and 117 from the treatment group. A summary of participants' demographic data is presented in Table 26.1.

Table 26.1 Participants' demographics between the treatment group and the comparison group.

Variables	Treatment group (*N*, %)	Comparison group (*N*, %)
Male	42, 35.9	78, 34.8
Female	75, 64.1	146, 65.2
White	87, 74.4	139, 62.1
Minority groups	29, 24.8	72, 32.1
Asian	1, 0.8	6, 5.9
First-generation	14, 12.0	33, 14.7
Non-generation	103, 88.0	191, 85.3

Note: *N* is the total number of the study participants who consented to the study and whose information was provided.

26.4.4 Research Questions

To assess the impact of the muddiest point activities, the study was guided by the following research questions. (1) Are there significant differences in performance between students exposed to the muddiest point activities infused with TFAs and their counterparts exposed to the TFAs alone in a General Chemistry 1 course? (2) How does viewing peers' responses on the muddiest point activities influence, if at all, students' learning experiences? We hypothesized that there will be statistically significant mean differences in performance between the treatment and the comparison group on the midterms and the final exam.

26.4.5 Implementation of the Muddiest Point Activities in the Treatment Group Through the Top Hat Technology

The Top Hat technology was specifically utilized for real-time in-class assessment in capturing learning barriers or areas of confusion in a General Chemistry course through the muddiest point activities. The muddiest point activities were implemented after the completion of each chapter, that is, before introducing a new chapter (medium-formative cycle). Implementation after each chapter was preferred rather than at the end of each lesson (short-formative cycle) to afford students additional moments for self-reflection on the material and to identify and address any existing gaps in learning after the instruction. Top Hat teaching technology was employed to provide students with the main concepts covered in each chapter to enhance reflection. For example, in soliciting the muddiest points from chapter 3, the following prompt was provided to students: "*Write the most confusing concept from Chapter 3. Note that in Chapter 3, we discussed the naming of ionic and molecular compounds, mole concept of compounds (determining moles, mass, and molecules of compounds), determining mass percent composition of elements in a compound, determining empirical and molecular formulas of compounds.*".[7] Next, students were asked to identify specific concepts, based on the provided topics or concepts, and anonymously type, using their electronic devices, the most confusing or difficult concepts.[7] The polled anonymous responses were displayed on the projector screens during the class. Students were also informed to "like" other students' responses that they felt were potentially confusing, but not instantly captured in their responses.[7]

The responses were analyzed and categorized according to frequency counts. Frequently identified muddy concepts were addressed by the course instructor during the subsequent class meetings. Furthermore, students received, through the learning management system, recorded video clips or PowerPoint presentations of instructor-solved problem sets for additional tutoring, and additional homework problems for practice.[7] Moreover, the muddiest concepts were addressed before introducing a new concept or chapter and were also intentionally integrated into the subsequent chapters throughout the semester by the instructor.[7] Three supplemental instruction

(SI) leaders, undergraduate students who previously taken the course successfully and who frequently attended the lecture sessions, worked closely with the instructor to develop practice problems toward addressing the frequently reported muddiest points, which were then implemented during the SI sessions in addition to general topics. The control group also had three SI Leaders who supplemented teaching on general topics only.[7]

26.4.6 Data Collection and Analysis

RQ1: Are there significant differences in performance between students exposed to the muddiest point activities infused with traditional formative Assessments (TFAs) and their counterparts exposed to the TFAs alone in a General Chemistry 1 course?

Quantitative data addressing this research question included the student' scores on three midterm exams and a final cumulative exam and participants' demographic data, such as gender, race/ethnicity, and first-generation college status. Each of the midterm exams was out of 100 points, whereas the final exam was out of 150. For the purposes of the study and ensuring consistency in the measured scale when comparing performance across the exams, the exams scores for each test was converted to 100%.

Independent *t*-tests were performed at $\alpha = 0.05$ to determine if there were any statistically significant mean differences in performance on the entry preparation scores, ACT math, three midterm exams, and the final exam between the treatment and comparison group. Effect sizes adjusted Cohen *d* for unequal sizes,[21] were also calculated for the mean differences that were statistically significant on the exams. Additionally, an independent *t*-test was also conducted to determine whether there were statistically significant mean differences in averaged midterm exam scores for the students' demographics (*e.g.*, gender, race and, first-generation status) between the treatment and the comparison group.[7]

RQ2: How does viewing peers' responses on the muddiest point activities influence, if at all, students' learning experiences?

These qualitative data were collected using an open-ended questionnaire. Specifically, participants were asked to describe how viewing the muddiest points reported by their peers influenced their learning experiences in the course. The NVivo qualitative analysis software[22] was employed. A total of 109 participants completed the open-ended questionnaire at the end of the semester. However, twelve of the participants' statements were irrelevant and therefore excluded from the analysis. A statement was deemed irrelevant when the participant did not respond to the question accordingly. For example, "N/A" and listing of the confusing concepts instead of descriptions of the impact of seeing the muddiest points reported by their peers were deemed irrelevant. An inductive, thematic approach was employed in the coding process,[20] which proceeded in two phases. In phase I, two researchers together coded about 20% of the 97 relevant statements. The codes were allowed to emerge with the coding focused mainly on the reported impact of seeing peers' responses. In phase II, one researcher continued with the analysis of the remaining responses using the same approach. Additionally, both researchers audited the identified codes from phase II against the raw data, for coherence and reliability. There were no discrepancies identified during the confirmatory audit phase.[23] Salient codes were developed into categories, which were further merged to generate themes.[20]

26.5 Results

This section describes key findings addressing the two research questions. We first describe results from the quantitative data addressing RQ1 followed by the qualitative findings addressing RQ2. We also provide representative participants' responses to support the qualitative findings.

26.5.1 RQ1

Independent *t*-test results showed that the treatment group scored significantly higher in the three midterm exams than the comparison group (see Table 26.2). For the final exam, the differences in performance were not statistically significant although the treatment group scored slightly higher than the comparison group. The mean differences were, however, associated with small effect sizes 0.2–0.4, adjusted Cohen's *d* for unequal sizes (0.2—small, 0.5—moderate, and 0.8—large).[18,21] Specifically, scores for the treatment group were higher by 6.1%, 4.8%, 5.5%, and 1.7% on Exam 1, Exam 2, Exam 3, and the final cumulative exam, respectively. These results indicate that the muddiest point activities and TFAs combination led to better performance than TFAs alone, suggesting that the muddiest point activities bolstered the performance of students in the treatment group.[7]

Further analysis of data by demographics showed that regardless of gender, ethnic minority groups, and first-generation status, students in the treatment group performed significantly better compared to their counterparts

Table 26.2 Independent *t*-test comparison mean scores between the treatment group and the comparison group on the exam scores.

Dependent variables	Treatment group			Comparison group					
	N	*M*	S.D.	*N*	*M*	S.D.	df	*t*	*p*
Exam 1	116	78.4	12.6	221	71.9	6.3	289	4.089	<0.001*
Exam 2	117	62.8	17.7	221	58.0	20.9	272	2.194	0.029*
Exam 3	116	49.2	19.3	217	43.7	20.6	248	2.412	0.017*
Final exam	117	50.1	15.2	214	48.4	18.0	274	0.869	0.386

Note: * = $p < 0.05$ and denotes the exam scores that were statistically significant between the two groups. *M* = mean, S.D. = standard deviation, and *N* is the total number of consented participants whose information was provided. Academic performance is measured by an average exam score out of 100%.

Performance by Demographics on the Averaged Midterm Exams Scores between the Treatment and the Comparison Group

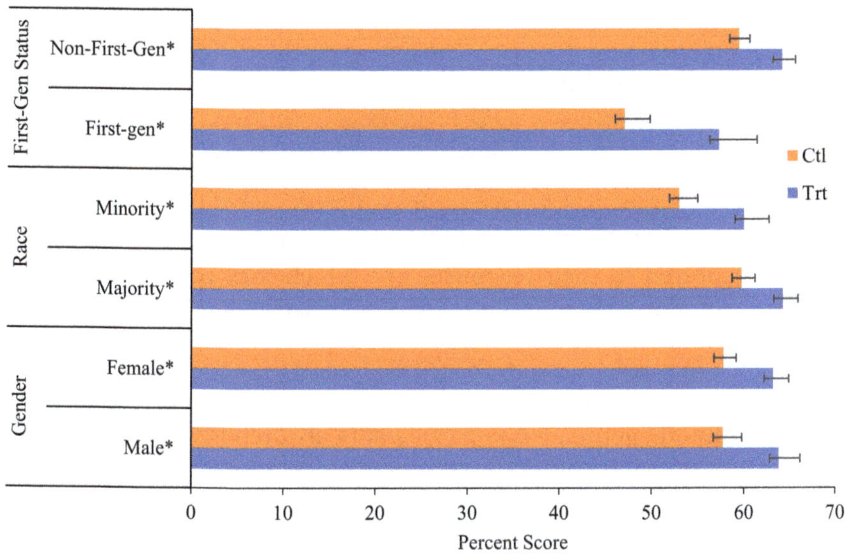

Figure 26.1 Independent *t*-test results on the averaged midterm exam scores between the treatment and comparison group based on demographics. *Indicates statistically significant mean differences, $p < 0.05$.

in the comparison group (see Figure 26.1). Specifically, male students in the treatment group performed significantly better ($M = 63.9$, S.D. = 14.9, $N = 42$), $t(99) = 2.018$, $p < 0.05$) than those in the comparison group ($M = 57.7$, S.D. = 18.3, $N = 78$). Similarly, female students in the treatment group ($M = 63.3$, S.D. = 15.1, $N = 75$, $t(219) = 2.309$, $p < 0.05$) performed significantly better than those in the comparison group ($M = 57.8$, S.D. = 17.4, $N = 146$). Based on generation status, results indicated that first-generation students in the treatment group ($M = 57.4$, S.D. = 15.9, $N = 14$, $t(45) = 1.986$, $p < 0.05$), performed significantly better than those in the comparison group ($M = 47.2$, S.D. = 16.2, $N = 33$). Additionally, non-first-generation students in the treatment group ($M = 64.4$, S.D. = 14.7, $N = 103$, $t(239) = 2.476$, $p = 0.014$), performed significantly better than those in the comparison group ($M = 59.6$, S.D. = 17.3, $N = 191$). Results based on race/ethnic groups indicated that the majority students in the treatment group ($M = 64.4$, S.D. = 14.7, $N = 87$, $t(206) = 2.120$, $p < 0.05$) performed significantly better compared to those in the comparison group ($M = 59.8$, S.D. = 17.6, $N = 139$) and so did the minority students—treatment group ($M = 60.1$, S.D. = 15.1, $N = 29$, $t(99) = 1.951$, $p < 0.05$) and comparison group ($M = 53.1$, S.D. = 16.8, $N = 72$). These results suggest the potential for the CAT activities to improve the performance of all students, including those from marginalized groups.[7]

26.5.2 RQ2

About 65% (63 participants) reported a positive influence of the CAT activities while 35% (34 participants) revealed no influence on seeing peers' responses on the muddy points. The emerging themes on the reported benefits are described as follows.

26.5.2.1 Theme 1: Identifying and Addressing the Learning Gaps

About 40% of participants (21 participants) mentioned that seeing their peers' responses helped them to identify potential areas of confusion not initially captured in their responses and which they could not recall. For example, some participants said: "This was helpful made me realize some things I didn't even realize I was confused about." ... "It reminded me of concepts that I also believed to be confusing that I may have forgotten to consider." These data suggest that the peers' responses further triggered other forgotten muddy points. Some participants also reported that seeing their peers' areas of confusion prompted them to revisit the concepts for better mastery—"Seeing these [peers' responses] made me go back and study these concepts to make sure I knew them well." ... "This was helpful because I knew what [the concepts] to focus on when studying." Overall, the peers' responses enhanced the students' desire to comprehend the concepts better.

26.5.2.2 Theme 2: Decreased Anxiety in the Course

Results revealed that 33% (18 participants) reported a decreased anxiety in the course. Specifically, the participants mentioned that they felt better knowing that they were "not alone" as their peers experienced moments of confusion. For example, some participants said: "It helped me feel better knowing I wasn't the only one confused about the concepts." ... "It helped me not feel alone when I struggled with [concepts] because I saw others struggle too." "Helped me identify that I wasn't alone since the points that many were confused on, I was also confused." These data suggest that seeing the most confusing concepts reported by their peers made the participants feel less anxious in the course.

26.5.2.3 Theme 3: Increased Confidence and Persistence in the Course

Results showed that about 17% (nine participants) reported increased confidence and persistence in the course. By realizing that their peers experienced moments of confusion, they developed more resilience and desired to press on, and not drop the course. For instance, some participants remarked: "Yes, I was impacted by this [seeing peers' responses] because it helped me see that I was not alone in my confusions which pushed me forward when I wanted to stop."... "[It] helped me to build confidence knowing I wasn't the only one struggling." Additionally, some participants developed confidence in asking questions as a result of seeing the reported muddy points by the peers—"[It] helped me feel better and more comfortable asking questions knowing others are confused too." This finding indicates that seeing the most confusing concepts reported by the peers bolstered the participant's confidence and persistence in the course.

Findings from the current study revealed that marginalized students in the treatment group, where Top Hat and muddiest point activities infused with the TFAs were implemented, performed better than their counterparts in the comparison group, in which TFAs alone were availed to students. This indicates the activities provided equitable access to learning for all students, including marginalized student populations, in the treatment group compared to the comparison group. Qualitative findings from the open-ended survey responses also showed reported benefits of seeing peers' responses through Top Hat.

26.6 Summary and Discussion

Results indicated statistically significant differences ($p < 0.05$) in performance scores between the treatment group and the comparison group in all of the three midterm exams. However, there were no significant differences between the two groups on the final exam even though the treatment group had a higher mean score. These finding contradict with Snead's (2017) that found no significant mean differences in performance in the general chemistry courses from muddiest point activities given after the end of each lesson.[24] Given that participants were required to reflect on two chapters or units, they were able to identify the areas of struggle with greater ease and prepare adequately for the midterm exams. In contrast, the final exam included question items covering 10 chapters. Therefore, broad learning objectives were assessed compared to the midterms which might explain the non-significant mean differences between the treatment group and the comparison group.[7] One explanation for the significant mean differences in the midterms may partly be due to scaffolding learning by focusing on smaller segments of the lesson units to uncover and close the learning gaps.[7] By scaffolding learning on smaller lesson units, the teachers can better determine exactly where students need help, where they succeed, and which supports are most effective, and enhanced instructional adaptations.[1,3] Additionally, the significant impact of muddiest point activities on the midterms can be attributed to the fact that the participants were given enough time to reflect and review concepts cover within the chapter units—medium assessment cycle[7] compared to if the muddiest point activities were given at the end of each lesson— short cycle assessment.[24]

Results based on demographics showed that the muddiest point activities afforded a positive learning experience for all students in the treatment group, regardless of their gender, first-generation status, and race/ethnicity. This might be partly due to the reinforced reflective learning, an important component of metacognitive regulation skills.[25,26] The current findings indicate that the muddiest point activities can improve student performance in general chemistry courses, including the marginalized student populations.

Qualitative findings revealed the benefits of sharing anonymous responses from peers on confusing or difficult concepts. Benefits included identifying and addressing gaps in learning, reduced anxiety in the course, and increased confidence and persistence in the course. Compared with previous semesters, taught by the same instructor, withdrawals from the course decreased by about 2% during the semester in which the muddiest point activities were implemented. For identification of and addressing knowledge gaps, it is worth noting that students' explanations stated that seeing what their peers reported helped remind them of some of the things they had forgotten. This finding further revealed that this activity was a good platform for students to share out their areas of confusion. While some students may be less inclined to discuss with peers during and outside of class,[27] Top Hat provided a platform for students to anonymously communicate the concepts they struggled with, and for the instructor to identify students' gaps in learning in real-time, as well as the frequently reported muddy points. The latter catalyzed the analysis of the reported muddy points and facilitated timely feedback in closing the gaps in learning. Overall, the integration of the muddiest point activity into the course through Top Hat provided a safe opportunity to ask questions about content material and provided the instructor with insights into the general patterns of student-reported muddy points.

Seeing the peers' responses on muddy points also made the participants feel better and calm. Chemistry anxiety includes discomfort regarding learning chemistry, assessments in chemistry courses, and fear of chemistry as a course.[28] Consequently, students perceive chemistry courses as difficult, which, in turn, negatively affects their performance in chemistry courses.[28] Findings also revealed reported increased confidence and persistence in the course as a result of the peer responses.

In conclusion, the results indicated significantly ($p < 0.05$) higher performance scores on all the midterm exams, favoring the muddiest point activities infused with TFAs. Furthermore, women and students from marginalized groups in the intervention group performed better than their counterparts in the comparison group.[7] Qualitative findings revealed benefits in seeing reported muddy points by peers in the course such as improved ability to identify and address the gaps in learning, decreased anxiety in the course, and increased confidence and persistence in the course. Some participants were motivated to continue in the course amidst the areas of struggle especially after realizing "they were not alone", as their peers were also experiencing similar struggles in the course.

Together, the quantitative and qualitative findings reveal the need to tap into digital assessment platforms such as Top Hat to improve learning experiences for all students and provide equitable access to learning (see also Chapter 23, which introduces another digital assessment platform, the 'inclusive online collaborative assessments in chemistry').[29] When successfully implemented, digital assessments can cultivate an inclusive learning environment, where diverse voices are heard and students' learning gaps can be identified and addressed in real-time.[7] In turn, this can narrow the opportunity gap in performance between marginalized and non-marginalized student groups.[7] Ultimately, this is pivotal in improving retention in chemistry and other STEM courses as well as broadening participation of marginalized groups in the future STEM workforce. Furthermore, findings from this study provide insights into future assessments and the use of digital assessments in chemistry courses. We contend that muddiest point activities can improve students' mastery of specific or targeted topics and foster a positive learning experience for all students. Additionally, we encourage instructors to consider soliciting student-generated responses on difficult or confusing concepts rather than constraining students to instructor pre-chosen topics, as part of the instructors' digital-topic specific pedagogical content knowledge (d-TSPCK), see Chapter 6 by Mavhunga and colleagues.[30] This expands the bandwidth in uncovering the unanticipated confusing chemistry concepts that are likely to be overlooked.

Acknowledgements

This work was in part supported through the Oklahoma State University Foundation, the Edward E. Bartlett Endowed Chair fund.

References

1. T. Angelo and P. K. Cross, *Classroom Assessment Techniques: A Handbook for College Teachers*, Jossey-Bass, San Francisco, 2nd edn, 1993.
2. P. Black and D. Wiliam, Developing the theory of formative assessment, *Educ. Assess.*, 2009, **21**(1), 5.
3. L. B. Nilson, *Teaching at its Best: A Research-based Resource for College Instructors*, John Wiley & Sons, 2016.
4. W. S. Harwood, The one-minute paper, *J. Chem. Educ.*, 1996, **73**(3), 229.
5. Y. He, Constructive Error Climate: A Classroom Assessment Technique in Science Classes Breadcrumb, *J. Coll. Sci. Teach*, 2020, **49**(4), 37–40.
6. D. B. King, Using clickers to identify the muddiest points in large chemistry classes, *J. Chem. Educ.*, 2011, **88**(11), 1485.
7. C. Z. Muteti, T. Kerr, M. Mwavita and J. M. Mutambuki, Blending muddiest point activities with the common formative assessments bolsters the performance of marginalized student populations in general chemistry, *Chem. Educ. Res. Pract.*, 2022, **23**(2), 452–463.
8. F. Mosteller, The 'Muddiest Point in the Lecture' as a feedback device, *On Teaching and Learning: The Journal of the Harvard-Danforth Center*, 1989, **3**, 10–21.
9. X. Chen, *STEM Attrition: College Students' Paths Into and Out of STEM Fields. Statistical Analysis Report. NCES 2014-001*. National Center for Education Statistics, 2013, https://nces.ed.gov/pubs2014/2014001rev.pdf, accessed May 2021.

10. A. V. Maltese and R. H. Tai, Pipeline persistence: Examining the association of educational experiences with earned degrees in STEM among US students, *Sci. Educ.*, 2011, **95**(5), 877.
11. National Science Foundation, National Center for Education Statistics: Women, Minorities, and Persons with Disabilities in Science and Engineering, 2019, https://ncses.nsf.gov/pubs/nsf19304/digest, accessed May 2021.
12. S. L. Eddy, S. E. Brownell and M. P. Wenderoth, Gender gaps in achievement and participation in multiple introductory biology classrooms, *CBE—Life Sci. Educ.*, 2014, **13**(3), 478.
13. S. Freeman, S. L. Eddy, M. McDonough, M. K. Smith, N. Okoroafor, H. Jordt and M. P. Wenderoth, Active learning increases student performance in science, engineering, and mathematics, *Proc. Natl. Acad. Sci. U. S. A.*, 2014, **111**(23), 8410.
14. K. M. Soria and M. J. Stebleton, First-generation students' academic engagement and retention, *Teach. Higher Educ.*, 2012, **17**, 673–685.
15. S. L. Eddy and K. A. Hogan, Getting under the hood: How and for whom does increasing course structure work? *CBE—Life Sci. Educ.*, 2014, **13**(3), 453.
16. S. W. Draper and M. I. Brown, Increasing interactivity in lectures using an electronic voting system, *J. Comput. Assisted Learn.*, 2004, **20**, 81–94.
17. D. Bruff, *Teaching with Classroom Response Systems: Creating Active Learning Environments*, John Wiley & Sons, 2009.
18. C. R. Ward, J. H. Reeves, and B. P. Heath, Encouraging active student participation in chemistry classes with a web-based, instant feedback, student response system, http://confchem.ccce.divched.org/sites/confchem.ccce.divched.org/files/2003SpringConfChemP2.pdf, accessed May 2021.
19. M. Heritage, *Formative Assessment: Making it Happen in the Classroom*, Corwin Press, 2010.
20. J. W. Creswell and J. D. Creswell, *Research Design: Qualitative, Quantitative, and Mixed Methods Approaches*, Sage Publications, 2017.
21. J. Cohen, *Statistical Power Analysis for the Behavioral Sciences*, Academic Press, 2013.
22. L. Richards, *Using NVivo in Qualitative Research*, Sage, 28 October 1999.
23. A. Tashakkori and C. Teddlie, *Mixed Methodology: Combining Qualitative and Quantitative Approaches*, Sage Publications Ltd, Thousand Oaks, CA, 1998.
24. L. P. Snead, *The Effect of Using the Muddiest Point Technique in a Large General Chemistry Class*, Drexel University, 2017.
25. J. M. Mutambuki, M. Mwavita, C. Z. Muteti, B. I. Jacob and S. Mohanty, Metacognition and active learning combination reveals better performance on cognitively demanding general chemistry concepts than active learning alone, *J. Chem. Educ.*, 2020, **97**(7), 1832.
26. C. Z. Muteti, C. Zarraga, B. I. Jacob, T. M. Mwarumba, D. B. Nkhata, M. Mwavita, S. Mohanty and J. M. Mutambuki, I realized what I was doing

was not working: the influence of explicit teaching of metacognition on students' study strategies in a general chemistry I course, *Chem. Educ. Res. Pract.*, 2021, **22**, 122.
27. R. W. Preszler, A. Dawe, C. B. Shuster and M. Shuster, Assessment of the effects of student response systems on student learning and attitudes over a broad range of biology courses, *CBE—Life Sci. Educ.*, 2007, **6**(1), 29.
28. B. B. Widanski and W. C. McCarthy, Assessment of chemistry anxiety in a two-year college, *J. Chem. Educ.*, 2009, **86**(12), 1447.
29. See Chapter 23: J. J. Stewart, in *Digital Learning and Teaching in Chemistry*, ed. Y. J. Dori, C. Ngai and G. Szteinberg, Royal Society of Chemistry, United Kingdom, 2023.
30. See Chapter 6: E. Mavhunga, M. Rollnick and D. van der Merwe, in *Digital Learning and Teaching in Chemistry*, ed. Y. J. Dori, C. Ngai and G. Szteinberg, Royal Society of Chemistry, United Kingdom, 2023.

CHAPTER 27

Embedding Feedback in Digital Learning Environments to Promote Learners' Thinking About Their Thinking in Chemistry

GWENDOLYN A. LAWRIE*

School of Chemistry & Molecular Biosciences, The University of Queensland, St Lucia, Queensland 4072, Australia
*E-mail: g.lawrie@uq.edu.au

27.1 Background

As students transition between high school and tertiary chemistry studies, they enter new and more complex learning environments in which it is assumed they will rapidly become autonomous in terms of time management skills, engagement in learning and completion of assessment. This transition requires a structured program of support including the provision of formative feedback to help learners transfer their prior knowledge and to enable their construction of new understanding.[1,2] Furthermore, tertiary learning environments have become increasingly complex through the introduction of blended or hybrid learning. The pivot into online teaching during the 2020 coronavirus (COVID-19) pandemic further amplified the need for

Advances in Chemistry Education Series No. 11
Digital Learning and Teaching in Chemistry
Edited by Yehudit Judy Dori, Courtney Ngai and Gabriela Szteinberg

Published by the Royal Society of Chemistry, www.rsc.org

well-scaffolded and resourced online learning environments as part of effective pedagogies and practices (see Chapter 4, which considers an example of this challenge).[23]

Online learning is not a new paradigm. Substantial progress has been made in recognising the elements and interactions that are required to enable active and constructive learning.[3] In particular, constructive interactions between students and content, students with peers, and students with teachers are regarded as highly desirable.[3] Online learning, enabled through the internet, involves either synchronous or asynchronous activities delivered entirely remotely or as part of a blended or hybrid environment. Learning interactions are typically facilitated through a combination of the learning management system, virtual learning environments, external platforms (often provided by textbook publishers) and custom-made digital technologies or tools.

One of the challenges for instructional design of effective online learning environments is to identify mechanisms of capturing learners' thinking, enabling provision of constructive feedback.[4] Constructive and active online learning environments can develop students' self-regulatory behaviours. In this chapter, findings from design-based research approaches reveal insights into instructional strategies for delivering effective technology-enabled feedback. Student perceptions of feedback reveal that feedback is only effective if students recognise it, value it and act on it!

27.2 Theoretical Lens

The provision of feedback, integrated with assessment, is core to student learning. Feedback is a complex construct, it is fluid in its definition dependent on the context, stakeholder perspective and intended purpose. Shute (p. 154) defined feedback as '*information communicated to the learner that is intended to modify his or her thinking or behavior for the purpose of improving learning*'.[5] Her focus on task level feedback aligned with the seminal work of Hattie and Timperley who recognised four levels of feedback.[6] Task level feedback involves performance and guidance related to a specific activity or assessment task, often involving criteria. Process level feedback relates to the way that learners engage in tasks in terms of processes, strategies and ways of modelling thinking that are required. Self-regulatory level feedback enables the learner to connect the process and outcomes of activities and tasks to inform self-monitoring, relational thinking and action towards meeting the required outcomes. Self-level feedback focuses on the characteristics of the individual often involving praise or normative information, the former can potentially have a negative impact or demotivate learners.[5,6]

In traditional classrooms, the transmissive model of feedback that focusses on what teachers do, has been replaced by one that focuses on the actions adopted by learners in response to receiving feedback, or their feedback literacy.[7] Feedback literacy is a student's capacity and readiness to make sense of comments and/or marks given for their performance then adjust thinking to improve future outcomes. Sustainable feedback practice involves '*closing the feedback loop*',[8,9] where feedback is conceptualised in terms of student agency in seeking, making sense of, and reflecting on their performance or their approaches to learning.

27.2.1 A Social Constructivist View of Feedback

Substantial research, over several decades, has established the foundations and guidance for instructors to engage learners with effective feedback to develop their feedback literacy.[5–11] Feedback becomes more constructive through a process involving dialogue between teachers and students. Nicol proposed the adoption of four effective characteristics that promote effective feedback dialogue.[10]

- Feedback should be tailored to students' needs: adaptive.
- Feedback should involve conversations or exchanges between students and teachers: discursive.
- Feedback should connect to an action that a student has taken related to a task or goal: interactive.
- Feedback should enable students to consider and intentionally adjust their thinking to aim for improved outcomes related to a future task: reflective.

Dialogic feedback is characterised by interactive oral or written communication between two or more people that relates to the quality of student work where interpretations and meanings are agreed, clarified and understood.[7] From a social-constructivist perspective, dialogic feedback can be delivered in multiple ways promoting interactions and dialogue between students and teachers as well as students with their peers.[12] Key aspects that support dialogic feedback mean that it should be information rich, involve elaboration, encourage reflection, encourage agency in learning, involve interactions and develop trust between students and teachers so further questions can be asked.[12] The format, formulation and timing of feedback is a critical consideration.

Prior to the widespread availability of technologies, written feedback was the primary format used by teachers to initiate discussion or reflection (usually in the form of indicating correct answers, grades or comments on work). This format continues to be valued by both teachers and students.[10]

There are now multiple different levels, categories and characteristics of feedback, summarised in Table 27.1 below, considered in research literature. Several have evolved in traditional learning environments, however technology-enabled feedback represents a new feedback paradigm which extends the potential modes available for sharing feedback with learners.[7]

Technology-enabled feedback places the emphasis on how students act on feedback (feedback literacy) rather than how teachers deliver feedback.[7] The affordances of this environment include the ability to provide just-in-time

Table 27.1 Levels, categories, and characteristics of feedback identified in research literature.

Levels of feedback[6]	Categories of feedback[7,10,12]	Characteristics of effective feedback dialogue[10]
Task	Verification/Corrective	Adaptive
Process	Dialogic (adaptive, discursive, interactive, reflective)	Discursive
Self-regulatory	Facilitative	Interactive
Self-level	Peer	Reflective
	Modelling	
	Self-regulatory	

and asynchronous feedback from multiple sources, so the teacher no longer represents the single source of feedback.[8] Examples of technology-enabled feedback include the use of clickers or polling tools in lectures where students receive immediate feedback, correction of misconceptions involving dialogic interactions with peers and teachers *in situ* and links to relevant course-based resources.[7] While just-in-time feedback is considered valuable, there is contrasting evidence that the practice of delayed feedback can enhance outcomes due to the spacing of information.[13] A potential affordance of technology-enabled feedback is flexibility in tailoring the timing of feedback delivery.

In asynchronous online environments, self-dialogue becomes more important and dialogic feedback can be delivered in combinations of written, verbal/audio or video formats. All three formats have been found to have equal impact on student achievement but notably video feedback offers the opportunity to build social presence and a sense of a community of inquiry.[14]

27.2.2 Chemistry Discipline-specific Feedback in Online Environments

In chemistry, conceptual understanding involves multiple levels of thinking which requires scaffolding by teachers through modelling and use of external representations.[15] It appears logical that discipline-specific feedback must be considered and embedded in learning environments. In addition, chemistry requires strong mathematical skills in problem-solving processes, therefore students will often tend to focus on rote-learning the steps in algorithmic processes that are provided in worked solutions rather than understanding the underlying concept. The chemistry teacher has a central role in terms of discipline-specific feedback particularly in modelling visuospatial skills and thinking to support students' development of representational competence.[16]

To date, the provision of technology-enabled, chemistry-specific feedback has been explored mostly related to quiz or homework platforms (see Chapter 24).[24] However, there are also good examples of active learning enabled through technology (flipped and hybrid classrooms) incorporating clicker questions, polling tools, quizzes and digital whiteboards.[16,17] Students can be engaged in predicting outcomes, drawing structures and peer discussions.

Online self-directed learning activities in chemistry in the absence of a teacher require thoughtful scaffolding to guide students in their engagement with concepts, chemistry representations and checking their thinking.[4] Simulations and dynamic interactive representations can be used to scaffold concept check questions through inquiry-based exploration prompted by feedback.[4]

27.3 Study Context and Inclusive Practice

In the synchronous, in-person learning environment, structured formal feedback is supplemented by informal incidental or experiential feedback (referred to as co-regulation) where students informally calibrate their thinking against their peers and the teacher's explanations.[18] In addition to immediately reviewing student work, teachers will 'read' the room in terms of students' body language and facial expressions so they can respond immediately. In online learning environments, opportunities for dynamic informal feedback are difficult to engineer; they need to be deliberately designed into the interactions that are facilitated online. Indeed, a recent study comparing online, and blended learning environments observed that a higher interaction frequency with the teacher increased the chance that the students adopted the teacher's regulatory strategies.[18] Figure 27.1 displays the different interactions that are important in online learning environments annotated with potential combinations and forms of feedback.[3]

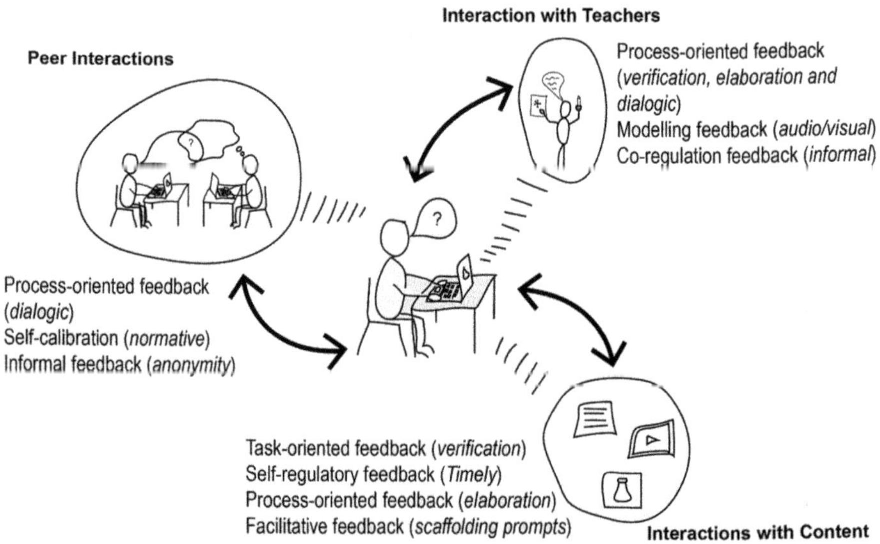

Figure 27.1 Sources and formats of feedback in online learning environments.

Self-assessment, reflection and self-regulation by learners can be enabled through the provision of multiple feedback events (different sources) and synthesis across multiple processes.[4,7,12] There is an opportunity to intentionally design technology-enabled formative feedback mechanisms that develop these skills in chemistry learners through provision of timely feedback. This chapter describes multiple interventions that aim to achieve this while responding to access and participation in learning.

In Australia, large first-year chemistry classes represent diverse cohorts based on the multiple pathways that students have transitioned from high school contexts, international contexts and their different stages of education. Inclusive practice includes the introduction of pedagogies and practices that provide students with flexibility and multiple means of engaging with learning and assessment.[1,2] The provision of multiple modes and categories of feedback, as part of the designing activities that widen equitable access and participation in online active learning, is one strategy to increase student confidence. Strategic embedding of formative feedback, with a focus on dialogic and peer feedback, has potential to support students in self-regulatory processing and subsequent actions to fill any perceived gaps in their understanding. These approaches are described further in this chapter including examples of practice.

27.3.1 Methods (Design-based Research in Action)

Application of a design-based research (DBR) methodology[19] supports the adaptation of a single course across multiple semesters involving collection, reflection on and response to empirical research data. In the context of this study, annual evaluation and reflection cycles resulted in the progressive introduction of scaffolding for provision of different forms of formative feedback in increasingly blended learning courses. The iterative evaluation cycles involving combinations of feedback strategies are summarised in Table 27.2, these interventions are described below.

The standard formats of feedback that arise from active learning lectures and laboratory practical sessions based on dialogic interactions between students, their teachers, tutors and peers are recognised in Table 27.3. The course included peer-assisted study sessions (PASS) led by senior undergraduate students. Exams, quizzes and problem-solving PASS worksheets were assigned as providing both verification and elaborative feedback based on correct answers, grades and worked solutions. Modelling feedback is integrated into lectures and PASS through the teachers' use of structural models and visualiser tools.

27.3.2 Feedback Interventions

The initial intervention in 2014 introduced online modules that were designed to engage students to consider their prior knowledge based on provision of formative feedback then directed them to self-regulated learning activities

Table 27.2 Schematic representation illustrating the different forms of feedback that were progressively incorporated in a first semester general chemistry course. Online activities were optional between 2014 and 2018 but were formally embedded as part of a blended learning course transformation in 2019. All learning activities were 100% online in 2020 due to the COVID pandemic.

Key
🟢 Dialogic 🔵 Peer ⚫ Facilitative 🟣 Self-regulatory 🟡 Modelling 🟠 Verification

Activity or Task	2014	2015	2016	2017	2018	2019	2020	2021
Lectures	🟢🔵🟡	🟢🔵🟡	🟢🔵🟡	🟢🔵🟡	🟢🔵🟡	🟢🔵🟡	🟢🔵🟡	🟢🔵🟡
Workshops	N/A	N/A	N/A	N/A	N/A	N/A	N/A	🟢🔵🟡🟠
Laboratory practical sessions	🟢🔵🟠	🟢🔵🟠	🟢🔵🟠	🟢🔵🟠	🟢🔵🟠	🟢🔵🟠	🟢🔵🟠	🟢🔵🟠
Online review modules	⚫🟣🟡	⚫🟣🟡	⚫🟣🟡	⚫🟣🟡	⚫🟣🟡	⚫🟣🟡	⚫🟣🟡	⚫🟣🟡
Peer-assisted study sessions	🟢🔵🟡🟠	🟢🔵🟡🟠	🟢🔵🟡🟠	🟢🔵🟡🟠	🟢🔵🟡🟠	🟢🔵🟡🟠	🟢🔵🟡🟠	N/A
Weekly computer quizzes	🟣🟠	🟣🟠	🟣🟠	🟣🟠	🟣🟠	🟣🟠	🟣🟠	🟣🟠
Blended learning tasks[a]	N/A	N/A	N/A	N/A	🟢🔵	🟢🔵🟣🟡	🟢🔵🟣🟡	🟢🔵🟣🟡
Midsemester exam	🟠	🟠	🟠	🟠	🟠	🟠	🟠	🟠
Final exam	🟠	🟠	🟠	🟠	🟠	🟠	🟠	🟠

[a]Involved student-generated explanations or representations shared online.

Table 27.3 Example references are provided to give insights into 2014 students' thinking.

Theme (properties)	Example responses
Verification Feedback	*Providing me with **solutions for the working out** of questions is a very helpful type of feedback. When it comes to handing in assignments and other online activities, I find it very helpful is feedback is given on **what I need to correct**. For example, in an equation question, having feedback on **what I did wrong** with the working out is very helpful.*

Table 27.3 (*continued*)

Theme (properties)	Example responses
References to seeking: • a correct answer • grades • worked solutions • more tests or quizzes.	*By giving students* ***the solutions to all the questions*** *that are asked in lectures, pass classes and past exams, which you have done to some degree. Some students obviously would use this to help work out the answers. The solutions would be really helpful, especially with calculations.* *Don't need feedback,* ***as long as the answers for CROM and PASS are posted and I'm getting the questions right****, that's enough positive feedback for me.*
Dialogic Feedback	*Have* ***one on one sessions with the teachers and talk about assessment and the area where u might need improvement****. Students could make appointments for these sessions as it is not plausible for teachers to have sessions with every student and this way only the ones that are really keen will make the effort.*
References to seeking: • one-on-one advice • PASS leader advice • being told what to do • verbal explanation.	*In person like PASS.* ***I can ask questions and get answers to things I didn't get.*** ***verbal explaining why the answer is incorrect*** *and explaining correct method to get the answer.*
Written Feedback	*Written feedback. It is* ***easier to understand, and we can review the feedback as much as we can****. It's better than audio feedback because we might miss out some words as they may speak too fast or maybe because of the accent that is difficult to catch.*
References to seeking: • written feedback • comments on work • emailed advice	*Written feedback is* ***useful for referring back to later*** *People giving you marks with written feedback on it,* ***kinda personalized****, you know? Like the labs, that was the best feedback in the course. The exam viewing sessions are bad cause it comes out of our time.*
Peer dialogic feedback	*I think the best form of providing feedback is peer studying. When you study together in a group,* ***it is easier to recognise your level of understanding and knowledge****. Thanks to that, we can study more or revise better.*
References to seeking: • self-calibration against other students • student explanations.	*I like being able to discuss concepts and problems in a small group* ***until we all understand them, because explaining things back and forth is really helpful*** *to my learning.*
Self-regulatory feedback	*I know what how I learn and what works for me, so the weekly CROMs and questions provided in lectures and PASS,* ***provided me with feedback which allowed me to identify gaps in my knowledge and understanding****.*

(*continued*)

Table 27.3 (*continued*)

Theme (properties)	Example responses
References to: • a link between feedback and thinking • intent to act on feedback • reflection on thinking	*I found mid-semester exam was a great indicator of how well I was going in the course and showed me that I* ***needed to change my studying habits and try harder*** *as I wasn't do as well as I thought I was.* *An example of where feedback was used effectively to support the way I learned that is already present in the course is the PASS questions as it enabled a week by week* ***evaluation of which concepts were struggled with and provided the ability to develop understanding in these certain areas.***

involving facilitative feedback in online modules. This initiative has been described in detail elsewhere.[1,2,4,9,17] These weekly chemistry review online learning modules (CROMs) diagnose high school conceptions using a concept inventory tool[20] linked to individual formative feedback rather than the correct answer (no course marks). Based on their answers to these concept questions, students were directed to self-regulated online learning activities situated in a dedicated folder in the learning management system (Blackboard™). Students were encouraged to complete these first two formative steps in the module prior to completing the summative online quiz (course marks) that tested the week's lecture concepts. Annual evaluation data was collected through an end of semester online questionnaire that captured demographic data, Likert-style scales and open response items. The student assessment of their learning gains (SALG) instrument[21] was adapted for this context and used to collect perception data. An incentive of 1% bonus marks was offered in 2014 providing a high response rate. Insights from two specific Likert-style items from the SALG scale are considered further in this chapter to explore students' perceptions of the extent to which learning activities and assessment helped their learning. The course structure of activities and assessment were retained between 2014 and 2018 with the only change being the quiz platform in 2015 (from Blackboard tests to an external provider, Sapling).

With an increasing focus on blended learning, a new task was introduced applying pedagogical design to increase peer interactions and involving embedded dialogic and reflective feedback. This online, collaborative peer discourse (CPD) activity was trialed in 2018 (hence was optional for students) but became linked indirectly to summative course marks in the following year. In 2019, funded through an institutional initiative to transform traditional courses into blended learning courses, the CROMs (online learning modules) were translated into the edX Edge online learning platform and formally embedded as weekly pre-lecture preparation.

Reflecting on the previous annual evaluation data, a need was identified to increase the value of interactions between students and their peers, hence 'Blended Learning Tasks' that applied technology-enabled peer discourse feedback were introduced. These tasks were also designed to strengthen the interactions between students and instructors as well as students with students in active lectures. For brevity, only sufficient details to establish the nature of feedback (dialogic, peer, self-regulatory and modelling) that is involved in these tasks are provided in this chapter.

- Collaborative peer discourse task (CPD): facilitated through a custom-made digital platform, integrated into Blackboard. Students are guided stepwise through a process of submitting an individual explanation of the real-world concepts that underpin a real-world phenomenon observed in a video. Then, as part of a group of four, students collaboratively and synchronously consider their different individual explanations to derive a consensus explanation. The final step involves reflection on thinking as a result of the group discussion. Students can engage with dialogic, peer and self-regulatory feedback through reflection.
- Digital noticeboard task (Padlet): delivered through the edX Edge platform, students are required to hand draw their representation of either a structure or chemical system, capture the drawing digitally and anonymously submit them to a Padlet wall. The lecturer reviews the class submissions to identify incorrect use of chemical representational formalisms and misconceptions. Students have access to the submissions of peers and can evaluate and calibrate their own thinking.[8] Dialogic elaboration and modelling feedback is introduced into the lecture involving the discussion of the correct representation of either the structure or chemical system.

All evaluation data have been collected through procedures that have been approved by the institutional ethics review committee.

27.4 Results and Discussion

27.4.1 Students' Perceptions of 'Useful' Feedback Across Blended Online and In-person Learning Environments

One of the enduring challenges for instructors is to raise their students' awareness of what constructive formative feedback 'looks like' and develop their agency in responding to feedback as part of learning. Traditionally, in most tertiary courses, students' perceptions of the level of feedback that they have received are captured through standardised end of semester institutional course evaluations. These scores are infamously difficult to improve.[10]

Students enrolled in a very large general (first year) chemistry course in 2014 were invited to complete an open response question as part of an online evaluation survey at the end of semester: 'What is the best form of providing you with feedback that will support the way you learn?'. This qualitative data ($N = 1155$) was thematically coded informed by levels and forms of feedback in Table 27.1. The distribution of feedback themes are displayed in

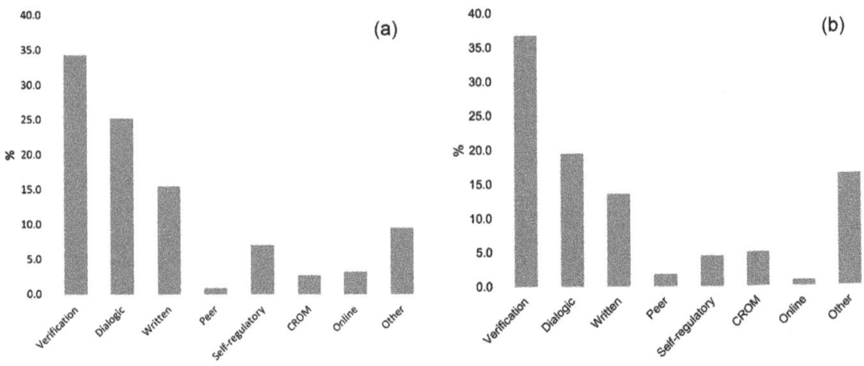

Figure 27.2 Distribution of themes that arose from coding students' perceptions of the best form of feedback provision that can support their learning collected through an open response survey question: (a) 2014 data ($N = 1155$), (b) 2015 data ($N = 220$).

Figure 27.2(a). This evaluation was repeated in 2015, without a bonus mark incentive, and responses to the same question ($N = 220$) coded with the same themes, displayed in Figure 27.2(b).

The relative distribution of themes categorising feedback that students valued were consistent across 2014 and 2015. Students tended to anticipate receiving corrective written feedback from their lecturers based on their high school experiences. This expectation was affirmed in Figure 27.2(a) where in 2014, 37% of students report a preference for verification feedback and/or corrective feedback, most referred to seeking worked answers for any form of course assessment (examinations, quizzes, PASS worksheets, laboratory quizzes), see examples in Table 27.3. In chemistry, worked solutions represent a traditional form of elaborative feedback (detailed explanations are not necessarily related to summative grades).[5]

Several students simply cited that they preferred '*online*' delivery of feedback, however there was a common perception that this actually related to the immediacy of feedback involving a mark or grade. This property is a recognised affordance of technology-enabled feedback as being 'just-in-time'.[7]

The second highest valued form of feedback (Figure 27.2) involved direct in-person conversations or explanations with either their lecturers, tutors or PASS leaders (senior undergraduate students). Students referred to seeking affirmation and clarification of concepts through dialogue (Table 27.3). Interestingly, conversations with other students in PASS sessions was not perceived to be a form of peer feedback by students—rather feedback provided by PASS leaders was universally valued perhaps reflecting student perceptions that these leaders had a greater authority than their own peers. Interestingly, there was frequent reference by students to indicate that they associated written feedback with longevity in terms of being a resource that could be returned to at any time during studies (Table 27.3).

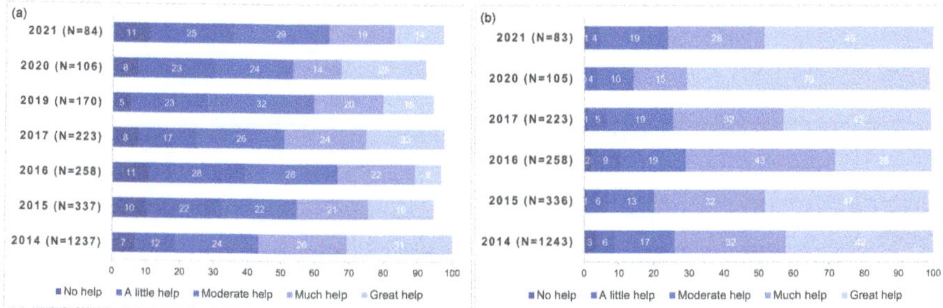

Figure 27.3 Response distributions for the SALG Likert-style item 'HOW MUCH did each of the following ASSESSMENT activities HELP YOUR LEARNING in chemistry': (a) feedback given in the CROM step 1 concept review questions; (b) the weekly quizzes.

Figure 27.3 displays the data from the SALG Likert-style scale item that captured how much the feedback that had originated from the chemistry review online modules (CROMs) had helped their learning. It is evident that formative feedback (designed to encourage students to reflect and immediately act on feedback) provided as part of the concept diagnostic questions (Figure 27.3(a)) was valued as being less helpful than the verification feedback provided from the associated summative online quizzes (Figure 27.3(b)) that focus on rehearsal of problem-solving. A limitation of this data set is the decrease in response rates annually; however, the patterns clearly indicate that students' perceptions of verification feedback was higher than formative, self-regulatory feedback. This bias is sustained between 2014 and 2021 indicating an intransigent perception.

27.4.2 Increasing the Value of Peer Interaction Feedback

As observed in Table 27.2, peer feedback was rarely cited as being valued even though the PASS sessions involved dialogic interactions between students. This very low recognition of the value of peer feedback prior to 2018, catalysed the introduction of technology-enabled peer feedback mechanisms as part of the transformation into a blended learning course in 2019. The aim was to integrate and enhance the connectivity between self-regulated learning and in-class activities. Student-generated submissions were collected and shared anonymously through two assessment activities: synchronous CPDs and Padlet noticeboards. Students were able to self-regulate in terms of how many tasks they opted to complete (there was a minimum of three out of five submissions required for full marks). Task scaffolding encouraged reflection on thinking involving self-calibration based on other students' submissions and explanations. Annual course evaluation surveys collected students' perceptions of to what extent these tasks had helped their learning (SALG item) (Figure 27.4). They were also asked to comment on how assessment had helped their learning during the semester in a linked open response field.

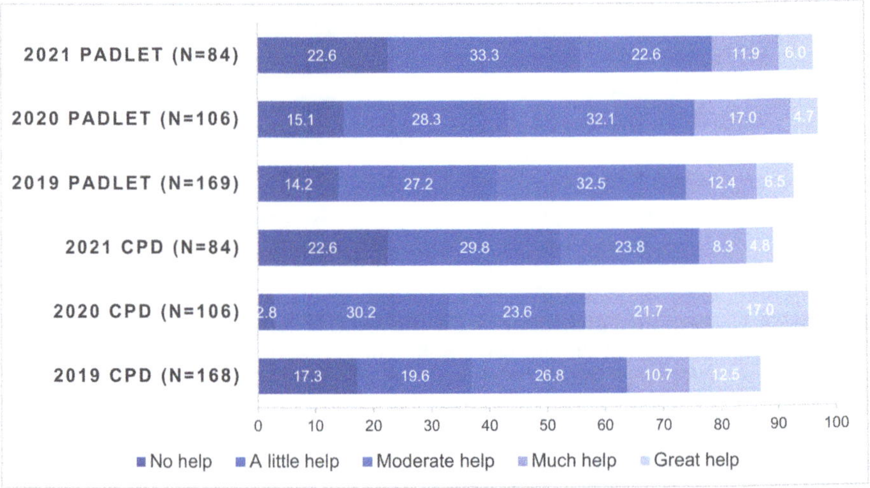

Figure 27.4 Response distributions for the SALG Likert-style item 'HOW MUCH did each of the following ASSESSMENT activities HELP YOUR LEARNING in chemistry'. Collaborative peer discussions (CPD) and a Padlet drawing task in blended (2019 and 2021) and online (2020) versions of the course are displayed.

The data presented in Figure 27.4 revealed that only <25% of students who responded perceived that these two assessment tasks had more than moderately helped their learning. One 2019 student commented *'I didn't find the [CPD chat] or Padlet tasks helpful at all, just very time consuming relative to their benefit'*. Another suggested it would be better to have instructors explain rather than their peers: *'[CPD chat] felt like a waste of time. Would have learned more if it was finished with a video explanation of what was going on by someone who knew'*. This finding reinforces the sentiments captured in the earlier perception data (Figure 27.2) in which only a low percentage of students identified as actively engaging in self-regulatory feedback. There have been multiple examples in open response comments linked to the perception data in Figure 27.4 from the questionnaire to indicate that the blended learning tasks had supported self-regulatory processes for many students, for example:

> 'There should be more [CPD chat] discussions for every topic because it really motivates me to learn the concept and also understand it enough for me to be able to explain it to another student'. [2019 student]

> 'The [CPD chat] discussions help solidify my understanding of topics since I was required to explain them to others'. [2019 student]

> 'The Padlet was great as even though if I got the answer wrong, I was forced to attempt the question and then see where I went wrong once the explanations released'. [2020 student]

Students' perceived value of peer feedback gained through the CPD task was also noted in the open comments in the questionnaire:

> 'I quite enjoyed the [CPD] sessions as it gave me a chance to discuss concepts with my piers (sic) so that I could further my understanding of them' [2019 student]
>
> '[CPD] discussions were extremely beneficial. It was a good way to link abstract concepts to real life. I found it good to discuss ideas with my peers through an anonymous way which helped with self-confidence.' [2019 student]

Finally, an increase in perceived value of the CPD task in helping learning in 2020 relative to other years (Figure 27.4) was observed in the data emphasizing the importance of student–student interactions in learning. The 2020 cohort had experienced the pivot into emergency response teaching due to COVID perhaps catalysing a greater need to engage and participate in peer interactions.

27.4.3 Further Insights on Feedback in Large Classes

It has been previously observed that students discriminate individualised (personal) feedback and generalised (whole class) feedback with the former having greater impact on their perceived value of feedback.[22] In this study, it appears that, regardless of task format, students tended to value individualised feedback more than generalised feedback and peer feedback was valued as lowest. However, many students recognised that it is difficult for instructors in large classes to be available to every student. Overall, it may be difficult to shift students' prior perceptions but there is no doubt that individual students are experiencing positive gains from different combinations of feedback, perhaps increasing their feedback literacy.

As a final thought, it is important to also consider that tasks that asked students to generate and share representations of their thinking provided the teachers with deeper insights regarding the diversity in student thinking. Teachers could reflect further on student outcomes and provide additional feedback during lectures to model explanations of correct thinking and point out alternate conceptions that arose. The feedback loop can then be closed as students apply these multiple sources of feedback at the next assessment opportunity.

27.5 Lessons Learned and Recommendations for Practice

The approaches described in this chapter have evolved across multiple annual cycles of design-based research in a single context involving intervention and evaluation, and as a result of this a picture has emerged. Our findings confirm that students apply a diverse range of approaches for seeking, engaging with, and acting on feedback to help their learning. Inclusive practice strategies widen access to participation in learning through delivery of feedback in multiple forms, modes and timing (both immediate feedback and allowing time for students to reflect on and act). Our students sought individual personalised feedback on their work and thinking, however they also recognised the challenges for instructors to achieve this in large classes. Several core recommendations for successful technology-enabled feedback practice have been developed based on these findings and are proposed in Table 27.4 below.

Carless and Boud[11] (p. 1323) suggest that technology-enabled feedback supports growth in student feedback literacy because it 'facilitates much-needed visibility for feedback processes, enhances students' responsibility by encouraging them to revisit feedback, and provides a means by which students can demonstrate to themselves and teachers that feedback loops are being closed'. Our chemistry education community needs more shared examples of chemistry-discipline specific, technology-enabled, effective feedback practice.

Table 27.4 Strategies for introducing feedback related to multiple interactions in online environments.

Learner–content interactions	Learner–teacher interactions	Learner–learner interactions
Provide immediate feedback where possible. This is highly valued by students and is an affordance of online learning.	Set expectations and scaffold feedback literacy skills through modelling to students how to act on feedback.	Include an online discussion forum where students can connect anonymously and exchange Q&A informally. Endorse and encourage peer support close to exams.
Avoid publishing numerical marks or grades in isolation from any feedback.	Provide generalised feedback where possible in contact sessions. Students are able to link this to their individual thinking.	Do not remove peer learning activities based on students' low value of them. Persist because students will benefit from peer feedback, even if it is not valued.
Use facilitative feedback, such as hints or prompts, that encourage students to persist in an activity or make another attempt at a task.	Provide multiple opportunities for students to seek in-person help from lecturers, tutors and peer leaders, online as well as in class.	Monitor and intervene if peer learning appears to involve incorrect thinking.
Use video explanations to increase the value of written feedback or worked solutions. Link to course resources and activities to encourage action.	Model "expert" thinking during explanations using digital visualisation and drawing tools.	

Acknowledgements

The contributions of the 2012 Office for Learning and Teaching Grant project team to the instructional design of the diagnostic instrument that underpinned the concept review modules is gratefully acknowledged. Funding of the 2019 transformation into a blended learning course through the University of Queensland UQ2U initiative is acknowledged. The contributions of students enrolled across multiple years in general chemistry for voluntarily sharing their perceptions data are also gratefully acknowledged.

References

1. G. Lawrie, A. Wright, M. Schultz, T. Dargaville, G. O'Brien, S. B. Bedford, M. Williams, R. Tasker, H. Dickson and C. Thompson, *Int. J. First Year High. Educ.*, 2013, **4**(2), 111.
2. G. Lawrie, A. Wright, M. Schultz, T. Dargaville, R. Tasker, M. Williams, R. Tasker, H. Dickson and C. Thompson, *Transforming Institutions: Undergraduate STEM Education for the 21st Century*, 2015, p. 399.
3. R. M. Bernard, P. C. Abrami, E. Borokhovski, C. A. Wade, R. M. Tamim, M. A. Surkes and E. C. Bethel, *Rev. Educ. Res.*, 2009, **79**(3), 1243.
4. M. A. Al Mamun, G. Lawrie and T. Wright, *Comp. Educ.*, 2020, **144**, 103695.
5. V. J. Shute, *Rev. Educ. Res.*, 2008, **78**(1), 153.
6. J. Hattie and H. Timperley, *Rev. Educ. Res.*, 2007, **77**(1), 81.
7. N. Winstone and D. Carless, *Designing Effective Feedback Processes in Higher Education: A Learning-focused Approach*, Routledge, 2019.
8. D. Boud and E. Molloy, *Assess. Eval. High. Educ.*, 2013, **38**(6), 698.
9. M. Yorke, *High. Educ.*, 2003, **45**(4), 477.
10. D. Nicol, *Assess. Eval. High. Educ.*, 2010, **35**(5), 501.
11. D. Carless and D. Boud, *Assess. Eval. High. Educ.*, 2018, **43**(8), 1315.
12. A. Espasa, R. M. Mayordomo, T. Guasch and M. Martinez-Melo, *Active Learn. High. Educ.*, 2019, 146978741989130.
13. A. C. Butler, J. D. Karpicke and H. L. Roediger III, The effect of type and timing of feedback on learning from multiple-choice tests, *J. Exp. Psychol.*, 2007, **13**(4), 273.
14. K. S. Taber, *Chem. Educ. Res. Pract.*, 2013, **14**(2), 156.
15. S. Padalkar and M. Hegarty, *J. Educ. Psychol.*, 2015, **107**(2), 451.
16. R. C. Jeske, J. A. Jones and C. L. Stanford, *In Active Learning in Organic Chemistry: Implementation and Analysis*, American Chemical Society, 2019, pp. 69–86.
17. G. A. Lawrie, M. Schultz, C. H. Bailey, M. A. Al Mamun, A. S. Micallef, M. Williams and A. H. Wright, in *Technology and Assessment Strategies for Improving Student Learning in Chemistry*, American Chemical Society, 2016, pp. 1–21.
18. J. Broadbent, S. Sharman, E. Panadero and M. Fuller-Tyszkiewicz, *Int. High. Educ.*, 2021, **50**, 100805.
19. F. Wang and M. J. Hannafin, *Educ. Technol. Res. Dev.*, 2005, **53**(4), 5.

20. M. Schultz, G. A. Lawrie, C. H. Bailey, S. B. Bedford, T. R. Dargaville, G. O'Brien, R. Tasker, C. D. Thompson, M. Williams and A. H. Wright, *Int. J. Sci. Educ.*, 2017, **39**(5), 565.
21. E. Seymour, D. Wiese, A. Hunter and S. M. Daffinrud, *In National Meeting of the American Chemical Society*, 2000, pp. 1–40.
22. A. Poulos and M. J. Mahony, *Assess. Eval. High. Educ.*, 2008, **33**(2), 143.
23. See Chapter 4: R. Iha, J. Luo, B. Lutes and G. Szteinberg, in *Digital Learning and Teaching in Chemistry*, ed. Y. J. Dori, C. Ngai and G. Szteinberg, Royal Society of Chemistry, United Kingdom, 2023.
24. See Chapter 24: R. Pullen, P. Rutledge, S. Schmid and S. Wilkinson, in *Digital Learning and Teaching in Chemistry*, ed. Y. J. Dori, C. Ngai and G. Szteinberg, Royal Society of Chemistry, United Kingdom, 2023.

CHAPTER 28

Introduction to Building Communities of Learners and Educators

M. H. TOWNS*

Purdue University, West Lafayette, IN 47906, USA
*E-mail: mtowns@purdue.edu

28.1 Building Communities of Learners and Educators during the Pandemic

The COVID-19 pandemic and ensuing rapid pivot to teaching online in March 2020 left many teachers and faculty scrambling. How were we to change the way we helped students learn chemistry from a face-to-face environment to an online environment? And with what technology? What options are available? Who in our network of chemistry teaching friends knows how to use them? What criteria do we use for choosing new tools in a matter of days? How do we teach laboratories in this online environment? How do we support student learning? How do we address equitable access to technology and not leave students behind? How do we create, sustain, and support a student community online? How do we support students' goals, hopes, and dreams through this troubling time? What are we going to do to meet these challenges and sustain our own mental health and wellbeing?

Perhaps part of the response to that formidable group of questions lies in the formation of communities of learners and educators, teachers and faculty

Advances in Chemistry Education Series No. 11
Digital Learning and Teaching in Chemistry
Edited by Yehudit Judy Dori, Courtney Ngai and Gabriela Szteinberg

Published by the Royal Society of Chemistry, www.rsc.org

members. Fortunately, the next three chapters focus on the formation of communities for learners and educators. There is no doubt that the formation of remote or online communities was thrust upon students and educators by the pandemic, and although we may have all been in the same storm (the pandemic) we were not all in the same boat. What is valuable in these three chapters is a thoughtful reflection upon how community was created that wisely draws upon the literature to guide choices or to analyze data gathered during the pandemic. All three chapters have insights for those eager to provide support—whether that is support in terms of learning chemistry at a distance through technology, or through content, pedagogical, and emotional supports by engaging in a community of educators with a common goal.

Additionally, each chapter addresses diversity and equity with a particular lens relevant to the author's context and country. What is held in common across the three countries of these author groups is a concern about and attention paid to, student access to the technological tools chosen to deliver instruction and support learning. There is evidence that educators turned to tools already well known to students such as Facebook or Discord.[1] However, to access even tools that are well known to students a device and reliable internet are required. Connectivity issues were rampant at the beginning of the pandemic, and remain so in many parts of the world, which only exacerbated educational inequities. At Purdue University there were efforts to support students who lived in areas without internet, and we had students driving to campus or local libraries to access the internet and do homework, quizzes, exams and access course materials while sitting in their cars or trucks. To note that these issues were a challenge to learning and student success is an understatement and there's little doubt that they differentially impacted students.

How were such challenges addressed by the chemical education community? The next three chapters provide information and insight into creating community for students and educators.

28.2 Using Technology to Build Community

Fung and Kim[2] from the National University of Singapore focus on building community with students and focus on student engagement in an online environment. One of the most pressing questions that immediately emerged in early 2020 was what technological tools to use. The choice of tools turned out to be critical in creating and sustaining community among students. To the great benefit of the chemical education community, Fung and Kim introduce and apply salient theoretical ideas from Kurt Lewin's field theory[3] and Wood's[4] notion of "behavioural friction" to this choice.

Kurt Lewin's field theory can be used to evaluate tools or platforms before they are adopted and implemented in a classroom setting.[3] The five elements are:

- Interactional synchronicity;
- Familiarity, convenience, and accessibility;

- Anonymity/pseudonymity;
- Distractions to learners;
- Instructional design and other miscellaneous features.

The chapter helpfully explains each element or criterion and relates it to other important aspects of building community online and influencing engagement. Wood's concept of behavioural friction is then brought in to help conceptualize how a person's actions and motivations may be restrained and/or constrained within this environment which also has a cultural context in which the student operates. Thus, actions may be restrained by behavioural friction meaning that there are forces which may impede a given action that a student could take. That "force" could be new technology, the difficulty in using the tool, or even the lack of desire to turn on a camera when joining a class. Thus, the activation barrier to using new technology to interact with peers or the instructor has been conceptualized by Wood as behavioural friction.

Student engagement is a prerequisite for building an online learning community. Fung and Kim note that significant research has been carried out that connects perceived psychological safety to student online participation.[5] Student engagement is also tied to the user experience and thus plays a key role in online learning. Importantly, they include ways of conceiving of the interaction between student engagement and platform effectiveness and note that when the platform is ineffective (*i.e.*, difficult to use, non-intuitive, has performance issues, *etc.*) and not optimized in a way that promotes engagement then peer interactions suffer.

Melding these theoretical perspectives together and leveraging the research literature provides new insights into not only choosing tools, but in considering the types of barriers to use conceptualized as behavioural friction. The chapter concludes with a brief evaluation of types of digital platforms using the criteria and ideas discussed in the chapter. It is a fresh and enlightening perspective on building student community as we continue to move forward in the pandemic, in an environment that has seen the biggest disruption to education in decades.

28.3 Building Community With Educators

In terms of building community with educators—teachers in secondary schools and faculty at the tertiary or university/college level—there are two chapters that report research related to the formation of community during the pandemic. The first is by Shauly, Shwartz, and Avargil[6] from the Technion—The Israel Institute of Technology and focuses on professional learning communities of teachers in secondary schools in Israel. The second is by DeKorver and Herrington[7] from Grand Valley State University in the United States of America and focuses on building a community of practice composed of university and college chemistry faculty. Both chapters rely heavily on theory to guide their research and in doing so provide well-grounded studies that illuminate actions taken to form community during the pandemic.

Shauly, Shwarz, and Avargil describe professional learning communities (PLCs) and their role in supporting teachers in Israel.[8] A PLC provides support to educators who teach a common subject by providing a framework to meet for discussions and the sharing of materials and methods to help students learn, thus improving their content knowledge, pedagogical content knowledge (PCK), student learning, and assessment techniques. In the study there are approximately 120 teachers, with 25 comprising a PLC of leaders, and 55 who participated in PLCs that functioned closer to their home area. Across this group there were teachers embedded in the PLCs who conducted their lessons in Hebrew and Arabic, which provided a different type of diversity lens to consider the transition to remote teaching and is reflective of the school system in Israel. The groups were set up to be longer term associations which helped teachers to develop and try practices, provided social and emotional support and reduced feelings of isolation, and provided a space to share and to create materials, techniques, and assessments.[9,10]

The authors use the literature to describe the nature of a PLC which is a longer-term grouping that provides support as teachers discuss innovations, challenges, how to teach certain concept, and methods of assessment. The authors also note that what happened in 2020 was not distance learning which is characterized by longer planning times and more thoughtful and reflective development of curricula. They specifically characterize the pivot that occurred in March 2020 as "emergency remote teaching." The challenge the teachers in the PLC faced were similar to many other teachers and faculty who reconceived teaching, learning, and assessment in a remote environment. Because they were teaching chemistry, there was the additional challenge of considering how to continue implementing and conducting laboratories in a remote environment. New technological tools needed to be assessed, learned, and implemented to support student learning and to conduct classes. As in other countries, there were questions of equity and access around technology which emerged. Finally, there was the challenge of sustaining mental health and well-being and remaining positive in an uncertain environment where personal and professional life became co-located during lockdowns. Adapting to this reality caused teachers to reach out for support from colleagues through the PLCs and through social media as documented in the chapter.

The chapter explores the challenges secondary teachers faced, it compares teachers work with the routine before the pandemic, and it describes the support aspects of the PLCs. One of the strengths of the chapter is the literature that the authors use to frame and inform. Beyond the rich discussion of PLCs, there is a discussion of the Refined Consensus Model for PCK[11] and how it can be used to further inform the analysis of the data and the findings that emerged.

Ultimately this study helps the chemical education community to understand the value of PLCs to teachers. Through analysis of a questionnaire and semi-structured interviews across the PLCs including teachers who spoke different languages, and teachers with a variety of years of experience in the classroom, the following findings emerged. Certainly, the critical role of

PLCs in supporting and sustaining Israeli teachers during the ERT was established. The community and network allowed teachers to turn to one another to further develop their content knowledge and PCK, it served as professional development, and it provided much-needed social and emotional support. They note that this was true for both Hebrew and Arabic-speaking teachers in the PLCs. This observation points to the importance of community as likely to be universal, in any culture. One of the unique solutions to the issue of continuing laboratory during the pivot to remote teaching came from the Israeli Ministry of Education, which shared an archive of recorded experiments. Given that many teachers and faculty around the globe recorded experiments and demonstrations to substitute for hands-on learning experiences, coordination at the national level through the Ministry of Education was a different solution than was found in other countries such as the US which lacks such a coordinating entity.

DeKorver and Herrington from Grand Valley State University in the United States provide the final chapter in this set. Their focus is on exploring the features of the strategies for teaching chemistry online (SFTCO) Facebook group that was created in March 2020 to support chemistry educators at the undergraduate level (primarily). This chapter extends their initial report on features of SFTCO after it had been in operation approximately two months.[1]

In the initial moments of formation, the goal of this group was to provide professional development and support for faculty engaged in the rapid pivot to online learning. DeKorver and Herrington note that there is a vast amount of literature on the characteristics of professional development for teachers and Herrington has a long history of engagement and research around the professional development of high school teachers.[12,13] In a review by Darling-Hammon, Hyler, Gardner they identified seven characteristics of effective professional development.[14] The experiences are content focused, incorporate active learning, support collaboration, use models of effective practice, provide coaching and expert support, build in time for feedback and reflection, and are of sustained duration such that faculty may have time to learn, practice, implement and reflect upon the new strategies that may change learning in their classrooms. A review of literature focusing on undergraduate STEM education that is related to professional development (essentially change strategies) revealed a strong social component to those who successfully enacted reforms.[15] This means that the collaborative aspects and sustained duration wherein faculty would have time to learn, practice, make use of feedback, implement, reflect, and iterate all the while having meaningful discussion with colleagues creates a noticeable positive effect in the success of engaging in revised practices.

While the previous chapter used a PLC model as a lens for exploring community, DeKorver and Herrington use a community of practice (CoP) model to investigate the SFTCO.[16] In their chapter they note the differences between the two models drawing attention to the distinction that CoPs have voluntary membership and are centred on a shared domain that is of interest to the community.

DeKorver and Herrington use the Facebook posts and replies as data and analyzed them to determine in what ways the SFTCO met the criteria for a CoP, to characterize how the content of questions and resources has changed over time, and to determine how the group engaged on matters of diversity equity and inclusion. The results are detailed and give the reader a sense of how the analyses allowed the researchers to address their research questions. The discussion of equity and inclusion is very illuminating characterizing the challenges faculty faced with students who did not have access to the internet in the early days of remote teaching. Perhaps indigenous to the US educational system the challenges of addressing student accommodations stemming from the Americans with Disabilities Act are also discussed. While the SFTCO community discussed these issues and challenges the authors note that often the discussions did not lead to specific solutions.

This chapter also makes use of the concept of technological pedagogical content knowledge (TPCK) where technological knowledge is melded together with PCK.[17] The concept of TPCK is especially salient in the pivot to online learning. In each chapter in this set educators had to consider what technologies to use and how to best use them to support student learning. Perhaps before the pandemic instructors had to consider how to support students in using the learning management system, or an online homework system. However, during the pandemic nearly all the curriculum was moved to some sort of platform. Thus, the inclusion of TPCK in this study as a framing device to guide analysis of the Facebook posts was illustrative. It emphasizes in CER the importance of being aware of a wide range of theoretical perspectives that can be used to inform and guide a research project.

The chapter ends with lessons learned from this research project and notes that the SFTCO CoP was critical in providing ideas about content (especially free resources), and pedagogical, technological, and emotional support.

28.4 Conclusions

What stands out across all three chapters focused on the building of communities of learners and educators in the pandemic is the way in which each set of authors chose to use theory to inform their perspectives, research design, data collection, and analyses. Each chapter provides the opportunity to learn from these theories and improve our own research and practices. Kurt Lewin's Field Theory and Wood's notion of "behavioural friction" are incredibly useful theoretical perspectives for those evaluating technology or seeking to move department discussions of technology beyond personal empiricism. The framing of PLCs in the chapter by Shauly, Shwarz, and Avargil and CoP in DeKorver and Herrington's chapter provides readers with two ways of conceptualizing communities of educators. The use of PCK, the refined consensus model of PCK, and the TPCK across these two chapters provide readers with not only a wealth of theoretical perspectives and knowledge, but also with excellent examples of how these ideas were used productively in studies. Truly, the theories used in these three chapters are a feast for readers.

References

1. B. DeKorver, A. Chaney and D. Herrington, *J. Chem. Educ.*, 2020, **97**, 2825.
2. See Chapter 29: Y. Kim and F. M. Fung, in *Digital Learning and Teaching in Chemistry*, ed. Y. J. Dori, C. Ngai and G. Szteinberg, Royal Society of Chemistry, United Kingdom, 2023.
3. B. Burnes and F. Cooke, *Int. J. Manag. Rev.*, 2013, **15**, 408.
4. W. Wood, *Good Habits, Bad Habits: The Science of Making Positive Changes that Stick*, Pan Macmillan, London, UK, 2019.
5. L. Roberts and C. Rajah-Kanagasabai, *Australas. J. Educ. Technol.*, 2013, **29**, 612–625.
6. See Chapter 30: A. Shauly, G. Shwartz and S. Avargil, in *Digital Learning and Teaching in Chemistry*, ed. Y. J. Dori, C. Ngai and G. Szteinberg, Royal Society of Chemistry, United Kingdom, 2023.
7. See Chapter 31: B. DeKorver and D. Herrington, in *Digital Learning and Teaching in Chemistry*, ed. Y. J. Dori, C. Ngai and G. Szteinberg, Royal Society of Chemistry, United Kingdom, 2023.
8. L. Orland-Barak. Discussion: Teacher Learning in Community: Premises, Promises and Challenges, in *STEM Teachers and Teaching in the Era: Professional Expectations and Advancement in the 21st Century Schools*, ed. Y. Ben-David Kolikant, D. Martinovic and M. Milner-Bolotin, Springer Nature, Switzerland AG, 2020, pp. 53–61.
9. P. R. Hallam, H. R. Smith, J. M. Hite, S. J. Hite and B. R. Wilcox, Trust and collaboration in PLC teams: Teacher relationships, principal support, and collaborative benefits, *NASSP Bull.*, 2015, **99**(3), 193.
10. R. Waldman and R. Blonder. A sense of community in a professional learning community of chemistry teachers: a study of an online platform for group communication, in *STEM Teachers and Teaching in the Digital Era: Professional Expectations and Advancement in the 21st Century Schools*, ed. Y. Ben-David Kolikant, D. Martinovic, and M. Milner-Bolotin, Springer Nature: Switzerland AG, 2020, pp. 11–140.
11. J. Carlson and K. R. Daehler, The refined consensus model of pedagogical content knowledge in science education, in *Repositioning Pedagogical Content Knowledge in Teachers' Knowledge for Teaching Science*, ed. A. Hume, R. Cooper, and A. Borowski, Springer Nature: Singapore Pte Ltd, 2019, pp. 77–92.
12. D. G. Herrington, K. Luxford and E. J. Yezierski, *J. Chem. Educ.*, 2012, **89**, 442.
13. D. G. Herrington and E. J. Yezierski, *J. Chem. Educ.*, 2014, **91**, 1368.
14. L. Darling-Hammond, M. E. Hyler and M. Gardner, *Effective Teacher Professional Development*, Learning Policy Institute, Palo Alto, CA, 2017.
15. C. Henderson, A. Beach and N. Finkelstein, *J. Res. Sci. Teach.*, 2011, **48**, 952.
16. E. Wenger, *Communities of Practice: Learning, Meaning, and Identity*, Cambridge University Press, Cambridge, UK, 1998.
17. P. Mishra and M. J. Koehler, *Teach. Coll. Rec.*, 2006, **108**, 1017.

CHAPTER 29

Bringing Back Learning Communities in the 21st Century

FUN MAN FUNG*[a,b] AND YONGBEOM KIM[c]

[a]Department of Chemistry, National University of Singapore, Singapore; [b]Institute for Applied Learning Sciences and Educational Technology (ALSET), National University of Singapore, Singapore; [c]School of Computing, National University of Singapore, Singapore
*E-mail: fun.man@u.nus.edu

29.1 Introduction

In this chapter, we propose a framework for creating student communities using the revised community of inquiry (RCOI) developed by Shea *et al.* (2012)[1] with concepts from Kurt Lewin's field theory[2] and psychological safety.[3] The urgent shift to remote learning in the COVID-19 pandemic has impacted universities and institutions worldwide, and it has brought up more concerns about educational inequalities. Students from low-income families often face greater challenges transitioning to remote learning, such as limited access to suitable devices and learning environments.[4] Furthermore, since students' main struggles in a digital classroom include staying motivated and isolation from instructors and other students, the lack of academic and wellbeing support means that students feel isolated or emotionally stressed when transitioning to online learning.[4]

Advances in Chemistry Education Series No. 11
Digital Learning and Teaching in Chemistry
Edited by Yehudit Judy Dori, Courtney Ngai and Gabriela Szteinberg

Published by the Royal Society of Chemistry, www.rsc.org

While not an institutional or systemic solution, we propose that instructors use digital methods and social media to deliberately create safe learning communities to alleviate student isolation in a digital classroom. Peer support can then be provided to address issues such as self-discipline, study motivation, and physical and mental well-being.[4] An online learning community can support students emotionally and assist them in their learning, helping students and closing the achievement gap.[4]

In our evaluation of the best practices in building online learning communities for students, we will use Kurt Lewin's field theory, as well as the concept of psychological safety to encourage student engagement and participation, a prerequisite for an online learning community.[2,3]

29.2 Bringing Back Online Learning Communities

29.2.1 Introduction to Learning Communities

To define a learning community, we look to the Revised Community of Inquiry (RCOI) framework. While Garrison *et al.* (1999) defines a Community of Inquiry as one that specifies three critical presences (Cognitive, Teaching, Learning),[5] Shea *et al.* (2012)[1] found that there is an additional, necessary requirement for communication among students, especially in an online setting.[5] Shea *et al.* (2012) proposes there are four critical 'presences' in a learning community: social, cognitive, teaching and learning presence.[1] Social presence is established when students communicate meaningfully with peers and develop interpersonal relationships. Cognitive presence refers to a continuous re-evaluation of course content, where students ask questions and learn in a constructivist way. Next, teaching presence refers to the facilitation of discussion, activities and conditions insufficient to establishing online learning communities. Notably, online courses required an additional aspect of active participation, motivation, or self-regulation, which they coined 'learning presence'. We summarise the four presences in Table 29.1.

Interestingly, the findings by Shea *et al.* (2012) also suggest that the learning presence is the greatest predictor of course performance, which perhaps

Table 29.1 A summary of all four presences in a (Revised) COI.[1,5]

Presence	Description
1. Social[5]	The ability for students to communicate with each other as fellow people, establishing interpersonal relationships.
2. Cognitive[5]	The ability for students to construct ideas and meaning *via* discussion.
3. Teaching[5]	The support of students' learning *via* (1) designing and organising course content and (2) facilitating discourse amongst students by faculty staff.
4. Learning[1]	The self- or co-regulation of students to actively learn and participate in courses.

suggests that the element of learner motivation may be the most important to learners.[1] We also feel that it is a key step in overcoming educational inequalities, as it directly relates to self-discipline and study motivation—two key factors that hinder students from historically underrepresented groups.[4]

Furthermore, it also appears to us that that the additional *learning presence* has only come to light in the context of online learning—indicating that learner motivation is a far greater challenge in online courses, and Chapter 3 in this book explains some of these challenges.[6] It seems to be insufficient to simply transfer lectures, tutorials and resources to an online LMS—the authors of Chapter 4 in this book describe some adaptations in their introductory chemistry course.[7] We posit that doing so simply corresponds to a superficial fulfilment of teaching and cognitive presences, but not learning nor social, as will be discussed in the next section.

29.2.2 Field Theory and Behavioural Friction

The basis for the evaluation of practices to create online learning communities will reference Kurt Lewin's field theory.[2] Lewin asserts that the behaviour of a person is determined by the person themselves and their surrounding cultural environment—that environmental cues act as contexts, or invisible forces on actions, restraining or driving behaviour.[2] Wood (2019) further expands on this with the concept of "behavioural friction", a term that refers to restraining forces that impede a given action.[8] Put simply, Wood states—make something inconvenient, and people will not do it. Conversely, to encourage behaviour, we should make the action 'frictionless'.

Wood's (2019) concept of behavioural friction is a powerful effect that shapes student behaviour. For instance, students in housing blocks tended to form friendships with neighbours, but did not take the trouble to do the same with those further away.[8] In a digital classroom, where students lack a common physical space to coincidentally run into peers, proactively reaching out to connect with peers represents a large increase in behavioural friction, resulting in little peer interactions and student participation. An example of this is when Danjou (2020) used private Facebook groups for their students, but noted a significant lack of student interactions, with some students remaining isolated throughout the academic semester.[9]

In summary, it is insufficient to simply know that students can send messages to each other; educators need to optimise the environmental cues around the actions such that students are most likely to be engaged. To the student, this 'user experience' determines the friction of using the classroom—a make-or-break factor in establishing an online community.

29.2.3 Psychological Safety

Depending on its features, a digital classroom can create an environment with either low or high psychological safety.[10,11] High levels of psychological safety is a pre-requisite for student engagement and participation, as it represents "one's emotional ability to express oneself in a virtual community without fear of negative consequences in relation to well-being, self-image, and status".[3] Online environments may inadvertently undermine psychological safety given the permanence of posts broadcasted to an entire cohort of students.[12] Roberts and Rajah-Kanagasabai (2013) found that the strongest predictors for student participation in an online forum are perceived psychological safety, self-consciousness and fear of negative evaluations.[11] It is also worth noting that both self-consciousness and the fear of negative evaluations are linked significantly to psychological safety, making it likely the most important factor in student participation.[11]

29.2.4 Creating Digital Classrooms for Learning Communities

When looking at potential environmental contexts and cues for an online learning community, there are two key aspects: the social context (student engagement), as well as the platform and its features (platform effectiveness). Ideally, both faculty and peers should initiate and support active conversations for communities, and platforms should be chosen that minimise the friction related to student participation. Considering these two aspects for online learning communities in education today gives rise to Figure 29.1, which highlights the importance of both student engagement and platform features.

The two aspects—student engagement and platform effectiveness, are mostly independent (see Figure 29.1), and hence instructors need to consider

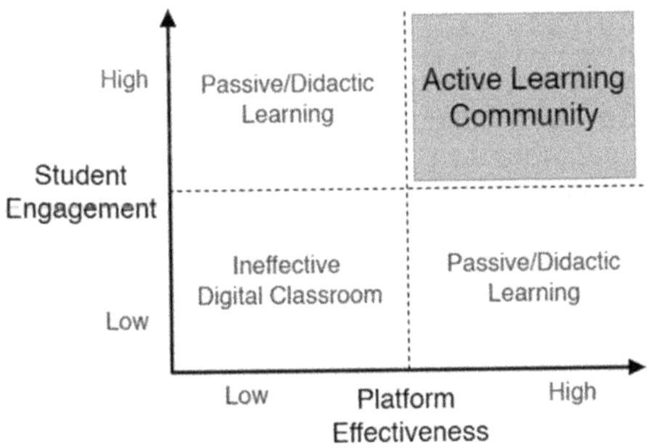

Figure 29.1 A figure of student engagement *vs.* platform effectiveness. The grey area represents the desirable outcome, where there is both a high level of student engagement from educators combined with effective platforms. The figure illustrates that the two are not competing but are independent and support each other to create a community. Without these, the classroom risks becoming didactic, or ineffective at conveying its material to students.

both aspects in developing an effective online community. Educators need to shape the online environment to one rich in discussions and inquiry, and this is complemented by platforms providing a frictionless experience to encourage the behaviour.

29.2.5 The Educator's Role in an Online Learning Community

First, educators have the most important role to fill in forming online learning communities. They initiate the participation and interactions within the space and have a significant impact on the psychological safety and online culture of the community. Shea (2006) found that students felt a stronger sense of community when instructors had a greater teaching presence, and actively created a risk-free environment by diagnosing misconceptions *via* discourse and affirming student contributions.[13] Joyner *et al.* (2020) further affirms the importance of the instructor, stating that an anchor for initial engagement, such as a shared goal for activity is important in forming student social interactions.[12]

The abovementioned anchor for initial engagement is paramount when getting a new student community off the ground. While an active student learning community could theoretically be self-sustaining in student participation, educators need to facilitate initial student engagement, for instance in providing a shared goal or activity, which is instrumental in building a community of learners—Chapter 3 in this book touches on this as well.[6] Furthermore, students also need a constant reminder to participate, without which the community risks becoming inactive, which further discourages participation. While group work and group projects are one way of generating initial engagement, in our experience, the best way is to generate discourse with open-ended discussion questions and engagement from peer tutors, inspired by the supplementary instruction (SI) model.

Alden (2017) describes the evolution of the SI model, which, at its core, is a programme targeted at historically difficult courses. In this program, peer tutors, called SI leaders, organise additional SI sessions where they provide students with problems and work to facilitate and prompt discourse on a topic.[14] Alden applied the model to an online classroom on Slack, Piapp and Facebook, observing the formation of peer study groups and support systems for the next few semesters.[14] In our CM1121 organic chemistry module, we engaged students with weekly questions for discussion on Discord, and observed healthy discourse among the students. In the end, we accumulated over a thousand messages sent over a 13-week duration, with even students coming up with their own questions for discussion.

Another way to provide the initial engagement is to use features such as quiz bots to give instant feedback as students interact. Ong *et al.* (2021) used a Telegram quiz bot on an environmental chemistry module to facilitate student interaction and discourse.[15]

In addition to providing the initial anchor for student engagement, instructors and tutors should also help students feel comfortable participating by affirming student contributions, as well as initiating and facilitating discussion and discourse by taking a constructivist approach. While the length of this section is relatively short, the importance of the educator, or even student tutors, to engage students in an online community cannot be understated—in our experience, it is unequivocally a prerequisite for any activity in an online space. We suggest that educators or tutors prepare special, open-ended questions to prompt discussions, remind them to log on frequently, and avoid answering queries by students immediately, if possible, instead directing them to the student body as an opportunity to facilitate discourse instead.

29.2.6 The Choice of Platform in an Online Learning Community

Beyond the direct role of the educator in enabling discussions, educators also need to choose platforms to best encourage participation, by increasing psychological safety and minimising behavioural friction in posting, messaging and connecting with peers. In doing so, we have identified five key features of platforms before they are adopted into a classroom setting.

The criteria are: (1) interactional synchronicity; (2) familiarity, convenience and accessibility; (3) anonymity/pseudonymity; (4) distractions to learners; and (5) features for instructional design. With the consideration of these features, we are sure that instructors can choose platforms best suited for their classroom.

29.2.6.1 Interactional Synchronicity

We follow the definitions of synchronicity given by Joyner *et al.* (2020).[12] Interactional synchronicity is a form of synchronicity where students can interact real-time with their peers, through elements such as messages, voice or video conferencing. This is in contrast to cohort synchronicity, where students simply follow the same cohort schedule.[12] From this section onwards, all mentions of synchronicity refer to interactional synchronicity.

Synchronicity is a pre-requisite to forming online student communities,[12] and any complementary design elements can be immensely helpful. It is worth noting that most social media platforms are synchronous, by the strictest definition—posts on forums are visible in seconds. But such platforms feel completely different from sending messages on instant-messaging platforms which *highlight* the presence of people on the other side of the screen. Notably, its importance is illustrated in increasing psychological safety and decreasing the behavioural friction associated with connecting with fellow students.

First, synchronicity increases student psychological safety by imitating the transient nature of real-time communication.[12] In a typical asynchronous online forum, students' fears of judgement and negative evaluations by their peers are amplified by the very nature of the forum. Messages are logged and often split into threads, which makes them easily searchable. Students, then, will be much more careful of their posts—avoiding mistakes, and often posting entirely to avoid (risking) attracting negative evaluations by the class, and even the instructors.

Synchronous platforms alleviate this problem by allowing older messages to be 'buried' by new messages, decreasing the likelihood of someone digging up a past mistake. To us, this also seems to be a good compromise: keeping older queries for potential references in the future, while also providing students a sense of comfort in the inconvenience of digging them up.† Then, especially in an active community, students will feel safer with the knowledge that their posts will not be under their peers' scrutiny indefinitely.

Various other synchronous features can also help students ease the difficulty in communication as well. Features that underscore synchronicity, such as the ability to see who is online, active and/or typing are key user features in such platforms. Going beyond the impermanence of messages, these features make it easier for students to connect with their peers, as they can instead focus on the few people online now instead of a big group of unknown user profiles. This highlights the presence of other students in the classroom, and these individuals also have a much higher chance of responding quickly, making it easier to connect. In short, in choosing a platform to host an online learning community, interactional synchronicity is a highly recommended prerequisite, and the best platforms have additional features to support real-time communication by highlighting online and active students in the moment.

29.2.6.2 Familiarity, Convenience and Accessibility

Next, platforms should minimise the behavioural friction to log on and participate in discussions—they can do this by being familiar, convenient and accessible.

The best platforms to use are ones that have already established themselves in the student population. For example, instant messaging platforms that students use and check every day to communicate with others. This is important, since students will likely have a habit of using the platforms regularly, significantly boosting student participation. For instance, if a learning community is hosted on Telegram, an instant messaging service, students who use Telegram regularly will inadvertently check said learning community as well and are more likely to respond to discourse regularly.

The alternative to this is platforms that actively remind students of discussions taking place in real-time. Instant messaging platforms do this by pushing notifications onto phones, but these reminders can come in

†Of course, tools such as searching come in handy here. However, we are mainly concerned with the 'friction' of digging up the messages, which becomes much higher.

many different forms, such as email notifications, or even summaries of recent discussions to contribute to, which allows platforms to become more convenient—students need not proactively check discussions to participate.

Without piggybacking on students' existing habits or regular notifications that helps students check into discussions daily, students only log into platforms out of obligation, perhaps to check for course updates and announcements and most may not participate in forums at all. This means that the number of students online at any given time will be far fewer, likely insufficient to sustain any form of conversation. Familiarity and convenience of social platforms is a clear prerequisite to sustaining student activity on social platforms, and by extension, an active community of inquiry.

Finally, the platforms should be convenient and accessible for students, as these qualities will significantly decrease the behavioural friction associated with using the platform itself. One consideration may be whether platforms can be used on multiple devices including smartphones, allowing students to participate on the go, as well as receive live updates and notifications. The mode of notifications themselves is also paramount. Live notifications on phones may prompt students to check its source immediately, but notifications in the form of emails may mean students only check them once a day or so—which may or may not align with the educators' goals. Platforms, as well, should not have any stability or performance issues. While this does seem like a low bar for any application, many students today on computers with lower performance may be dissuaded from participating in or using the platform at all if encountering such issues—potentially exacerbating inequity issues for some students. If this is a significant concern, perhaps lightweight applications such as messaging apps that take up less bandwidth may be best-suited for the particular course. To encourage students to use platforms, they should be able to log into and use the platforms frictionlessly.

Therefore, educators need to consider a platform's familiarity, convenience and accessibility to students before deciding on one. While these seem like relatively unremarkable points, they could have a significant impact on the level of student participation in an online learning community. Educators should look to ensure that it is easy, convenient, or already a habit for students to log onto these platforms, as well as consider qualities such as availability on devices, performance and stability before making a decision.

29.2.6.3 Anonymity/Pseudonymity

Whether a platform is anonymous or not also has a significant impact on student participation, since anonymity is significantly associated with perceived psychological safety by students, allowing students to communicate more openly and express feelings and views on sensitive issues.[11] It encourages students to ask questions, clarify expectations and results in increased participation from students who are not active communicators in the classroom, with students expressing that anonymity/pseudonymity "freed themselves from others' eyes and evaluations" and "from making mistakes in front of others".[10]

Furthermore, anonymity in student platforms has other beneficial side effects. Notably, there is no significant difference in the likelihood of responding to a message nor the perceived credibility of a message with respect to anonymity.[11] It has also been posited to help students focus on the content of posts, rather than its author.[10]

The detrimental effects of identification on psychological safety can also explain, among many other reasons, why many institutional student forums and learning management systems (LMSs) have little student participation except when such activities contribute towards their academic grade. Student profiles are often present for both students and lecturers, and the loss of anonymity represents a risk of judgement by both peers and faculty. Especially for students from historically underrepresented groups, anonymity is also a way to remove their marginalised personas during discussions, where the emphasis is on content, not the authors.

Despite its advantages, full anonymity on online forums also has its downsides. It potentially enables a wide range of negative behaviours from students such as negative, disruptive postings and attacks on other students without repercussions. Educators may even find great difficulties in implementing anonymity, especially in courses that engage in learning analytics, where online behaviour is tracked and analysed. As such, anonymity may not be ideal for online learning communities.

In practice, the compromise that educators can opt for is pseudonymity, where every student can choose a pseudonym such that the user remains anonymous. Most social media platforms support such a feature, where they allow for the creation of fictitious profiles, but pseudonymity is likely best supported by platforms such as Discord that actively encourage nicknames, or student forums built by institutions themselves, where such features can be added.

Educators who engage in learning analytics could even associate students' names to their pseudonyms for learning analytics, either by a back-end authentication solution or by simply asking the students to identify themselves privately, which will still preserve psychological safety between peers. Furthermore, in practice, platforms like Reddit and Discord have specific and strict ground rules that if breached, would result in messages being removed, warnings and eventually being banned from the community itself. In our previous experience using Discord in our CM1121 organic chemistry course, while we saw little negative behavior from students, we were fortunate to have a team of moderators to constantly moderate the activities on Discord. Even current students could act as moderators, alerting the instructor if discourteous messages are shown.

While anonymity or pseudonymity is not a prerequisite for an online learning community, it is a definite consideration for educators in their choice of media platforms as having the feature could potentially be a silver bullet for the problem of psychological safety in a digital classroom. Lastly, it may also be worth noting that anonymity may hamper students from forming connections.

29.2.6.4 Distractions to Learners

Distractions in social media platforms have always been a key concern for their adoption in the classroom, and for good reason. A platform to facilitate collaborative learning and learning communities should not lead students away from the relevant learning materials with a single click or tap. In short, a platform chosen for the classroom should allow for a separation of students' social and educational personas.[14]

Different social media platforms will do this to different degrees. For instance, Instagram and TikTok blend all content into a singular feed, and the user has little control of what is shown. While this blend may serve well for the purposes of science communication and initial engagement to the topic, it also means that learners will constantly face the urge to check feeds and click on other distractions. For instance, Alden (2017) used Piapp and Facebook, which has a singular news feeds for private groups and thus allows the instructor to keep the learning materials separate from other non-educational content.[14] Platforms such as Discord, Reddit, and Slack also provide clear compartmentalisation, using different 'servers' or 'workspaces', which serves to separate the educational environment from social or entertainment spaces.

This criterion, notably, is one that educators will have to balance with familiarity, convenience and accessibility (Section 29.2.3.2). It is understandable that many students use social media for entertainment and extensions of their social persona, which then serves as distractions as students are tempted to check their latest news feeds. Nevertheless, instructors should consider avoiding platforms that mix content from different sources freely without any user control, thus avoiding such problems—though, this may be dependent on the type of community and assessments instructors wish to set up.

29.2.6.5 Features for Instructional Design

Other than the five features and criteria mentioned before, educators also should consider other features that can significantly help a student learning community or consider including these features in their student forum. Every platform has its own features that shape the nature of its interactions, and by extension, any online classroom on it. These features should be considered in relation to the course and its pedagogical tools, as well as their potential in instructional design, or shaping the learning experience for students.

One key example of this is the "Spoiler function" in Discord, Telegram and other discussion boards. The function allows students to censor images and text in messages posted in the chat, which can then be revealed by clicking on the message itself. In our Discord server for students taking an introductory organic chemistry module, the feature was vital, allowing students working through the problem more slowly to solve and discuss the problem in the chat without accidentally seeing their peers' solutions. This resulted in constant discussions around every problem, where students can join in any time to solve problems before looking at their peers' solutions.

As different platforms have different quirks and unique features, educators should be on the lookout for the potential of instructional design in any such features, or even consider its implementation in their own student forum. Many seemingly minor features can fundamentally reshape an online community. For instance, karma or reputation-based platforms such as Reddit, where users are rewarded by 'likes' or 'upvotes' may incentivise students to post for 'reputation'. Gamified platforms such as Coursemology incentivise student participation with levels and experience points. Collaborative workspaces such as Miro Board that allow students to draw structures in real-time can immensely support an online organic chemistry classroom.[16] Bots in platforms such as Telegram can provide an endless source of engagement for student discussions.[15] Shared multimedia or digital tools can provide equitable resources and lab experiences remotely,[17–21] and we feel that educators can use the myriad of features available to shape their online learning communities.

29.2.6.6 Summary

Whether choosing a platform or building one to host an online learning community of students, educators have a plethora of considerations. While the criteria for evaluations outlined in this chapter provide a guideline, educators should keep in mind the fundamental ideas for increasing student participation: increasing perceived psychological safety, as well as minimizing behavioural friction for students to participate. When choosing/building a platform, educators should always keep in mind the course and their goals for the classroom environment and evaluate the platform in reference to them. For instance, chemistry courses that require drawing of structures may benefit from platforms that allow free-hand drawing, or gamification can promote participation, should the course be adaptable to such a system.

A summary of the suggested guideline in evaluating online platforms is shown below (see Table 29.2).

29.2.7 Evaluation of Different Types of Digital/Social Platforms

In this section, platforms that could be used in digital classrooms will be evaluated using the criteria set forth earlier. These platforms can be categorized loosely under 'social media'. Namely, we evaluate six different types of social media platforms—instant messaging, social networking, video conferencing, image sharing, collaborative workspaces and channel-based

Table 29.2 A summary of the suggested criteria, and their accompanying descriptions for educators to consider when choosing online platforms.

Criteria	Description
1. Interactional synchronicity	• Always choose live, synchronous platforms when possible. • Consider synchronous design elements: users online, typing, *etc.*
2. Familiarity, convenience and accessibility	• Use platforms that students are already using and check regularly or send regular notifications to students. • Try to choose platforms available on mobile phones. • Do not use platforms with stability or performance issues.
3. Anonymity/Pseudonymity	• Try to adopt platforms with anonymity/pseudonymity • Anonymity/pseudonymity can give safety to marginalised students.
4. Distractions to learners	• Never choose platforms where educational and other content mixes freely.
5. Features for instructional design	• Consider unique features in relation to the learning community and course.

messaging platforms. Both the categorisation of these platforms are our own opinions, and their evaluation is not backed by any empirical studies—they are simply our suggestions.

29.2.7.1 Instant Messaging

Instant messaging refers to apps such as WhatsApp and Telegram, often used as a replacement to short message services (SMS). These apps are synchronous with a single feed, and many students use these apps as a means of communication This means that they have pre-existing habits of using these apps, which makes them a suitable choice for digital classrooms. From our personal experiences, instant messaging platforms tend to have a more 'casual' atmosphere, which may be beneficial or detrimental, depending on the educator's preference.

However, it is also notable that some instant messaging apps may not allow for anonymous/pseudonymous users, and the ease of switching over to different chats mean that distractions still persist. Furthermore, apps such as Telegram allow for the implementation of bots, a potential tool for student engagement, as well as the ability to hide messages under 'spoilers' or attach images that can greatly lubricate discussions. Finally, if instructors are to participate in instant messaging groups/platforms, it also could mean that their personal data (*e.g.*, phone number, Telegram username) is shared with students. All in all, instant messaging apps are a suitable choice for digital classrooms, but due to their lack of anonymity/pseudonymity, instructors may have to play a larger role in the initial engagement of students.

29.2.7.2 Social Networking

Social networking platforms refer to platforms that allow users to interact with others and find people with similar interests, for instance, Facebook or Twitter. These platforms have been previously used before in higher education. For instance, Danjou (2020) used Facebook in a digital classroom by using its 'private groups' feature.[9] Social networking platforms encourage student engagement with their familiarity, synchronicity posts and notifications with their availability on phones. It may also be worth looking at the potential 'reactions' of the platforms, as positive affirmations in the forms of 'likes' or 'upvotes' may encourage students to participate more.

However, hosting a learning community on such platforms also presents its challenges. First, many social networking platforms have little means to filter or compartmentalise content, and at best, distractions are only a click away. Furthermore, as most students have pre-existing profiles on these platforms, pseudonymity is not an option, potentially lowering students' perceived psychological safety. As such, we note that while social networking platforms are certainly an option, instructors need to play a much larger role in increasing psychological safety and student engagement before an online learning community can be formed.

29.2.7.3 Video Conferencing

Next, there are video conferencing apps, such as Zoom, Skype or Unhangout that many digital classrooms employ. In the context of forming online learning communities, the most glaring limitation of such platforms is the lack of interaction—synchronous or asynchronous—outside of scheduled lesson or tutorial times and are insufficient in sustaining student communities. It is also worth noting that video conferencing platforms may exacerbate the educational inequalities that arise due to a technology gap—students may not have equipment such as webcams for the sessions, they may not be able to find a suitable place for online discussions at home, or their internet bandwidth may not be sufficient for the app. We suspect that these deficiencies can somewhat be alleviated *via* the use of a secondary platform: by sharing notes, clips of recorded lectures and encouraging meaningful discourse. However, digital equity still may remain a concern for instructors or institutions that are heavily reliant on conferencing platforms for lecture delivery, *etc.* Re-playable, asynchronous lectures have been a key strength in online learning,[7] and we believe that video conferencing should be supported by other platforms that allow students to message, ask questions and connect with each other.

29.2.7.4 *Image Sharing*

Image sharing sites such as Instagram and Snapchat, that facilitate sharing images and videos, have also popped up as a possible medium of instruction. For instance, Lim *et al.* (2017) used Instagram and Snapchat in laboratory teaching to allow students to access course-related content at their convenience, ultimately helping students to observe mistakes and understand the experiments better.[22]

While many of these platforms do have instant messaging features built into them, they are often overlooked and underused by students. Hence, such sites are incapable of supporting an online learning community by themselves, due to a lack of features for synchronous interactions beyond scheduled classroom times. Should such platforms be used, we believe that interactions using images and videos should be carefully designed by instructors to build rapport and prompt discussions.

29.2.7.5 *Collaborative Workspaces*

Collaborative workspaces have also emerged as solutions for digital classrooms, facilitating new types of digital student interactions. Google Drive and MS Office 365 applications allow students to work together, and Miro Board, a collaborative virtual whiteboard, allows students to draw shapes and molecules together.

However, many of these platforms do not support instant messaging features. While students can chat *via* comments, it could pose a considerable challenge for instructors to get students to log onto the platform regularly, especially given inconsistent notifications for comments. In our experience using Miro Board for our CM1121 introductory organic chemistry module, an alternative messaging platform was still needed for students to ask questions and have discussions. However, we also noted that Miro Board facilitated these conversations, as students were simply able to screenshot or link drawings to the platform, making it easier to communicate.

29.2.7.6 *Forums*

Lastly, forums, such as LMS forums, Discord, Slack, Microsoft Teams, Piazza and even Reddit have been used in higher education. These platforms often support a set of conversations in a private workspace, with persistent channels or threads of discussions as chat groups. Some let users call and message other students privately.

In the context of a digital classroom, these platforms provide the benefits of instant messaging and video conferencing platforms, as well as some other advantages. Multiple channels help to organise discussions on the platform, unlike those with single news feeds. Furthermore, since different communities are separated by the distinction of workspaces and servers, there is a separation of educational and social content.

Many of these platforms are available on phones, and some (such as Discord and Slack) support pseudonymity as well, with miscellaneous features such as "spoilers" elaborated on previously. However, one drawback is that both instructors and students may face a lack of familiarity with these platforms. Therefore, it is imperative that educators choose platforms that allow them to repeatedly remind students to participate, engage them regularly, and even integrate coursework. All in all, channel-based messaging platforms combine the benefits of many different platforms, making them largely suitable for most types of learning communities—provided that an adequate amount of student engagement is given.

29.2.8 Conclusion

In conclusion, creating online student communities is a big step instructors can take in their classrooms to alleviate educational inequalities in their classrooms. Establishing learning presence will be key to helping students stay motivated and disciplined in courses, and a student community can provide them peer support and supplementary help should they fall behind. This will also be key to supporting disadvantaged students, who may face more difficulties in their transition to a digital classroom.

To do so, educators need to maximise psychological safety and minimise the behavioural friction in participating, and they have two roles to play. First, it is imperative that they provide an anchor for student interactions—we recommend this *via* providing open-ended discussion questions and peer tutors to engage students. Second, they need to carefully consider the best platform to host their online classroom.

For platform choice, we identify five key features: (1) interactional synchronicity; (2) familiarity, convenience and accessibility; (3) anonymity/pseudonymity; (4) distractions to learners; and (5) features for instructional design. Out of the various platform choices, we generally recommend either instant messaging platforms or forums. They each have their pros and cons—the former allows for a more casual and potentially safer environment amongst students, while the latter is more flexible and allows for integration with coursework but requires greater student engagement. Again, this is up to the educator. A summary of our suggestions is available in Table 29.3.

We hope that our suggestions will not be regarded as the rule for online classrooms but inspire educators instead to design the best online classroom for their course.

Acknowledgements

The authors acknowledge Han Jia Yi for his time and effort reviewing our contribution.

Table 29.3 A summary of our evaluation of social media platforms for building online communities in digital classrooms.

Platform	Evaluation	Examples
1. Instant messaging	• Platforms have notifications and pre-existing users. • Promotes a casual atmosphere. • Little anonymity/pseudonymity. • Available on phones.	• Telegram • WhatsApp • GroupMe
2. Social networks	• Platforms have notifications and pre-existing users. • Reactions can promote participation. • May introduce distractions to students. • Available on phones.	• Facebook • Twitter
3. Video conferencing	• Difficult for students to interact after class hours. • May exacerbate the technology gap—bandwidth, camera, mic, *etc.* • Ideally should be supported by a secondary platform.	• Zoom • Skype • Unhangout
4. Image sharing	• Opens different modes of instruction and student engagement (*e.g.*, bite-sized content on Instagram). • Not built to facilitate discussions. • Ideally should be supported by a secondary platform	• Instagram • Snapchat • YouTube
5. Collaborative workspaces	• Different features (*e.g.*, drawing) can facilitate communication between peers. • Can support some discussion with thread/comment feature. • Students need constant reminder to access platform. • Ideally should be supported by a secondary platform	• Miro Board • Jamboard • Google Drive • MS Office 365
6. Forums	• Is the most flexible platform for different classrooms—lots of alternatives. • Chats are organized—easier to search through questions. • Easier to integrate with LMS/coursework. • Some support anonymity, "spoilers", synchronous elements for participation. • Students may be unfamiliar; need extra engagement/reminders to participate.	• LMS • Piazza • Discord • Reddit

References

1. P. Shea, S. Hayes, S. U. Smith, J. Vickers, T. Bidjerano, A. Pickett, M. Gozza-Cohen, J. Wilde and S. Jian, *Internet High. Educ.*, 2012, **15**, 89–95.
2. B. Burnes and B. Cooke, *Int. J. Manage. Rev.*, 2013, **15**, 408–425.
3. Y. Zhang, Y. Fang, K.-K. Wei and H. Chen, *Int. J. Inf. Manage.*, 2010, **30**, 425–436.

4. R. Raaper and C. Brown, The Covid-19 pandemic and the dissolution of the university campus: implications for student support practice, *J. Prof. Cap. Community*, 2020, **5**(3/4), 343–349.
5. D. R. Garrison, T. Anderson and W. Archer, *Internet High. Educ.*, 1999, **2**, 87–105.
6. See Chapter 3: F. M. Ho, K. Ibraj and N. Graulich, in *Digital Learning and Teaching in Chemistry*, ed. Y. J. Dori, C. Ngai and G. Szteinberg, Royal Society of Chemistry, United Kingdom, 2023.
7. See Chapter 4: R. Iha, J. Luo, B. Lutes and G. Szteinberg, in *Digital Learning and Teaching in Chemistry*, ed. Y. J. Dori, C. Ngai and G. Szteinberg, Royal Society of Chemistry, United Kingdom, 2023.
8. W. Wood, *Good Habits, Bad Habits: The Science of Making Positive Changes that Stick*, Pan Macmillan, 2019.
9. P.-E. Danjou, *J. Chem. Educ.*, 2020, **97**, 3168–3171.
10. T. Miyazoe and T. Anderson, *J. Educ. Technol. Soc.*, 2011, **14**, 175–187.
11. L. D. Roberts and C. J. Rajah-Kanagasabai, *Australas. J. Educ. Technol.*, 2013, **29**(5), 612–625.
12. D. A. Joyner, Q. Wang, S. Thakare, S. Jing, A. Goel and B. MacIntyre, in *Proceedings of the Seventh ACM Conference on Learning@ Scale*, 2020, pp. 15–24.
13. P. Shea, *J. Asynchronous Learn. Networks*, 2006, **10**, 35–44.
14. E. Alden, *J. Chem. Educ.*, 2017, **94**, 2007–2009.
15. J. S. H. Ong, P. R. Mohan, J. Y. Han, J. Y. Chew and F. M. Fung, Coding a Telegram Quiz Bot to Enhance Content Mastery during COVID-19, *J. Chem. Educ.*, 2021, **98**(8), 2699–2703.
16. B. J. M. Ng, J. Y. Han, Y. Kim, K. A. Togo, J. Y. Chew, Y. Lam and F. M. Fung, *J. Chem. Educ.*, 2021, **99**, 708–714.
17. See Chapter 19: S. Bernholt, J. Eckhard, M. Rodemer, A. Langner, G. Asmussen and N. Graulich, in *Digital Learning and Teaching in Chemistry*, ed. Y. J. Dori, C. Ngai and G. Szteinberg, Royal Society of Chemistry, United Kingdom, 2023.
18. See Chapter 18: D. Donnelly-Hermosillo, L. Gerard and M. C. Linn, in *Digital Learning and Teaching in Chemistry*, ed. Y. J. Dori, C. Ngai and G. Szteinberg, Royal Society of Chemistry, United Kingdom, 2023.
19. See Chapter 21: M. Milner-Bolotin and V. Milner, in *Digital Learning and Teaching in Chemistry*, ed. Y. J. Dori, C. Ngai and G. Szteinberg, Royal Society of Chemistry, United Kingdom, 2023.
20. See Chapter 20: R. Pullen, A. Motion, S. Schmid, S. George-Williams, S. Wilkinson and S. Leach, in *Digital Learning and Teaching in Chemistry*, ed. Y. J. Dori, C. Ngai and G. Szteinberg, Royal Society of Chemistry, United Kingdom, 2023.
21. See Chapter 16: J. Krajcik, in *Digital Learning and Teaching in Chemistry*, ed. Y. J. Dori, C. Ngai and G. Szteinberg, Royal Society of Chemistry, United Kingdom, 2023.
22. R. R. X. Lim, A. S. Ang and F. M. Fung, in *Teaching and the Internet: The Application of Web Apps, Networking, and Online Tech for Chemistry Education*, ACS Publications, 2017, pp. 37–53.

CHAPTER 30

Supporting Chemistry Teachers in Emergency Remote Teaching—The Role of Professional Learning Communities (PLCs)

A. SHAULY[a], G. SHWARTZ[a] AND S. AVARGIL*[a]

[a]Faculty of Education in Science and Technology, Technion—Israel Institute of Technology, Israel
*E-mail: savargil@technion.ac.il

30.1 Introduction

At the beginning of 2020, the COVID-19 pandemic began, and lockdowns were imposed in an effort to slow its spread. Schools were compelled to urgently divert to distance learning defined as emergency remote teaching.[1] This differs from other forms of distance learning, which are planned and may include face-to-face activities. Emergency remote teaching presented a variety of challenges associated with teachers' professional knowledge of learning, teaching, and assessment, as well as personal difficulties addressing the uncertainty imposed by the pandemic. Adapting to the new reality led teachers to seek support that enabled them to implement innovative teaching methods and tools.[2,3]

Advances in Chemistry Education Series No. 11
Digital Learning and Teaching in Chemistry
Edited by Yehudit Judy Dori, Courtney Ngai and Gabriela Szteinberg

Published by the Royal Society of Chemistry, www.rsc.org

One channel of providing support is through professional learning communities (PLCs) where teachers with a common background meet regularly for discussions and joint learning. By sharing classroom experiences and practices, this platform aims to improve teachers' knowledge, students' learning, and assessment of learning.[4] The relationships formed within PLCs differ from other professional development programs for several reasons: (1) the PLC format is a long-term program, enabling support for teachers during the implementation groundwork of new practices in class;[5] (2) relationships formed among teachers encourage mutual aid, improve teachers' wellbeing, and reduce feelings of isolation; and (3) the experience of all members creates shared practice and knowledge.[6–8]

This research was conducted in the setting of a PLC network in Israel that operates from the Technion—Israel Institute of Technology and seeks to support the development of chemistry teachers. This network includes two types of PLC: local communities of teachers called "Chemistry Teachers' Community" (CTC), and "Leading Teachers' Community" (LTC) that includes senior teachers who lead CTCs. The LTC is run by coordinators: senior teachers and academic personnel.[9,10] Therefore, these communities are comprised of various partners holding different positions.

The PLCs in this research include members who speak and teach in different languages, mirroring the diversity in the Israeli school system.[11] Studies of STEM teachers' PLCs in Israel showed that participation in PLCs influenced the professional connections among teachers, and set the groundwork for teachers to consult with their peers regarding class activities, enabling them to share their class experiences, and to receive support and recognition of their work.[9,10,12] The inclusion of teachers from various sections of the Israeli school system enabled us to gain insights into the influence of diversity within the PLCs on teachers' perceptions of the support offered by the community and its importance to them.

Since PLCs operate continuously, the research goals of this chapter are to understand how teachers' work changed compared to their routine before the pandemic and learn about the support aspects PLCs provided. The role of PLCs in supporting teachers coping with remote teaching during the COVID-19 crisis was examined from the perspective of PLC members and included a comparison to non-member teachers. The study considered ethnic differences among chemistry teachers in Israel. The inclusion of all these perspectives enhanced the understanding of PLCs' contribution to teachers during the crisis. Based on this knowledge, practical recommendations were derived to improve support and communication in teachers' PLC and to manage the introduction of educational changes more effectively, whether planned or in response to an emergency.

Thus, the research questions we asked were:

1. How did teachers work change compared to their routine before the pandemic?
2. What were the support aspects the PLCs provided?
3. How was the support within PLCs affected by diversity among members?

30.2 Theoretical Framework and Literature Review

30.2.1 The Refined Consensus Model for PCK

The refined consensus model (RCM) for PCK[13] was used here to describe the challenges and support aspects of chemistry teaching during emergency remote teaching. The RCM entails a description of knowledge components for science teaching, as well as the complex layers of knowledge and teachers' experiences in a particular setting. Three levels of PCK are included in the RCM: collective PCK, *i.e.*, knowledge shared among a group of teachers, personal PCK, which is attributed to unique knowledge of each individual teacher, and enacted PCK which is the knowledge used in class for planning instruction and carry it out to attain student outcomes.[13]

According to the model, the basis for teachers' knowledge includes knowledge of assessment, pedagogy, content knowledge of subject matter, students, technology, and curriculum. The component of collective PCK is shared accepted public knowledge for teaching of a specific scientific discipline. Collective PCK can either be established based on academic professional literature or be developed by educators and teachers in any educational setting.[13] Some examples include developing together a learning sequence for a specific science subject or a representation for a science content subject.

Personal PCK refers to the personal level, meaning that every teacher has his or her own unique knowledge, skills, and beliefs. In the RCM model, personal PCK, which is the unified knowledge and skills of each teacher, and enacted PCK, which represents the teaching practice in class are distinguished.[13]

The current study focused on three aspects of the RCM model: first, aspects of the model's basis, especially pedagogical knowledge dealing with class management, use of technology, assessment, and knowledge of students; second, the collective PCK shared by PLC members; and lastly, the contribution of the PLC activities to the personal PCK formation of community members.

30.2.2 Professional Learning Communities Frame and Characteristics

PLCs bring together teachers with a common background who meet regularly for discussions, collective learning, and reflection. This platform aims to enrich teachers' knowledge, improve students' learning, and assessment.[4] Personal relationships formed within PLCs distinguish them from other PD programs by encouraging mutual assistance and improving teachers' well-being.[6–8,14] PLCs can bring together teachers of the same school.[14] Other communities bring together teachers from different schools who teach a common subject.[7,10,12]

A PLC is based on a constructivist approach, in which members work towards common goals.[14,15] PLCs promote knowledge exchange, problem exploration, and creative solutions; this fosters a sense of commitment, accountability,[4] and collective learning.[14,15]

According to Wenger and colleagues,[8] there are two key characteristics to PLCs: "community"—a closed group focused on a shared identity and goals, and "network"—an open set of connections and relationships among people. Therefore, collaboration and building relationships is a feature unique to PLC activities. Knowledge is built in communities through planned structured activities and discussions within the group. The fact that the closed community includes people with similar views increases willingness to ask questions and share information.[16]

30.2.3 STEM Teachers' PLCs

Chemistry teachers' PLCs in Israel operate according to a network model, described by Levy *et al.*,[12] as the "fan model"—"teacher-leaders participate in a PLC led by a team from the academic institute, while they simultaneously lead regional PLCs of high-school physics teachers all over Israel."[12] The fan model is applied to the PLCs of chemistry teachers described in this chapter.

Studies of science teachers' communities in Israel included an investigation of the activities introduced in PLC meetings and changes in classroom practices, for example, the teachers' use of learner-centered approach practices.[10] These studies showed that PLCs influenced the professional connections among teachers, enabling a platform for consultation and sharing experiences from class and between meetings.[9,10,12] These relationships help reduce feelings of isolation among those who are the only science teacher in their school and give novice teachers an opportunity to build their professional identity.[10]

Another example is a program dedicated to teachers who teach research and design in the Netherlands' high schools;[17] its goals were to add to the knowledge concerning integration of research and design practices and advancement of teaching methods. Results showed that following the PLC discussions, teachers formed a shared view of the subject matter. Teachers prepared class activities, implemented them, and shared their experience with the PLC members. This advanced teachers' skills by enriching personal PCK and adding to the groups' collective PCK.[17]

30.2.4 Teachers' Challenges During Emergency Remote Teaching

The change to work in an emergency remote teaching mode during the COVID-19 crisis required simultaneous adaptations in all aspects of the teachers' work: from changes in pedagogy and assessment to practical difficulties of using technological tools and communicating with students.[1] Since emergency remote teaching was adopted globally, this crisis offered a unique opportunity to study this transition.

Challenges of chemistry teachers in Israel were studied by Rap and colleagues[3] during the first weeks of the crisis. The authors divided teachers' reported needs into three categories: the need to master the use of educational technical tools, the need to adopt different teaching and assessment tools, and the inability to conduct laboratory experiments and demonstrations. During COVID-19 crisis, teachers were also concerned with other issues such as equitable teaching and well-being of students.

Collaboration between teachers was facilitated by other online resources. Teaching materials and guidance for using remote teaching tools were found in Teachers' WhatsApp group conversations, in chemistry teachers' web sites, and open sources on the internet.[3] Facebook groups were used to build a repository of teaching materials according to members' request, and for development of on-line teaching materials.[2,18]

30.3 Methodology

In this mixed methods study, qualitative and quantitative data collection were carried out simultaneously but separately, and analyzed together to provide an in-depth interpretation of the data.[19]

30.3.1 Research Setting

The need for continuous learning by in-service teachers has led to the establishment of a chemistry teacher PLC network in Israel. A network arranged according to the "fan model"[12] started at the Technion, Israel Institute of Technology in fall of 2018. This network consists of the coordinators of the leading teachers' community (LTC) that seek to support the development of a chemistry LTC and monitor regional 'close to home' chemistry teacher communities (CTCs). The LTC's coordinators provide the leading teachers a model for leading the CTCs. The LTC and CTCs are tailored according to the unique professional needs of chemistry teachers.

Both LTC and CTC met biweekly, face-to-face or by Zoom during lock downs. During meetings, community members discussed high-quality, innovative methods of teaching of chemistry, and the challenges they presented, while empowering and strengthening their professional identity. They discussed common difficulties encountered in teaching and studying chemistry and collaborated to suggest ways to address them in class. Teachers share a common vision of providing students with meaningful learning, deepening conceptual understanding of chemistry, enhancing students' positive attitudes towards chemistry, and increasing their motivation.[9]

In CTC meetings, leaders presented newly developed activities introduced in the LTC meetings, community members shared experiences from their classrooms for the benefit of others, presented their needs, and discussed pedagogical issues.[4,9]

30.3.2 Research Tools

In this study we used two research tools to collect data: interviews, and a questionnaire. These research tools were selected in order to learn the perspectives of PLC members and non-members on their remote teaching experience during lockdowns, in comparison to routine work, and the support they needed and received.

30.3.2.1 Interviews

Semi-structured interviews were conducted with the following groups of participants: coordinators of the LTC, LTC members, CTC members, and chemistry teachers not participating in a PLC. The interviewees' group was selected based on availability and reflects the diversity of communities' members. It includes participants from the different chemistry teachers' communities, two Arab speaking teachers, teachers with various teaching experience and age, and coordinators both from academia and high school. Although the questions were planned, we maintained flexibility and enabled variations and personalization of the interview protocol.[20] All interviews were audio-recorded, transcribed, and qualitatively analyzed. The goal of the interviews was to: (a) characterize the teaching process in class *vs.* teaching in emergency remote mode, (b) identify challenges teachers encountered, and (c) examine the role of PLCs during the crisis.

30.3.2.2 Questionnaire

A questionnaire was developed to understand teachers' strategies for coping with emergency remote teaching and the challenges they faced. It included four sections: (1) participants' demographic and personal data; (2) pedagogical aspects concerning teaching, assessment, and maintaining personal relationships with students; (3) technological challenges, and (4) characterizing teachers' work environment. Closed- and open-ended questions gave responders a chance to reflect on the 2019–20 school year. Face validity was achieved by consulting with senior teachers and science education researchers, and some items were rephrased to achieve better clarity. The questionnaire was electronically administered to chemistry teachers in Israel between November 2020 and February 2021. Statistical measures such as internal consistency and analysis of factors were carried out and are reported in the following sections of this chapter.

30.3.3 Research Participants

Research participants were comprised of different partners.

- LTC coordinators—these are experienced chemistry teachers who also work in academia, involved in research concerning chemistry education, or as teacher supervisors in the Ministry of Education.

- LTC that includes 21 members—most with more than seven years of teaching experience. The community joins members from various factions of the Israeli society, a third of them Arabic speakers.
- The Haifa CTC community includes 23 members—two of them are Arabic speakers and their teaching experience ranges from three to over 31 years of teaching.

The questionnaire was answered by 122 teachers. Most responses, 67.2%, were from PLC members. The other responses were from teachers who were not PLC members. In the questionnaire, personal data such as years of experience in teaching, age, and society, were collected. Teachers from the Jewish community, described as "Hebrew speakers", represented 57.4% of respondents. Arab and Druze teachers described as "Arabic speakers", were 44.6% of respondents. Teachers' experience ranged between two to over 40 years of experience. Figure 30.1 shows the distributions of teaching experience in years based on Huberman,[21] and the teachers' age distribution.

30.3.4 Method of Analysis

30.3.4.1 Thematic Analysis

Thematic analysis was used to analyze the texts of transcribed interviews and answers to open-ended questions in the questionnaire. This process allowed for a sensitive, insightful, and rich exploration of the text, exposing structures and underlying patterns. Our work followed the steps detailed by Braun and Clarke.[22] After an initial review of the responses to the open questions,

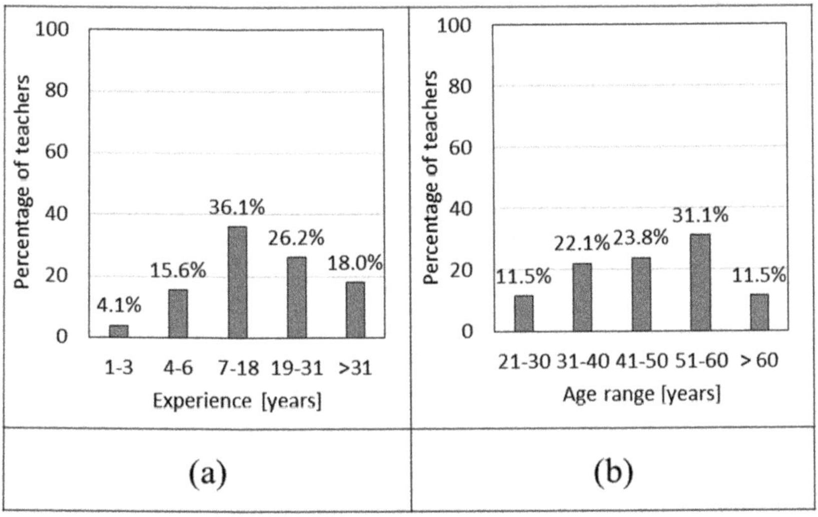

Figure 30.1 Characteristics of responders to the questionnaire ($N = 122$): (a) teaching experience; (b) age range.

and reading interview transcripts, an initial set of codes was formed. The set included the various types of teachers' knowledge, according to the conceptual framework, and aspects typical to chemistry teaching, (*e.g.*, hands-on learning). During coding, additional codes were defined, to categorizes all topics in the texts. For reliability, sample texts were coded separately by different researchers. Differences were discussed, and codes were clarified by a negotiated agreement approach.[23] Finally, codes were grouped into themes, considering the theoretical framework and the comparison with the results obtained from the quantitative analysis.

30.3.4.2 Statistical Analysis

The questionnaire included two types of closed-ended questions—Likert-type items and multiple-answer questions, and open-ended questions. Likert-type items had a five-level scale. Teachers were asked to relate to the differences in various aspects of teaching before and during the lockdowns.

Exploratory factor analysis was used to explore factors of the questionnaire's items. Exploratory factor analysis groups the items listed in the questionnaire into groups based on their calculated variance, without making assumptions on the number of factors or their nature. This grouping simplifies the interpretation of results, by reducing the number of variables to a lower number of factors discussed.[24] Table 30.1 presents the factors, exemplary items for each factor, and Cronbach's α values. Cronbach's α values, which represent the internal consistency within the data, indicate the reliability.

In the multiple-answer questions, teachers were allowed to mark more than one answer; responses were coded using thematic analysis and given the same weight. For example, in relation to using technological tools, teachers could mark multiple support sources, *e.g.*, prior knowledge, chemistry teachers support, school support, *etc.* The categories of responses to open

Table 30.1 Factors concerning teaching, assessment, and maintaining personal relations with students.

	Factor	Exemplary item	Cronbach's α
1	Teaching high-order thinking skills	"Students posing question during class session"	0.762
2	Assessment of learning	"Teacher ability to identify unclear topics among students"	0.752
3	Teacher's knowledge of students' learning characteristics	"Students' motivation for learning"	0.767
4	Teacher–student relationships in the learning environment	"The respect students have for the teacher"	0.742

ended questions were coded using the same set of codes used in the thematic analysis of the texts, and verified following discussion and agreement between two researchers, to ensure reliability.

30.4 Results

30.4.1 Characteristics of Chemistry Teachers' Work in Emergency Remote Teaching Mode Compared to Routine

To answer the first research question, how did teachers' work change compared to the routine before the pandemic, teachers were asked to compare their teaching experience during emergency remote teaching *vs.* their usual routine, in the aspects of teaching, assessment, and maintaining relationships with students. Responses to 18 items were collected using a five-level Likert scale ranging from 1 "almost impossible during COVID-19 crisis" to 5 "significant improvement due to COVID-19 crisis". Four factors were identified: (1) teaching high-order thinking skills; (2) assessment of learning; (3) teacher's knowledge of students' learning characteristics; and (4) teacher–student relationships in the learning environment. Figure 30.2 shows that assessment was the most difficult aspect of teaching in emergency remote teaching mode, as claimed by teachers (mean score = 1.89). On the contrary, teacher–student personal relationships, in class or on-linc, was rated a little above 3 (mean score = 3.1), meaning that teachers felt there was almost no difference between teaching during emergency remote teaching *vs.* their usual routine.

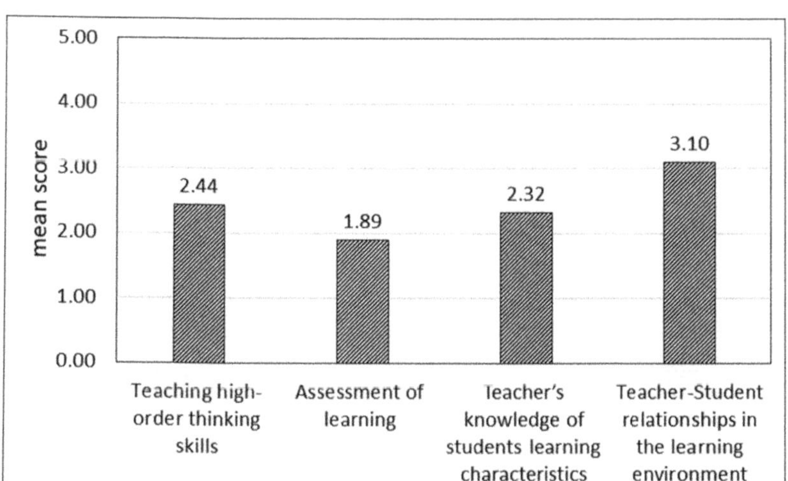

Figure 30.2 Mean score for factors dealing with pedagogical aspects during emergency remote teaching ($N = 122$).

Within the factors mentioned above, we wanted to further understand the challenges and advantages of teaching chemistry during the COVID-19 crisis, and to compare them to the teacher's regular routine. We analyzed responses to open-ended questions in the questionnaire receiving a total of 208 statements, of which 52% indicated challenges and 48% referred to advantages. Figure 30.3 presents the percentage of statements, divided into advantages or challenges, for each factor; categories dealing with teachers' work outside the classroom were mapped and grouped separately.

The first factor, teaching high-order thinking skills was mentioned both as a difficulty and an opportunity. The biggest difficulty was in the category of "hands-on-learning" (shown in a separate category in Figure 30.3) which included comments on the inability to work in the lab and conduct demonstrations. Other difficulties were related to instruction of scientific skills, two examples are:

"*[When teaching] Scientific topics, students acquire skills, which are difficult to pass on in remote teaching*"; and "*[topics of] stoichiometry and energy [were difficult], students have difficulties with complex calculations*".

Advantages were attributed to students who worked independently, practicing problem solving. For example:

"*The improved aspect of teaching was practice of multiple-choice questions. Each student gets their own time without interruptions from others.*"; and "*An increase of on-line assignments significantly improved understanding*".

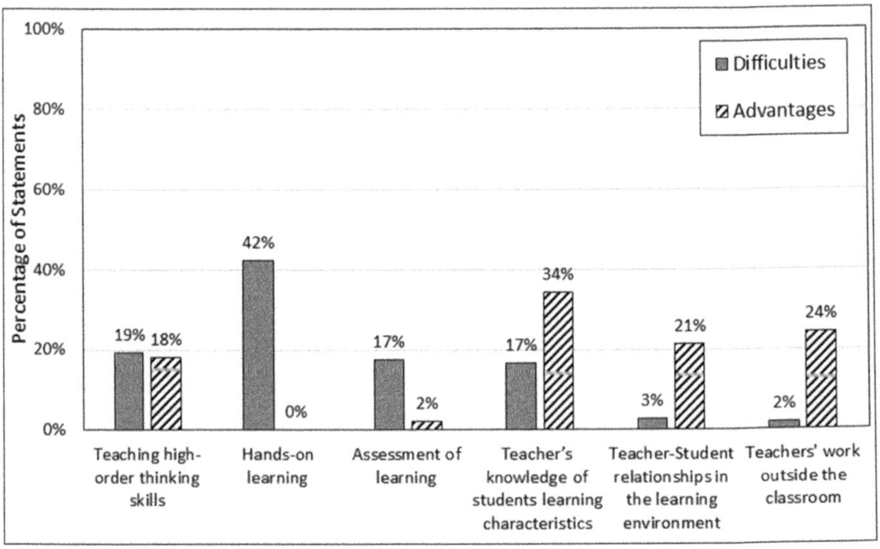

Figure 30.3 Number of extracts relating to themes concerning pedagogical aspects (*N* difficulties = 109; *N* advantages = 99).

In alignment with Figure 30.2, assessment was found to be a challenging aspect. 17% of statements indicated that teachers struggled with comprehending whether students understood the topics presented during class, monitoring the process of problem solving by students, the reliability of exams, and adapting exams to digital format. An experienced teacher explained during her interview:

> *"[In Zoom] they learned and solved well; the average grade was over 85. When we returned to class, I noticed they had figured out how to work with online worksheets, probably together, to turn in for good grades. But it was far from representing their true knowledge".*

The factor "teacher's knowledge of students' learning characteristics" included two aspects—self-regulated learning (SRL) and responding to individual students. SRL, attributed to individual work of students, was mentioned by many teachers. One teacher wrote:

> *"There is a significant shift of responsibility for learning to the student.... A student can choose to be more, or less, attentive and involved in the learning process".* Another teacher wrote: *"Focus on independent learning and self-regulated learning was definitely improved...".*

Another aspect of SRL mentioned by the teachers was a change in the traditional role of the teacher in imparting knowledge, as one teacher described:

> *"Distant learning enables the student to learn to be responsible for the learning process. Teachers' role changes from being the source of knowledge to a facilitator, a partner in process of building students' knowledge."*

Working with small groups or individual students was assisted by technological tools. Zoom breakout rooms were described as more effective compared to groups in a traditional classroom:

> *"The good thing I found was working in groups on assignments or lab reports. Moving among breakout rooms enables giving attention to every team without interruptions, making it possible to push them forward as a group."*

Teachers referred to the use of Zoom as a platform for private tutoring:

> *"Giving private lessons in Zoom [is an advantage]. There are less disturbances so everyone who wants to learn and has a computer can study better".*

Opportunities inherent to remote teaching, which benefit individual students even without direct interaction with a teacher, were noted:

> *"There are students who usually don't concentrate in class because they talk .... Distant learning was an improvement because they were attentive away from*

their friends". Another statement was: "*Some students thrive on-line: those with ADHD, Autism Spectrum Disorder, behavior problems, or a preference to learn alone and manage time independently*".

The last factor, 'teacher–student relationships in the learning environment, included issues of class management. Teacher reported that coping with behavior problems was simpler in on-line mode.

Teachers reported that their work outside of the classroom became simpler. Teaching from home saved time and enabled flexibility in scheduling and attending meetings. The advantage of working from home was described favorably in an interview with a teacher who is a mother to young children:

"*You don't need to run around, all meetings are on Zoom, ... talking to parents, all by Zoom, it's perfect. You don't need to leave home... In terms of daily planning, it is much better.*"

Available technology was also listed as an advantage, as one teacher detailed:

"*There is no need to look for a computer, projector or connection to the internet, all is available*".

30.4.2 PLC Support of Chemistry Teachers Working in Emergency Remote Teaching Mode

Aspects of support, which were the topic of the second research question, were addressed in the questionnaire by the multiple answer question "How did you cope with the technological challenges during emergency remote teaching?". List of answers included: prior knowledge, independent learning, support of family and acquaintances, chemistry teachers' support, PLC meetings, school support, *etc.* A total of 541 responses were collected, since teachers were allowed to give multiple answers, 73% of them from teachers in PLCs. Two aspects were examined: the source of support and the type of support. Sources of support were divided into three categories. One was personal sources, *i.e.*, family, friends, teacher's previous knowledge, and abilities; the others were support from chemistry teachers, and support of schools and the Ministry of Education. Type of support included categories for teachers who preferred to rely on their personal abilities, accept help individually from colleagues, or receive support in a group setting.

A comparison of preferred support source revealed that non-members were self-reliant, while PLC members preferred to receive support from other chemistry teachers (Figure 30.4(a)). This result was significant ($p < 0.05, \chi^2(2) = 8.563$). A higher percentage of PLC members favored group support compared to non-members, who relied on their personal abilities (Figure 30.4(b)). This tendency was not significant ($p > 0.05, \chi^2(2) = 4.155$).

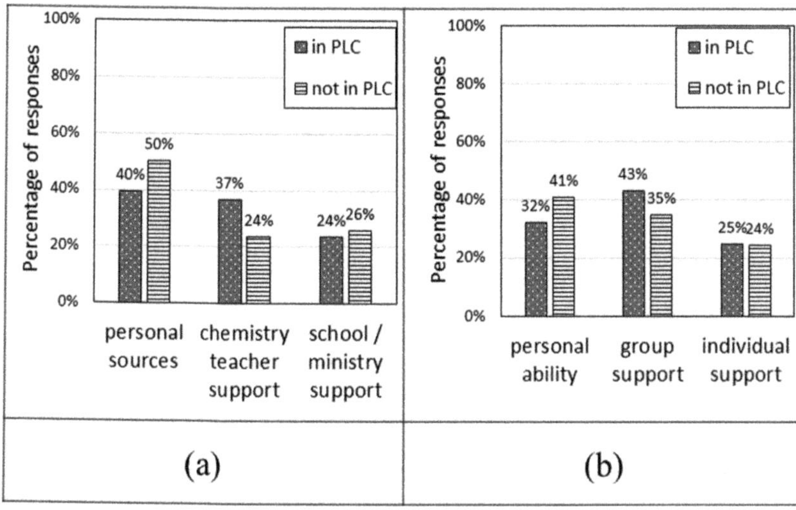

Figure 30.4 Support aspects preferred by PLC members (N = 394) and teachers non-members (N = 147): (a) source of support; (b) type of support.

On a professional level, ideas and teaching materials shared in the community were the basis for planning lessons and class activities, for example:

"The community took us a step forward, sharing ideas and [activities] ... Suddenly [teachers] became professional and shared a variety of things they created. Then they communicated it, building a process with us, like we were pupils; it was excellent".

The rapid transition to remote teaching was an emotional task. Sharing within the community reduced the feeling of isolation and supported teachers. The authors of Chapter 31 in this book report similar results. PLC members greatly valued the assistance they received by the community, and wanted this framework to continue.[2]

30.4.3 Diversity Within Chemistry Teachers' PLCs

The last research question dealt with the diversity among teachers. Among chemistry teachers in Israel, the diversity in spoken language is a central issue. Hebrew speakers are the majority; Arab speakers are bilingual. The PLCs described in this research include speakers of both languages. The spoken language in community discussions and activities is Hebrew. Thus, Arabic speakers faced an extra challenge of translating teaching materials. One Druze teacher explained:

"In every assignment I make adjustments... I always changed something or adapted it to the people I'm teaching."

Translation is a time-consuming task; working in a community enables teachers to share translations. In PLC with a majority of Arabic speaking teachers, materials were developed in Arabic, as a member of CTC said:

> *"I [usually] translate... because there was a majority of Arab teachers, we asked the [PLC] leaders to do the work in Arabic, so we will have a collection of activities, and they allowed it."*

Having access to a collection of activities is important to Arabic speaking teachers. The last quote was an important testimony to the role of chemistry teachers' PLCs in promoting equity, since Arabic speaking teachers found answers to their needs within a diverse PLC. An experienced teacher guide who has been involved in developing remote teaching materials said that gaps in translation have decreased in recent years. In her opinion, this was important for Arabic speaking teachers in the first weeks of emergency remote teaching.

Another noticeable difference between chemistry classes in different schools in Israel is the typical class size—in Arabic speaking schools, classes are typically large, whereas in Hebrew speaking schools, classes are smaller. In spite of this, all teachers were active in discussions. One example was a discussion about a set of four simple experiments on acid–base reactions. Teachers suggested different settings—presenting demonstrations *vs.* having students move between areas in the lab; using Google Forms or paper and pencil worksheets to guide students during activity, to suit the variety of classes they teach. The community addressed the specific needs of Arabic speaking schools and Hebrew speaking schools, discussed, and made valuable suggestions which enhanced the feeling of inclusion.

The relationships among PLC members were a source of support and inspired a feeling of unity in spite of diversity. This was evident in both communities studied here. One teacher who independently found solutions to difficulties, still valued the support of the group:

> *"...I think that only sharing within the community, with teachers, felt better, because we are all the same, all share the same challenges, and this is OK. I did not use tools suggested in the community to connect with my students".*

Other personal issues, such as working from home, the need for family support, or challenges of teachers who were parents of young children, were described by teachers belonging to all parts of society.

30.5 Discussion

Teaching in emergency remote teaching mode presented diverse challenges to teachers. In answer to the first research question, we concluded that assessment of students' learning and knowledge were the main concern of teachers, and they doubted the trustworthiness of assessment tools and their ability to formatively assess their students while teaching. Yet, teachers mentioned the benefits of monitoring and strengthening students' SRL and taking responsibility of the learning during this time. Difficulties in conducting experiments, included in the hands-on learning category, is a challenge unique to chemistry teaching, and needed a unique solution. Sharing an archive of recorded experiments was initiated by the Ministry of Education and was found helpful.

In regard to support (research question 2), results showed that PLC members preferred the setting of group support within the community. PLCs gave teachers the opportunity to form connections with other chemistry teachers. Since answers to the questionnaire showed a significant difference between PLC members and non-members regarding the preferred source of support, we concluded that chemistry teachers' PLCs had an important role in support and fulfilling teachers' needs during lockdowns.

Two of the knowledge gaps were mapped with respect to the RCM.[13] One was the basis of teachers' knowledge (*i.e.*, assessment knowledge, technical knowledge, and knowledge of students). The second was the pool of class activities and teaching techniques, included in collective PCK. Collaboration advanced PLC members towards filling these gaps. Sharing information, guidance, and useful tips about technological tools strengthened their technological knowledge. Sharing teaching activities, developed especially for online use, enriched collective PCK. Some teachers reported this was essential for their teaching online. Enrichment of collective PCK of members of a teachers' community, as a result of sharing teaching materials among members is described in Chapter 31 of this book.[2] This exposure to a variety of novel tools and activities was a direct contribution to both aspects of PCK. Inner components of PCK in the model, personal PCK and enacted PCK, draw from this pool of communal knowledge, according to the RCM. Reports of teachers basing class activities on what they learned during the crisis indicate that aspects of personal PCK and enacted PCK were enhanced. This agreed with indications that participation in teacher communities results in changes of class practices.[17,25] Therefore, we concluded that all aspects of PCK were improved.

PLC were not the only source of know-how in these fields. Teachers reported that they were able to acquire the needed skills individually and in private consultations with colleagues. These skills and knowledge add to the basis of knowledge of specific teachers, add to personal PCK and affect their class work, enriching enacted PCK. In our opinion, the independent work of teachers had a smaller contribution to collective PCK, although teachers reported limited sharing of information and solutions with colleagues.

The essence of chemistry teachers' PLCs was to enable collaborative work of teachers for the benefit of all members. According to the conceptual framework defining PLCs,[8,14] the relationships formed among community members were the basis for joint learning and professional development. Views of teachers included in the results section described cooperation and sharing of professional assets in communities. The significance of relationships in the community found here was also identified by other researchers as summarized by Vangrieken and colleagues.[25]

The communal relationships formed between Hebrew and Arabic speaking teachers in the communities were an integral part of the support provided by the PLC. Although Arabic speaking teachers have needs which differ from their Hebrew speaking colleagues and need to adopt materials shared in PLCs to their classroom, they found the community to be an important and useful source of support and materials during emergency remote teaching. The PLCs were also important to Arabic speaking teachers since it created connections among them and enabled sharing materials in Arabic. For Arabic speaking teachers in PLCs, the opportunity to consult with and share and receive help from community members of all sections of society was regarded in higher importance compared to the work related to translation and adoption of teaching materials written in Hebrew. The importance of PLCs was also reflected in teachers' statements describing emotional support given in PLCs, and in the similarity found in preference of speakers of both languages concerning source and type of support. This agreed with the conclusion of Hallam *et al.*,[6] who wrote on the importance of trust within a teachers' community.

30.6 Conclusion and Recommendations

This study showed the importance of chemistry teachers' PLCs network in supporting chemistry teachers in Israel who worked in emergency remote teaching mode during the COVID-19 crisis. In time of crisis, the preferred support sources for PLC members were their PLC colleagues and other chemistry teachers. In reference to the RCM, joint learning and sharing in the community contributed to teachers' professional knowledge, enhancing their professional development, collective and private aspects of PCK. Emotional support, which benefited their well-being, was also related to membership in PLC. This was evident for both Hebrew and Arabic speaking teachers.

It should be noted that PLCs were not the sole source of support for chemistry teachers. Results showed that teachers found support and professional assistance with colleagues in their school, family, and friends. Other teachers reported facing the challenges of working in emergency remote teaching mode alone, using their prior knowledge and gathering needed information and guidance on their own.

The biggest challenge, unique to chemistry teaching, *i.e.*, combining hands-on learning in teaching and assessment was confronted by extensive efforts made by teachers all over Israel with coordination by the Ministry of Education. It emphasizes the importance of cooperation among teachers, inherent to teacher communities.

As PLCs were important in times of crisis and were found to be a platform for developing professional knowledge among teachers, we recommend expanding the network of chemistry teachers' PLCs and making the communities accessible to more teachers. A focus should be given to discussion on learning assessment, especially on formative and summative assessment in on-line teaching. The PLC platform should be used to increase the amount of teaching materials available in Arabic and give Arabic speaking teachers easy access to them, on-line communication is well suited for this purpose.[2,26] Since hands-on learning is an important part of teaching chemistry, we also recommend using PLCs to discuss and develop methods to meet this challenge, for example by expanding the pool of home experiments which may be performed by students.

References

1. C. B. Hodges, S. L. Moore, B. B. Lockee, A. M. Bond and A. Jewett, in *Radical Solutions for Education in a Crisis Context: COVID-19 as an Opportunity for Global Learning*, ed. D. Burgos, A. Tlili and A. Tabacco, Springer Science and Business Media Deutschland GmbH, Singapore, 2021, pp. 37–51.
2. See Chapter 31: A. B. DeKorver and B. D. Herrington, in *Digital Learning and Teaching in Chemistry*, ed. Y. J. Dori, C. Ngai and G. Szteinberg, Royal Society of Chemistry, United Kingdom, 2023.
3. S. Rap, Y. Feldman-Maggor, E. Aviran, I. Shvarts-Serebro, E. Easa, E. Yonai, R. Waldman and R. Blonder, *J. Chem. Educ.*, 2020, **97**, 3278–3284.
4. L. Orland-Barak, in *STEM Teachers and Teaching in the Era: Professional Expectations and Advancement in the 21st Century Schools*, ed. Y. Ben-David Kolikant, D. Martinovic and M. Milner-Bolotin, Springer Nature Switzerland AG, Chapel Hill, NC, 2020, pp. 53–61.
5. M. M. Kennedy, *Rev. Educ. Res.*, 2016, **86**, 945–980.
6. P. R. Hallam, H. R. Smith, J. M. Hite, S. J. Hite and B. R. Wilcox, *NASSP Bull.*, 2015, **99**, 193–216.
7. J. E. Hamos, K. B. Bergin, D. P. Maki, L. C. Perez, J. T. Prival, D. Y. Rainey, G. H. Rowell and E. VanderPutten, *Sci. Educ.*, 2009, **18**, 14–24.
8. E. Wenger, B. Trayner and M. de Laat, *Ruud Moor Cent.*, 2011, **18**, 1–60.
9. R. Waldman and R. Blonder, in *STEM Teachers and Teaching in the Digital Era: Professional Expectations and Advancement in the 21st Century Schools*, ed. Y. Ben-David Kolikant, D. Martinovic and M. Milner-Bolotin, Springer Nature Switzerland AG, Chapel Hill, NC, 2020, pp. 11–140.
10. B. S. Eylon, Z. Scherz and E. Bagno, in *STEM Teachers and Teaching in the Digital Era Professional Expectations and Advancement in the 21st Century Schools*, ed. Y. Ben-David Kolikant, D. Martinovic and M. Milner-Bolotin, Springer Nature Switzerland AG, Chapel Hill, NC, 2020, pp. 65–90.
11. N. Resh and N. Blass, in *The Palgrave Handbook of Race and Ethnic Inequalities in Education*, ed. P. Stevens and A. Dworkin, Palgrave Macmillan, Cham, 2019, pp. 631–694.
12. S. Levy, E. Bagno, H. Berger and B. S. Eylon, in *STEM Teachers and Teaching in the Digital Era: Professional Expectations and Advancement in the 21st Century Schools*, ed. Y. Ben-David Kolikant, D. Martinovic and

M. Milner-Bolotin, Springer Nature Switzerland AG, Chapel Hill NC, 2020, pp. 159–184.
13. J. Carlson and K. R. Daehler, in *Repositioning Pedagogical Content Knowledge in Teachers' Knowledge for Teaching Science*, ed. A. Hume, R. Cooper and A. Borowski, Springer Nature Singapore Pte Ltd, Singapore, 2019, pp. 77–92.
14. S. M. Hord, *Action Res. Communities*, 2009, **30**, 40–43.
15. L. Stoll, R. Bolam, A. McMahon, M. Wallace and S. Thomas, *J. Educ. Change*, 2006, 7, 221–258.
16. S. E. Booth, *J. Educ. Comput. Res.*, 2012, **47**, 1–31.
17. T. E. Vossen, I. Henze, M. J. De Vries and J. H. Van Driel, *Int. J. Technol. Des. Educ.*, 2020, **30**, 295–320.
18. A. K. Fackler and C. M. Sexton, *Electron. J. Res. Sci. Math. Educ.*, 2020, **24**, 5–13.
19. J. W. Creswell, V. L. Plano Clark, M. L. Gutmann and W. E. Hanson, in *Advanced Mixed Methods Research Designs*, ed. A. Tashakkori and C. Teddlie, SAGE PublicationsSage CA, Thousand Oaks, CA, 2003, pp. 159–196.
20. L. K. Barriball and A. While, *J. Adv. Nurs.*, 1994, **19**, 328–335.
21. M. Huberman, *Teach. Coll. Rec.*, 1989, **91**, 31–57.
22. V. Braun and V. Clarke, *Qual. Res. Psychol.*, 2006, **3**, 77–101.
23. F. M. Watts and S. A. Finkenstaedt-Quinn, *Chem. Educ. Res. Pract.*, 2021, **22**, 565–578.
24. A. G. Yong and S. Pearce, *Tutorials Quant. Methods Psychol.*, 2013, **9**, 79–94.
25. K. Vangrieken, C. Meredith, T. Packer and E. Kyndt, *Teach. Teach. Educ.*, 2017, **61**, 47–59.
26. C. L. Blitz, *Can Online Learning Communities Achieve the Goals of Traditional Professional Learning Communities? What the Literature Says*, 2013.

CHAPTER 31

Strategies for Teaching Chemistry Online: A Community of Educators for the COVID-19 Pandemic and Beyond

BRITTLAND DEKORVER*[a] AND DEBORAH HERRINGTON[a]

[a]Chemistry Department at Grand Valley State University in Allendale, MI, USA
*E-mail: dekorveb@gvsu.edu

31.1 Contributions to Chemistry Education Research (CER)

The COVID-19 pandemic not only plunged chemistry instructors worldwide into the ocean of teaching as a novice.[1] It also resulted in a sense of isolation as instructors began carrying out all of their professional duties from the privacy of their home. Unfortunately, this experience of isolation was not new to chemistry instructors and the CER community.[2,3] The group "Strategies for Teaching Chemistry Online" (SFTCO) was initially formed to address the former issue: to fulfil an immediate professional development need as instructors navigated the difficulties of converting their classes to an online delivery. However, within a year of its creation, it was clear that the group was

Advances in Chemistry Education Series No. 11
Digital Learning and Teaching in Chemistry
Edited by Yehudit Judy Dori, Courtney Ngai and Gabriela Szteinberg

Published by the Royal Society of Chemistry, www.rsc.org

also confronting the problem of isolation, serving as a means of connecting people. Several members of the group noted this unforeseen function of the group and expressed their desire for the continuation of the forum beyond the culmination of pandemic-induced online teaching. This chapter will characterize the group and situate its activity as a community of practice (CoP), with practical implications for teaching as well as professional development (PD).

There is a tremendous amount of literature on characteristics of effective PD programs. A recent review of 35 methodologically rigorous studies demonstrating a positive link between teacher PD, teaching practices, and student outcomes identified the following critical characteristics: a focus on content; active learning; support for collaboration; models of effective practice; coaching and expert support; feedback and reflection; and sustained duration.[4] Yet, the vast majority of PD opportunities engaged in by high school or college level chemistry faculty, if they engage in any at all, tend to be short term, one-off experiences that do not allow for collaboration, coaching and expert support, mutual sharing of ideas, and feedback and reflection. This, in combination with the fact that often such PD opportunities are ones that are prescribed or required as opposed to ones that instructors choose to address their own particular needs, may help explain why these short-term PD experiences have little to no effect on teaching practices. A review of literature focused on change in undergraduate STEM programs found that successful change strategies were the ones based on emergent goals and shared purpose.[5] Furthermore, engaging in social interactions with other teachers and experts while attempting to enact instructional reform was the key difference between teachers who did and those who did not transform their classroom instruction as a result of a reform effort.[6]

The previous chapter discussed how professional learning communities (PLC), one recognized PD mechanism, supported chemistry teachers during the transition to remote instruction caused by the COVID-19 global pandemic.[7] This chapter will examine a different framework for a professional organization, the CoP, and use it as a lens for describing the role of SFTCO in the chemistry education community.

31.2 Theoretical Framework

31.2.1 CoP

Though PLCs and CoPs have some overlapping features, there are some distinct differences. Table 31.1 summarizes some of the key differences between these two types of communities. In contrast to a PLC, membership in a CoP is voluntary and centres on a shared domain of interest or need. Outcomes are more individualized in a CoP, based on the needs and goals of the individual member. Though CoPs are more organic and may appear to have fewer defining features than PLCs, Wenger identifies three characteristics that are crucial for CoPs: domain, practice, and community.[8]

Table 31.1 A summary of the key differences between PLCs and CoPs.

Element	PLC	CoP
Formation	Formed around a common issue of student needs that is based on data	Formed around a shared domain of interest (for SFTCO this was chemistry teaching)
Group membership	Often have assigned roles and is expected as an educator or faculty member	Voluntary participation and membership
Activities	Collective inquiry—this often involves identifying best practices in addressing the identified issue and collecting and analyzing data/student work to both clarify the problem and evaluate the efficacy of the implemented solution. The focus of the activities is on effecting change to address identified student needs	Members engage in a variety of activities including sharing of information, requesting information or feedback, discussing issues, and joint problem solving within the domain of interest. There is no one single issue that is the focus of these activities
Outcomes	Collective solution to a problem that has been assessed based on data	Individual improvement—deepening knowledge and expertise through ongoing interactions with community members

A CoP has an identity in its shared domain of interest. For SFTCO, a shared interest and expertise in the teaching of chemistry was the domain. Another key element of a CoP is that its members are practitioners. This differentiates a CoP from a community of interest (*e.g.*, people interested in collecting coins or a book club). The ultimate goal of the members of the CoP is to improve their practice through the sharing of resources, experiences, and ways of solving problems. In the case of the SFTCO group, the community was initially formed around the need for many people to transition their chemistry courses and labs online in a very short period of time. The one key area that PLCs and CoPs overlap is in the collaborative work of members. Such collaboration requires the development of a sense of trust and support among the members of the community.

For the study we describe in this chapter, we focused specifically on how the SFTCO group functioned as a CoP to support its members in developing knowledge and expertise in effectively teaching chemistry in an online environment. As the motivation for participation in a CoP is to meet the needs of the individual participants, to sustain a CoP it is important that the key emphasis or emphases of the CoP can and do shift as the needs of the members change. The framework we used to examine the ways in which the SFTCO group functioned as a CoP in meeting the needs of members as they transitioned instruction online and how the domain of the group shifted over time was technological pedagogical content knowledge (TPCK).[9]

31.2.2 (T)PCK

The concept of pedagogical content knowledge (PCK), introduced by Shulman, contends that effective teaching requires knowledge about the content discipline and how to best teach that content.[10] Furthermore, PCK is discipline specific. The ways in which chemistry teachers think about transforming or representing content to make their content accessible to their students are necessarily different from how a math teacher may do the same in their discipline. PCK has become a valuable framework for teacher preparation and development.

More recently, rapidly developing and changing technologies being used in education, especially true in the recent global shift to remote instruction, has compelled the inclusion of technology in the PCK framework. As good chemistry teaching requires more than just knowing the chemistry content, effective online chemistry teaching requires more than just knowing how different technologies work. It requires knowledge of how to best use different technologies to support the pedagogies that are most effective in supporting student learning of chemistry. This knowledge domain is known as TPCK, and just as PCK is the unique intersection of content knowledge and pedagogical knowledge related to a particular discipline, TPCK is the unique intersection of content, pedagogical, and technological knowledges.[9] As shown in Figure 31.1, there are also pairwise intersections between the three elements.

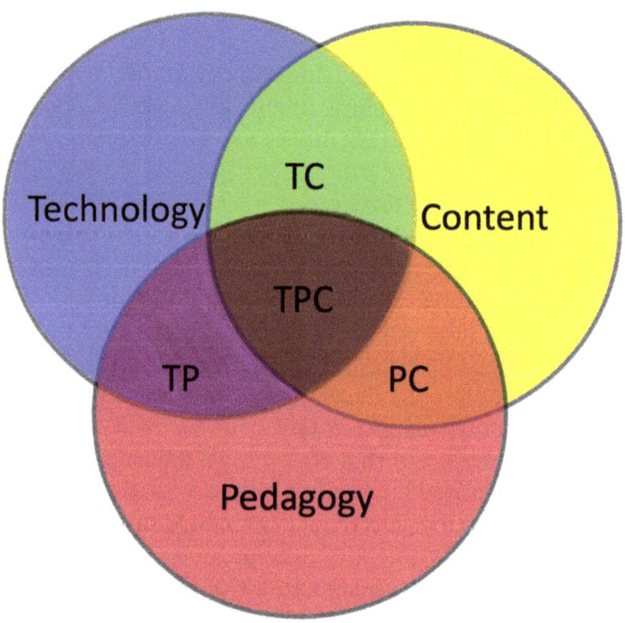

Figure 31.1 The overlap of knowledge of content, pedagogy, and technology. Technological pedagogical content knowledge exists at the intersection of all three. Reproduced from ref. 12 with permission from American Chemical Society, Copyright 2020.

Though content knowledge in chemistry is obviously different from that in history or sociology, even within the domain of chemistry there are differences in the content knowledge for teaching general chemistry as compared to that required to teach organic chemistry. Pedagogical knowledge in chemistry includes knowledge about specific teaching practices used in the classroom as well as knowledge about students, how they learn, and how to assess that learning. Finally, technological knowledge encompasses knowledge about how to use standard technologies such as textbooks in addition to things like learning management systems (LMS), assessment software, or simulations.

In thinking about how technology knowledge interacts with content and pedagogical knowledge, consider how the introduction of animations and simulations has changed the ways we can represent particle interactions for students (TCK). Or, how we can best use video conferencing software breakout rooms to support group collaborations in online classes (TPK). The interaction of all three is TPCK.

It is important to recognize the dynamic nature of the relationships in the TPCK framework, where a change in one dimension requires changes in the other two. One example in the shift to remote instruction was the inability of many students to perform in-person laboratory experiments. The changes in technology (in this case the lab equipment) required reconsideration of lab content in terms of what important lab content and skills could be taught online (*e.g.*, moving away from physical lab skills towards process skills like data analysis), as well as what instructional methods would best support the learning of this content. Thus, using TPCK as a framework to conduct a qualitative analysis of the posts from SFTCO provided a picture of how the focus of the group shifted as the needs of the community changed.

31.3 Research Questions

As the digital learning in chemistry was propelled to the primary concern of chemistry educators around the globe in March 2020, the SFTCO Facebook group was created to attempt to fill the need for chemistry instructors to learn new technologies and pedagogies. More than a year after its creation, we seek to understand the role that the group played at the outset of the COVID-19 pandemic and the role it continued to play thereafter. Particularly, we hope to answer these questions.

1. In what ways does "Strategies for Teaching Chemistry Online" meet the criteria of community, practice, and domain in order to function as a community of practice?
2. How has the domain, *i.e.*, the content of questions and resources posted by members shifted over time?
3. How has this group addressed topics of diversity, equity, and inclusion in their discussions?

31.4 Participants and Setting

The dataset was curated from a community of chemistry instructors within a Facebook group. Facebook groups are a virtual platform, created by a specific user with the purpose of uniting people who share a particular interest. In this case, the group "Strategies for Teaching Chemistry Online" was created by one of the authors to facilitate sharing information specific to teaching chemistry during the emergency transition to remote teaching in the spring of 2020. Membership in the group grew quickly, with over 3000 members joining in the first three months of the group's existence, sharing thousands of posts and comments. Members consist of secondary and post-secondary instructors, primarily from the United States (3600 or 74%) with more than 100 members from Canada and Brazil, and fewer than 100 members from each of the many other nations represented in the group. At the time of this writing, there were 4890 members, with membership increasing at a rate of approximately 50 new members per month.

31.5 Methods

As evidence of the group's engagement as a *community*, statistics about the group participation and activity were downloaded as Excel files from Facebook's group analytic features. Activity data were recorded by Facebook. These consisted of a list of the threads posted, the comments added to each thread, and the number of reactions each post and comment received. These activity data served as a proxy for engagement, with active participation by group members indicated by not only the posts and comments in reply, but the number of "reactions" (*e.g.*, likes, "wow", "angry", *etc.*) on each contribution. Due to the public nature of the posts, an Institutional Review Board determined this study was exempt from obtaining consent. All exemplar quotations reported herein have been paraphrased to protect the anonymity of the authors.

To determine the extent to which community content could be characterized as *practice*, the parent posts contributed by members were downloaded as Excel files *via* Grytics, a social media analytics application. The posts were separated according to month and then imported into MAXQDA qualitative data analysis software. Content of the posts were coded according to two separate schemes. Most applicable to this discussion is the deductive coding using the TPCK framework. Posts were categorized as pertaining to technology, pedagogy, (chemistry) content, or some combination of the three. Examples of posts for each code are provided in Table 31.2. Given that in the first three months after the group's creation members had generated over 1300 posts and 18 000 comments, it was determined that analyzing the data at the level of the parent post was most feasible. As the parent post directed the content of the subsequent comments in that thread, categorizing the data at the parent post level allowed us to determine how the general focus of the group's domain shifted over

Table 31.2 TPCK coding scheme applied to SFTCO Facebook Group posts.

Code	SFTCO examples
Technology (T)	How do I record a screen capture of my iPad?
Technology and pedagogy (TP)	I give weekly group quizzes and I need to set different time limits for different groups (for accommodations). Any idea how I do that in canvas?
Pedagogy (P)	Here's an interesting article about cheating on exams: [link]
Pedagogy and content (PC)	Does anyone have a general chemistry lab or demo for enthalpy that they love?
Content (C)	Is zinc considered a transition metal?
Technology and content (TC)	What is your favourite online periodic table?
Technology, pedagogy, and content (TPC)	I'm looking for something to teach molecular orbitals to my pchem students, but I need it to be accessible from their own devices. Has anyone had success with WebMO?
No code assigned	We've made it through another week. Take time for yourself this weekend and keep being awesome!

time. Further, not all posts had the same number of comments for a variety of reasons (*e.g.*, not many people with expertise in a particular area or questions were more closed or open), thus coding at the comment level could have skewed the characterization of the overall domain of interest towards content of the posts that had more comments. However, if we were unsure about how to characterize a parent post, the comments related to that post were reviewed.

The development of the coding scheme involved the SFTCO posts from the first ten weeks of the group's existence, with at least two researchers independently coding each post using the TPCK framework, comparing codes, and then negotiating any disagreements.[11] The findings reported in this chapter used this coding scheme to analyze the posts from the first two months the group had been in existence (March and April 2020) and those same two months a year later (March and April 2021). This time frame was selected so that analysis would include the initial emergency transition to remote teaching, as well as a later period after instructors had time to make more thoroughly considered decisions about their teaching. To be consistent with respect to the periodic fluctuations in various institutions' academic calendars, the same two months (March and April) were selected for the latter timepoint. For this analysis, posts from each month were initially coded by at least one researcher, and then the coded posts discussed among all researchers until consensus was achieved. If a post did not clearly fit into the TPCK framework, it was not coded. For March and April 2020, a total of 203 posts were coded and for March and April 2021, a total of 183 posts were coded. An analysis of how those posts were divided within the different codes (Table 31.2) is presented in Section 31.5.

31.6 Main Findings

31.6.1 CoP

As described earlier, a CoP must have these defining features: community, practice, and domain. In this section, we will present evidence that SFTCO satisfies these criteria and functions as a CoP for its members.

31.6.2 Community and Practice

The first indication of community is the number of group members, which at the time of this writing was over 4800. However, membership in a group is not enough to constitute a community. The members have to engage with each other in activities, discussions, helping each other, and sharing resources over an extended period of time, allowing them to develop relationships. These relationships are critical in providing a feeling of belonging which is central to community. McMillan and Chavis define a sense of community as "a feeling that members have of belonging, a feeling that members matter to one another and to the group, and a shared faith that 'members' needs will be met through their commitment to be together."[11] Of the 4800 members, 52% were "active", defined as having contributed to content within the past week and shown in Figure 31.2.

In the first 13 months of the group's existence, over 4000 posts were created with over 70 000 comments for an average of 16 comments per post. This is evidence of a high level of engagement, with many members participating and a great deal of discussion and communication among the members.

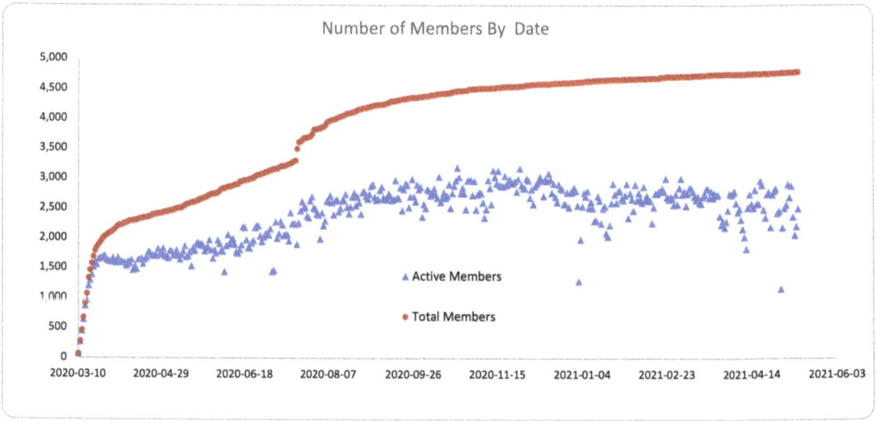

Figure 31.2 The number of SFTCO members who are active (participated in the group in the past week) as compared with total number of members from the group's creation in March 2020 to May 2021.

In addition to active participation, another important aspect of community and developing relationships is support and trust. This was also observed to be present in SFTCO in analyzing the topics of the ten highest rated posts from the period of 1 March 2021 to 30 April 2021. These posts were selected by ranking all types of posts in the date range by Facebook's "engagement" rating, which is the number of comments multiplied by two added to the number of "interactions" (like, wow, care, *etc.*). For example, a post with 10 likes and two comments would have an engagement score of 14. After filtering for the posts with the highest engagement during the date range, the results were coded for TPC content according to the coding scheme described in Table 31.2. Surprisingly, one year after the emergency transition to remote teaching, none of the posts that garnered the highest ratings were focused on the technology, content, or practice of teaching chemistry. Instead, they were humorous, supportive, or cathartic posts that were tangentially related to chemistry teaching. Many were memes or specific interactions with students cast into a humorous light. Other posts communicated frustration with students' level of engagement or expectations. None of the top posts were intended to begin a conversation about teaching chemistry. This is not inconsistent with our original analysis of the top-rated posts from the first ten weeks of the SFTCO group, where a high proportion of top-ranked posts there, too, did not pertain to TPCK.[12] However, after a year, an even larger proportion of the top-ranked posts fell into that category of "no TPC code". The high ratings of these posts indicate that group members engage not just to fulfil their professional needs, but their personal and emotional needs as well. One of these posts specifically stated the information was being shared because "*SFTCO has become part of my support.*" This sentiment of SFTCO serving as a support community and the phrase "*hive-mind*" were frequently expressed or used. This provides evidence that SFTCO acts as a community where members trust and support each other.

Members of a CoP also engage in activities within their domain and seek to improve their practice. Wenger-Trayner described a variety of activities that serve to help CoP members develop their practice.[8] Examples of posts from SFTCO are provided as evidence that its members are seeking to develop their practice, summarized in Table 31.3.

31.6.3 Domain

The TPCK coding scheme was used to interpret the group's domain. In the first two months of the group, technology was included in the content of the majority of the posts (Figure 31.3). One year later, the percentage of posts including technology decreased, with a large increase in the posts that were coded as just "content" or "pedagogical content". This illustrates how the focus of SFTCO shifted to meet the needs of its members. Moving from the initial need for technological know-how during the emergency transition to remote teaching towards more general aspects of teaching chemistry, such as how we should define a hydrogen bond and teach intermolecular forces, as

Table 31.3 Examples of CoP activities noted in the SFTCO Facebook group.

Activity	SFTCO examples
Problem solving	Can we discuss how people are handling assessment in this online environment?
Requests for information	Does anyone have an assignment on stoichiometry?
Seeking experience	Has anyone else had students who are not able to upload files to Blackboard for question answers? Can you tell me how you solved this?
Reusing assets	I have developed a biochem online lab for kinetics. I'm happy to share, so drop a comment below if you want me to send it to you
Coordination and synergy	Is anyone teaching inorganic chemistry in the fall and interested in collaborating to develop materials?
Discussing developments	Does anyone have experience using the new Canvas quiz tool? What do you think?
Mapping knowledge and identifying gaps	So far, I have done the following for my exam, randomized questions and answers, only showing one question at a time, using question pools, using random numbers in calculation questions, giving students a time limit, having them sign an academic integrity statement, is there anything else I should be doing?

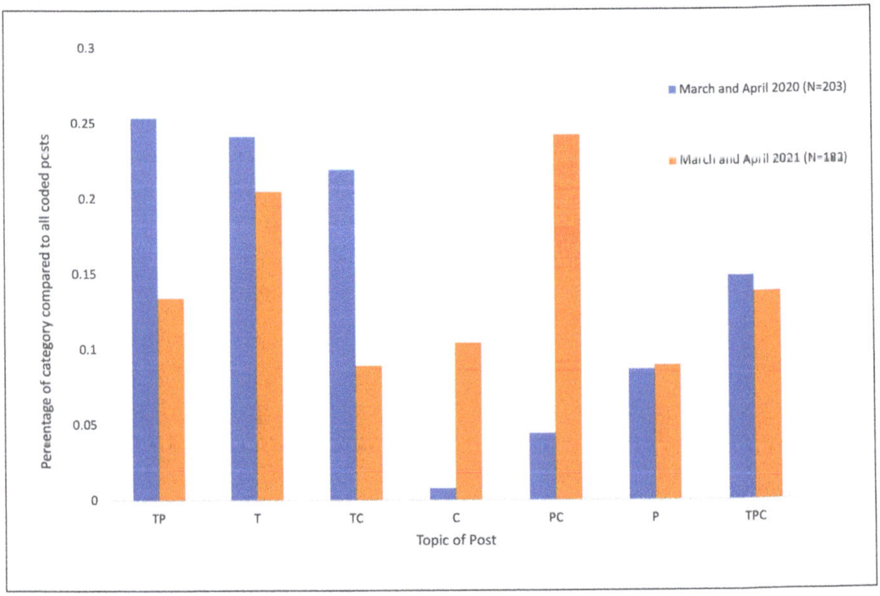

Figure 31.3 A comparison of TPCK coded posts in SFTCO, from the emergency transition to remote teaching to one year later.

instructors became more comfortable with teaching online and began thinking about returning to the classroom. Still other discussions had very little to do with teaching at all and were solely directed at an appreciation or understanding of chemistry. There were even a number of open position postings and inquiries as group members searched for colleagues or employment.

As these shifts in topics occurred, members would sometimes check their post for acceptability, often with a phrase such as "This isn't really about teaching online, but..." As these "off topic" posts became more frequent, the number of members offering an apologetic preface decreased. 14 months after the group's inception, a member questioned the future of the group, whether it would continue as instruction moved back to face-to-face. Many members responded with their hopes that the group would continue, regardless of the instructional delivery method. One user commented "*I don't care about the name; I want the ideas and the community.*" At this point, the group's name was shortened to "Strategies for Teaching Chemistry," a change which was suggested by a group member and enacted by the author/group administrator. These discussions signify that not only did the members acknowledge a shared domain, but that the community's domain was negotiable and broadened over time from *teaching chemistry during a transition to online teaching* to *teaching chemistry*. This is a hallmark of a CoP, where the joint enterprise and domain is driven by the members of the group, and not assigned by a higher authority.

31.6.4 Equity and Inclusion

One topic that received much participation, but often suffered from a lack of resolution, was how to ensure that the teaching was accessible to all students. For example, during the early days of the transition to remote teaching, there was a discussion about methods for tailoring curricular materials for use on platforms available on smartphones to ensure that students without reliable internet or technology access did not get left behind. These discussions about access to connectivity and technology frequently generated more questions and challenges than solutions. Several months later, there were more posts about diversity, primarily asking for resources to include a more diverse representations of science in their curricula, so students see themselves represented in the portrayals of science in their classrooms.[12] These conversation threads no doubt arose from the unique intersection of the new equity challenges encountered by the shift to remote instructions, the blatant evidence of racial inequities across the United States, and the lack of content-specific resources to support chemistry instructors in incorporating these types of inclusive pedagogies. An analysis of posts on this topic was conducted to investigate this need further (Figure 31.4).

To start, a lexical search of the posts from March–December 2020 for the terms: inclusion, equity, accessibility, accommodations, diversity, low-income, free, and access, as well as several derivatives of these words. This was the time frame when instructors were making choices for their teaching,

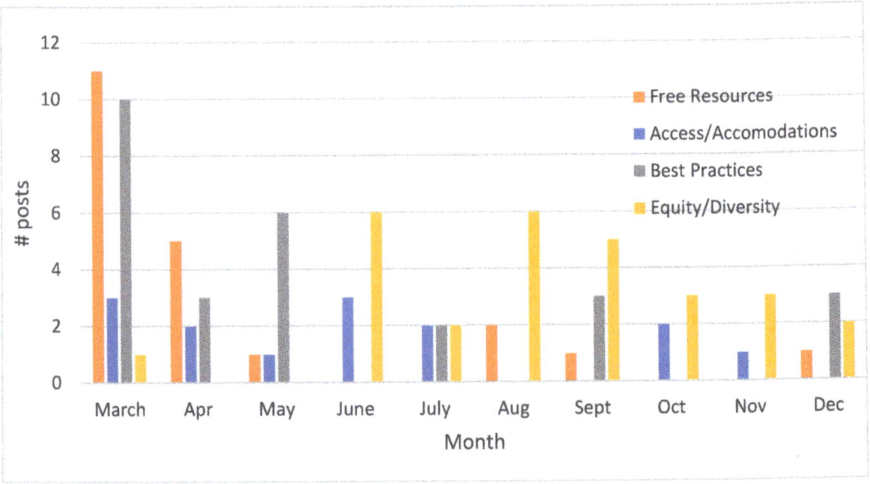

Figure 31.4 The number of posts in SFTCO focused on equity, diversity, and accessibility issues from the group's creation in March 2020 to December 2020.

first in the context of the emergency transition and then planning for the next term fully taught online. The analysis indicated that initially the community was primarily focused on sharing information about materials that were being offered for free (*e.g.*, "*Chem 101 is offering use of their platform for free until the end of the term*") and what kinds of things people were doing to provide equitable access to students who may not have access to reliable internet or technologies, or students who had additional demands such as work, childcare, *etc.* to balance as a result of the pandemic (best practices). There was also a greater focus in the community during the first few months on how to best make disability support accommodations for students in the online environment (access/accommodations). However, as we entered the summer months when people were more comfortable with the technology needed to get their courses online, develop course materials, and had some ideas about how to tackle assessment, the focus of the posts shifted towards things like how we can incorporate more diverse perspectives and better represent the diversity of scientists and their contributions in our chemistry courses (equity/diversity). This was a specific manifestation of the general shift in the group's domain, with a heavier emphasis on technology at the group's inception, gradually shifting to include topics that did not pertain strictly to online teaching. In addition to the familiarity with technology gained in the early months, instructors may have had fewer students during the summer months, allowing them to focus on more abstract principles of Diversity, Equity, and Inclusion (DEI). The group's deeper shift toward social justice may have also reflected the current events, with widespread protests across the United States in the wake of the murder of George Floyd. It should be noted that even though the majority of the group's members were from a nation experiencing upheaval due to racial protests, the total number of posts regarding DEI was very small compared to the total number of posts during this time. It is unknown whether the smaller degree of activity surrounding these issues indicates unwillingness or prioritization of other needs (*e.g.*, recovering from the emergency transition to remote teaching or preparations for online teaching during the upcoming term).

31.7 Lessons Learned

Changing teaching practice is difficult. Real and meaningful change requires not just implementing new teaching materials or using new teaching practices, but also a change in underlying beliefs about teaching and learning such that they are in alignment with the new materials or practices.[13,14] This difficulty is evidenced by the fact that despite the vast amount of research regarding best-practices in teaching and learning, the majority of college science courses still primarily rely on lecture instruction,[15] and decades of K–12 science education reform efforts have failed to produce notable changes.[14] Changing beliefs about teaching and learning, the precursor to real change, requires instructors to be aware of their beliefs and for those beliefs to be challenged.[16] This is especially difficult if we are isolated in our teaching and efforts to change our teaching, something that is possible even when we have other colleagues teaching in the same content area at our institution. Facebook groups had been observed as a place of informal professional development for teachers prior to the COVID-19 pandemic.[17–19] In light of the urgent need for professional development demanded by Emergency Remote Teaching (ERT), SFTCO served as a CoP for the members of the chemistry teaching community, providing materials as well as content, pedagogical, and technological support to meet the immediate needs of its members as they shifted their instruction online. Factors that contributed to the group's success as a CoP during this challenging time were its flexibility in domain, member-driven content, and presence of emotional support from the community. The high degree of participation in community support posts indicates the importance of this feature of the CoP. Though, as we described earlier, CoPs and PLCs have notable differences, in their chapter in this book Shauly, Shwartz, and Avargil[7] highlight the ways that PLCs have provided teachers with similar support.

However, the shift in the focus of the group illustrates that this CoP may serve as an ongoing source of PD as members find the time and space to grapple with more challenging pedagogical changes. The high level of community engagement indicated by the relatively large numbers of comments suggests that this group can challenge ideas and provide the support necessary to enact change. This is illustrated by the more recent discussions within the SFTCO group around the topics of equity and inclusion. The fact that many of these discussion threads did not end in satisfactory resolution or represent the very beginning of a conversation (identification of resources as opposed to how best to use such resources) indicates that chemistry instructors continue to want and need professional development support in this area, and that more attention should be given to this area in both research and practice.

Acknowledgements

The authors thank Ariel Chaney for her work on the initial stages of this project. They also appreciate the generous contributions of time and expertise about teaching chemistry made by the many members of SFTCO.

References

1. A. K. Fackler and C. M. Sexton, *Electron. J. Res. Sci. Math. Educ.*, 2020, **24**(3), 5–13.
2. D. G. Herrington, R. D. Sweeder, P. L. Daubenmire, C. F. Bauer, S. L. Bretz, D. M. Bunce, J. H. Carmel, R. Cole, B. K. DeKorver, R. M. Kelly, S. E. Lewis, M. Oliver-Hoyo, S. A. C. Ryan, M. Stains, M. H. Towns and E. J. Yezierski, *J. Chem. Educ.*, 2019, **96**(3), 393–397.
3. L. Padwa, A. M. Kelly and K. Sheppard, *J. Chem. Educ.*, 2019, **96**, 2383–2392.
4. L. Darling-Hammond, M. E. Hyler and M. Gardner, *Effective Teacher Professional Development*, Learning Policity Institute, Palo Alto, CA, 2017.
5. C. Henderson, A. Beach and N. Finkelstein, *J. Res. Sci. Teach.*, 2011, **48**, 952–984.
6. J. P. Spillane, *J. Curric. Stud.*, 1999, **31**, 143–175.
7. See Chapter 30: A. Shauly, G. Shwartz and S. Avargil, in *Digital Learning and Teaching in Chemistry*, ed. Y. J. Dori, C. Ngai and G. Szteinberg, Royal Society of Chemistry, United Kingdom, 2023.
8. E. Wenger-Trayner and B. Wenger-Trayner, Communities of Practice. A brief introduction, 2015, https://wenger-trayner.com/wp-content/uploads/2015/04/07-Brief-introduction-to-communities-of-practice.pdf, retrieved 7 February 2022.
9. P. Mishra and M. J. Koehler, *Teach. Coll. Rec.*, 2006, **108**, 1017–1054.
10. L. S. Shulman, *Educ. Res.*, 1986, **15**, 4–14.
11. D. W. McMillan and D. M. Chavis, *J. Community Psychol.*, 1986, **14**, 6–23.
12. B. DeKorver, A. Chaney and D. Herrington, *J. Chem. Educ.*, 2020, **97**, 2825–2833.
13. M. Fullan, *The New Meaning of Educational Change*, Teachers College Press, New York, 4th edn, 2007.
14. T. Burner, *Forsk. Forandring*, 2018, **1**, 122–134.
15. M. Stains, J. Harshman, M. K. Barker, S. V. Chasteen, R. Cole, S. E. DeChenne-Peters, M. K. Eagan, J. M. Esson, J. K. Knight, F. A. Laski, M. Levis-Fitzgerald, C. J. Lee, S. M. Lo, L. M. McDonnell, T. A. McKay, N. Michelotti, M. S. Palmer, K. M. Plank, T. M. Rodela, E. R. Sanders, N. G. Schimpf, P. M. Schulte, M. Smith, M. Stetzer, J. Stewart, B. V. Valkenburgh, E. Vinson, L. K. Weir, P. J. Wendel, L. B. Wheeler and A. M. Young, *Science*, 2018, **359**, 1468–1470.

16. R. Ceglie and V. Olivares, Representation of Diversity in Science Textbooks, in *The New Politics of the Textbook*, ed. H. Hickman and B. Porfilio, Sense Publishers, Rotterdam, 2012, p. 49.
17. N. Mansour, *Int. J. Environ. Sci. Educ.*, 2009, **4**, 25–48.
18. H. Bett and L. Makewa, *Asia-Pac. J. Teach. Educ.*, 2020, **48**, 132–136.
19. C. S. Bissessar, *Aust. J. Teach. Educ.*, 2014, **39**, 121–135.

Subject Index

References to figures are given in *italic* type.

Bold keywords are holistic ones

www.ingramcontent.com/pod-product-compliance
Ingram Content Group UK Ltd.
Pitfield, Milton Keynes, MK11 3LW, UK
UKHW021826300926
9009IPUK00003B/6

9 781839 165238